Current Strategies for the Biochemical Diagnosis and Monitoring of Mitochondrial Disease

Current Strategies for the Biochemical Diagnosis and Monitoring of Mitochondrial Disease

Special Issue Editor

Iain P. Hargreaves

MDPI • Basel • Beijing • Wuhan • Barcelona • Belgrade

MDPI

Special Issue Editor
Iain P. Hargreaves
John Moores University
UK

Editorial Office
MDPI
St. Alban-Anlage 66
Basel, Switzerland

This is a reprint of articles from the Special Issue published online in the open access journal *Journal of Clinical Medicine* (ISSN 2077-0383) from 2017 to 2018 (available at: http://www.mdpi.com/journal/jcm/special_issues/mitochondrial_disease)

For citation purposes, cite each article independently as indicated on the article page online and as indicated below:

LastName, A.A.; LastName, B.B.; LastName, C.C. Article Title. *Journal Name* **Year**, *Article Number*, Page Range.

ISBN 978-3-03897-240-2 (Pbk)
ISBN 978-3-03897-241-9 (PDF)

Contents

About the Special Issue Editor

Iain P. Hargreaves went to work as a postdoctoral researcher at the Institute of Neurology, University College of London, London, UK where he helped to establish the Neurometabolic Unit which specialises in the diagnosis of metabolic disorders, especially mitochondrial and neurotransmitter disorders. He worked in the Neurometabolic Unit for 21 years until he left to take up a lecturing position in biochemistry/clinical biochemistry at Liverpool John Moores University, Liverpool, UK. He has retained an honorary consultant clinical scientist and senior lecturer position within the Neurometabolic Unit and Institute of Neurology, respectively, and he is still actively involved in the mitochondrial diagnostic service. Interests: Defects in mitochondrial energy metabolism and coenzyme Q10 biosynthesis together with biochemical treatments of mitochondrial dysfunction.

Journal of
Clinical Medicine

MDPI

Editorial

Biochemical Assessment and Monitoring of Mitochondrial Disease

Iain P. Hargreaves [1,2]

1   Department of Pharmacy and Biomolecular Science, Liverpool John Moores University, Byrom Street,
    Liverpool L3 5UA, UK; i.hargreaves@ucl.ac.uk; Tel.: +44-151-231-2711
2   Neurometabolic Unit, National Hospital, Queen Square, London WC1N 3BG, UK

Received: 26 March 2018; Accepted: 27 March 2018; Published: 29 March 2018

Mitochondrial respiratory chain (MRC) disorders have a multifaceted clinical presentation and genetic origin. The adage, "any symptom, any organ or tissue, any age of presentation, any mode of inheritance", coined by Munnich and colleagues in 1992 [1] highlights the challenges faced in diagnosing these complex disorders, which requires a multidisciplinary approach involving the results of the clinical, histological, genetic, and biochemical investigations. In the biochemical context, the first-line investigations to determine evidence of a MRC disorder in patients are by the assessment of plasma or cerebral spinal fluid (CSF) lactate levels [2]. However, these determinations lack specificity and sensitivity, and a 'normal' result does not exclude the possibility of an underlying MRC disorder. An elevated plasma alanine level, an indicator of cellular pyruvate accumulation, has also been suggested as an appropriate marker of MRC dysfunction with an absolute level >450 µM being utilized as factor to determine the likelihood of mitochondrial disease, according to the Nijmegen diagnostic protocol [3]. However, an elevated plasma alanine level may only be present during a relapse in symptoms, and therefore a 'normal' plasma alanine level does not exclude an underlying MRC disorder [2]. Urine organic acid analysis may reveal evidence of elevated lactate, Krebs cycle intermediates, or 3-methylglutaconic acid in some patients with MRC disorders; however, these metabolites may only be present if the patient is acutely symptomatic and be absent during periods of stability [2]. However, the diagnostic utility of urine organic acid analysis in mitochondrial disease is supported by the study of Alban et al. (2017) [4], which reported an abnormal urine organic acid profile in 82% of patients with muscle MRC enzyme deficiencies. Nonetheless, renal immaturity is an important factor to consider, and an abnormal urine organic acid profile in a patient less than one year of age should be interpreted with extreme caution [2].

The diagnosis of mitochondrial disease is impeded by the paucity of reliable surrogate markers of MRC dysfunction presently available to select in preference to an invasive skeletal muscle biopsy, which is required for spectrophotometric enzyme assay. However, the hormone-like cytokine, serum fibroblast growth factor-21 (FGF-21), which is involved in the intermediary metabolism of carbohydrates and lipids, has been suggested as a potential reliable biomarker of MRC dysfunction [5]. In addition, the growth differentiation factor-15 (GDF-15) has also been identified as a potential marker of mitochondrial disease [6]. Although, at present, there are still concerns about the sensitivity of FGF-21 for detecting MRC disease in non-myopathic patients, and GDF-15 is regarded to have superior sensitivity but lower specificity [7]. A study by Morovat et al. [8] has indicated that although serum FGF-21 determination may have diagnostic utility in mitochondrial disease, it may prove more useful in monitoring disease progression and the effects of therapeutic intervention. Furthermore, the combined use of serum FGF-21 determination with urine organic acid analysis has also been suggested to improve the diagnostic value of either test used in isolation [4]. Surprisingly, however, the combined assessment of both FGF-21 and GDF-15 in adult patients with mitochondrial disease was not found to improve the diagnostic value of the individual tests [7]. An elevated plasma creatine level has also recently been suggested as a potential biomarker of mitochondrial disease; however, in view of the number of variables that may influence the circulatory level of this compound, its diagnostic value requires careful consideration [9].

The determination of coenzyme Q_{10} (CoQ_{10}) in plasma or blood mononuclear cells (BMNCs) appears to be of diagnostic utility in identifying patients with a deficit in the level of thisisoprenoid [10]. However, although this determination can't distinguish between primary or secondary CoQ_{10} deficiencies, it identifies an important subset of mitochondrial patients that may respond to CoQ_{10} supplementation [11]. The use of BMNC's also offers a means to directly assess MRC enzyme activities in patients with suspected mitochondrial disease, although a 'normal' result does not exclude the possibility that a defect may be expressed in other tissues [12].

The assessment of oxidative stress is also an important consideration in the context of mitochondria disease; although not a diagnostic parameter, it can provide important information about disease pathophysiology as well as the therapeutic efficacy of antioxidant strategies. The intracellular redox status of the antioxidant, reduced glutathione (GSH), as indicated by the ratio of GSH to its fully oxidised form, GSSG in white blood cells or BMNCs, may offer an appropriate surrogate for this evaluation [13].

Overall, at present, due to the lack of reliable validated biomarkers or surrogates for evaluating evidence of MRC dysfunction [14], spectrophotometric assessment of MRC enzyme activities in a skeletal muscle biopsy or tissue from the disease-presenting organ if accessible is still considered the 'Gold Standard' biochemical method for diagnosing patients with MRC disorders. The status quo is set to exist until more effort and funding can be centered on identifying appropriate biomarkers that fulfill all criteria required to have diagnostic utility for detecting MRC disorders.

Conflicts of Interest: The author declares no conflict of interest.

References

1. Munnich, A.; Rustin, P.; Rotig, A.; Chretien, D.; Bonnefont, J.P.; Nuttin, C.; Cormier, V.; Vassault, A.; Parvy, P.; Bardet, J.; et al. Clinical aspects of mitochondrial disorders. *J. Inherit. Metab. Dis.* **1992**, *15*, 448–455. [CrossRef] [PubMed]
2. Haas, R.H.; Parikh, S.; Falk, M.J.; Saneto, R.P.; Wolf, N.I.; Darin, N.; Wong, L.-J.; Cohen, B.H.; Naviaux, R.K. The in-depth evaluation of suspected mitochondrial disease: The Mitochondrial Medicine Society's Committee on Diagnosis. *Mol. Genet. Metab.* **2008**, *94*, 16–37. [CrossRef] [PubMed]
3. Wolf, N.I.; Smeitink, J.A. Mitochondrial disorders: A proposal for consensus diagnosis criteria in infants and children. *Neurology* **2002**, *59*, 1402–1405. [CrossRef] [PubMed]
4. Alban, C.; Fatale, E.; Joulan, A.; Ilin, P.; Saada, A. The relationship between mitochondrial respiratory chain activities in muscle and metabolites in plasma and urine: A retrospective study. *J. Clin. Med.* **2017**, *10*, 6. [CrossRef] [PubMed]
5. Tyynismaa, H.; Carroll, C.J.; Raimundo, N.; Ahola-Erkkilä, S.; Wenz, T.; Ruhanen, H.; Guse, K.; Hemminki, A.; Peltola-Mjøsund, K.E.; Orešič, W.T.M.; et al. Mitochondrial myopathy induces a starvation-like response. *Hum. Mol. Genet.* **2010**, *19*, 3948–3958. [CrossRef] [PubMed]
6. Kalko, S.G.; Paco, S.; Jou, C.; Rodríguez, M.A.; Meznaric, M.; Rogac, M.; Jekovec-Vrhovsek, M.; Sciacco, M.; Moggio, M.; Fagiolari, G.; et al. Transcriptomic profiling of TK2 deficient human skeletal muscle suggests a role for the p53 signalling pathway and identifies growth and differentiation factor-15 as a potential novel biomarker for mitochondrial myopathies. *BMC Genom.* **2014**, *15*, 91. [CrossRef] [PubMed]
7. Davis, R.L.; Liang, C.; Sue, C.M. A comparison of current serum biomarkers as diagnostic indicators of mitochondrial diseases. *Neurology* **2016**, *86*, 2010–2014. [CrossRef] [PubMed]
8. Morovat, A.; Weerasinghe, G.; Nesbitt, V.; Hofer, M.; Agnew, T.; Quaghebeur, G.; Sergeant, K.; Fratter, C.; Guha, N.; Mirzazadeh, M.; et al. Use of FGF-21 as a biomarker of mitochondrial disease in clinical practice. *J. Clin. Med.* **2017**, *6*, 80. [CrossRef] [PubMed]
9. Ostojic, S.M. Plasma creatine as a marker of mitochondrial dysfunction. *Med. Hypotheses* **2018**, *113*, 52–53. [CrossRef] [PubMed]

10. Yubero, D.; Allen, G.; Artuch, R.; Montero, R. The value of coenzyme Q_{10} determination in mitochondrial patients. *J. Clin. Med.* **2017**, *6*, 37. [CrossRef] [PubMed]
11. Rodríguez-Aguilera, J.C.; Cortés, A.B.; Fernández-Ayala, D.J.; Navas, P. Biochemical assessment of coenzyme Q_{10} deficiency. *J. Clin. Med.* **2017**, *6*, 27. [CrossRef]
12. Hargreaves, I.; Mody, N.; Land, J.; Heales, S. Blood mononuclear cell mitochondrial respiratory chain complex IV activity is decreased in multiple sclerosis patients: Effects of β-interferon treatment. *J. Clin. Med.* **2018**, *7*, 36. [CrossRef] [PubMed]
13. Enns, G.M.; Cowan, T.M. Glutathione as a redox biomarker in mitochondrial disease—Implications for therapy. *J. Clin. Med.* **2017**, *6*, 50. [CrossRef] [PubMed]
14. Finsterer, J.; Zarrouk-Mahjoub, S. Biomarkers for detecting mitochondrial disorders. *J. Clin. Med.* **2018**, *7*, 16. [CrossRef] [PubMed]

Journal of
Clinical Medicine

MDPI

Article

Blood Mononuclear Cell Mitochondrial Respiratory Chain Complex IV Activity is Decreased in Multiple Sclerosis Patients: Effects of β-Interferon Treatment

Iain Hargreaves [1,2,*], **Nimesh Mody** [1,3], **John Land** [1] and **Simon Heales** [1,4,*]

[1] Neurometabolic Unit, National Hospital, Queen Square, London WC1N 3BG, UK;
n.mody@abdn.ac.uk (N.M.); j.land@ucl.ac.uk (J.L.)

[2] Department of Pharmacy and Biomolecular Science, Liverpool John Moores University, Byrom Street, Liverpool L3 5UA, UK

[3] Institute of Medical Sciences, University of Aberdeen, Scotland AB24 3FX, UK

[4] UCL Great Ormond Street Institute of Child Health, University College London, London WC1E 6BT, UK

* Correspondence: i.p.hargreaves@ljmu.ac.uk (I.H.); s.heales@ucl.ac.uk (S.H.);
Tel.: +44151-231-2711 (I.H.); +44203-448-3818 (S.H.)

Received: 1 February 2018; Accepted: 18 February 2018; Published: 20 February 2018

Abstract: Objectives: Evidence of mitochondrial respiratory chain (MRC) dysfunction and oxidative stress has been implicated in the pathophysiology of multiple sclerosis (MS). However, at present, there is no reliable low invasive surrogate available to evaluate mitochondrial function in these patients. In view of the particular sensitivity of MRC complex IV to oxidative stress, the aim of this study was to assess blood mononuclear cell (BMNC) MRC complex IV activity in MS patients and compare these results to age matched controls and MS patients on β-interferon treatment. Methods: Spectrophotometric enzyme assay was employed to measure MRC complex IV activity in blood mononuclear cell obtained multiple sclerosis patients and aged matched controls. Results: MRC Complex IV activity was found to be significantly decreased ($p < 0.05$) in MS patients (2.1 ± 0.8 k/nmol $\times 10^{-3}$; mean $\pm$ SD] when compared to the controls (7.2 ± 2.3 k/nmol $\times 10^{-3}$). Complex IV activity in MS patients on β-interferon (4.9 ± 1.5 k/nmol $\times 10^{-3}$) was not found to be significantly different from that of the controls. Conclusions: This study has indicated evidence of peripheral MRC complex IV deficiency in MS patients and has highlighted the potential utility of BMNCs as a potential means to evaluate mitochondrial function in this disorder. Furthermore, the reported improvement of complex IV activity may provide novel insights into the mode(s) of action of β-interferon.

Keywords: mitochondrial respiratory chain; complex IV; blood mononuclear cells; multiple sclerosis; β-Interferon

1. Introduction

Multiple sclerosis (MS) is an inflammatory demyelinating disease of the central nervous system (CNS), in which cytokines and other inflammatory mediators are raised [1]. To date, the exact cause of MS has still to be fully elucidated, but it is believed to result from an abnormal response of the immune system to one or more myelin antigens in the CNS, such as components of the myelin [2]. The disease is characterized by an accumulation of macrophages and lymphocytes in the CNS leading to demyelination and destruction of neuronal axon [3]. These areas of demyelination are known as plaques and contain areas of gliosis and inflammation in most cases [4].

MS has a heterogeneous clinical presentation with symptoms including impaired vision, fatigue, spasms and paralysis of a number of muscle systems [5]. There are five basic types of MS of which relapsing remitting (RR) is the most common [6]. In RR-MS patients the disease develops in a discursive manner with symptomatic and asymptomatic phases. Over time, RR-MS patients may develop chronic lesions that result in irreversible axonal damage and loss, resulting in the conversion of RRMS to secondary progressive MS (SPMS).

Although MS is traditionally considered to be an autoimmune disease, neurodegeneration has also been implicated in disease progression [7]. One of the major factors that are responsible for neurodegeneration in MS is thought to be mitochondrial respiratory chain (MRC) dysfunction with evidence of impaired MRC complex I (NADH: ubiquinone reductase; EC: 1.3.5.1), III (Ubiquinol: cytochrome reductase; EC: 1.10.2.2.), and IV (Cytochrome c oxidase; EC: 1.9.3.1) activities being reported in post mortem cerebral tissue from MS patients, as well as in experimental autoimmune encephalomyelitis [8–10]. Although the cause of the MRC dysfunction in MS has still to be fully elucidated, oxidative and nitrosative stress are thought to be contributory factors [10]. Reactive oxygen species (ROS), such as superoxide (O_2^-) and reactive nitrogen species (RNS: nitric oxide (NO); peroxynitrite; $ONOO^-$) are generated during neuro-inflammation in MS and have been implicated, by our research and that of others, as mediators of demyelination and axonal injury [11–14]. Although the inflammatory environment in demyelinating plaques is conducive to the generation of ROS, activated lymphocytes and macrophages also release a host of pro-inflammatory cytokines, such as interferon gamma (IFN-g), which results in an upregulation inducible nitric oxide synthase (iNOS) activity within the CNS and a concomitant increase RNS generation [15]. ROS and RNS are able to induce MRC dysfunction by causing oxidative damage to mitochondrial DNA, mitochondrial membrane phospholipids, and/or the protein subunits of the enzymes [16]. The continued inflammatory process in the CNS of MS patients coupled to the impaired immune regulation results in high circulatory levels of RNS [17], which may have the potential to impair MRC function in peripheral tissue. Although few studies have assessed this phenomenon, a study by Kumleh et al. [18] reported evidence of impaired skeletal muscle MRC complex I activity in a small cohort of MS patients. Although it is difficult in living MS patients to accurately determine evidence of cerebral MRC dysfunction, the presentation of mitochondrial dysfunction in systemic tissue may provide an appropriate surrogate for this evaluation. The liberation of a skeletal muscle biopsy, which is considered the "gold standard" for MRC enzyme determination [19] would be relatively invasive and may not always be possible. However, the use of blood mononuclear cells (BMNCs) may provide an alternative relatively low invasive means to assess the evidence of MRC dysfunction in MS patients. Furthermore, in view of the relatively small amount of biological material afforded from a BMNC preparation it would not be possible to assess the activity of all the MRC enzyme complexes. However, in view of the particular susceptibility of MRC complex IV activity to RNS induced inactivation [19], assessment of the activity of this enzyme complex may therefore be judicious in MS patients. Furthermore, this tissue may also be informative with regards to the efficacy and mode of action of therapeutic agents, such as β-interferon, which may slow disease progression.

The purpose of this study was therefore to determine BMNC complex IV activity in MS patients and compare these results to age matched controls and MS patients receiving therapy in the form of β-interferon.

2. Experimental Section

2.1. Reagents

All of the reagents were analytical grade and obtained Sigma Aldrich Chemical Company (Poole, Dorset, UK). PD_{10} column used in the preparation of reduced cytochrome c for complex IV spectrophotometric enzyme assay were obtained from Amersham Pharmacia (St. Albans, Herts, UK).

2.2. Patients

Patients were diagnosed and consented by a consultant neurologist at the National Hospital, Queen Square, London, UK, according to the guidelines of Poser et al. [20]. They were divided into two groups:

(1) Patients not receiving β-interferon treatment. This group consisted of seven patients (male:female = 4:3). Six patients were aged between 30–39 years and one female patient was aged 65 years.

(2) Patients receiving β-interferon treatment. This group consisted of four patients (male:female = 3:1). Patients were aged between 26-36 years. Patients were selected and received β-interferon in accordance with the National Institute for Health and Care Excellence (UK) guidelines.

For this study, a control group of 24 healthy volunteers (aged 32–55 years, male:female = 9:15) were used.

2.3. Blood Mononuclear Cell (BMNC) Preparation

BMNCs were isolated from between 3–10 mL of lithium heparin blood by use of the ACCUSPIN system–Histopaque-1077 (Sigma-Aldrich, Poole, Dorset, UK). BMNCs were suspended in phosphate buffered saline, pH 7.2, and stored at –70 °C until analysis.

2.4. Spectrophotometric Enzyme Assays

Enzymatic determinations were undertaken at 30 °C using a Uvikon XL spectrophotometer (Northstar, Leeds, UK).

MRC complex IV activity was measured by the potassium cyanide sensitive oxidative of reduced cytochrome c at 550 nm, according to the method of Wharton and Tzagoloff [21].

To account for the mitochondrial enrichment of the preparations used, activity of the mitochondrial marker enzyme, citrate synthase (CS) (EC 2.3.3.1) was evaluated. This was determined according to the method of Shepherd and Garland [22] by the formation of 5-thio-2-nitrobenzoic following the incubation BMNCs with acetyl-CoA, oxaloacetate, and 5,5-Dithiobis-(2-nitrobenzoic acid). 5-thio-2-nitrobenzoic absorbs at 412 nm.

Complex IV activities were expressed as a ratio to CS activity (k/nmol) to take into account the mitochondrial enrichment of the BMNCs [23].

CS has units of activity of nmol/min/mL. Complex IV has units of activity of k/min/mL since the activity of this enzyme is expressed as a 1st order rate constant. Therefore, when complex IV activity is expressed as a ratio to CS activity the units are: k/min/mL divided by nmol/min/mL = k/nmol.

2.5. Protein Determination

Protein was determined according to the method of Lowry and colleagues [24] using bovine serum albumin as a standard.

2.6. Statistical Analysis

Statistical analysis was performed using one-way analysis of variance (ANOVA) followed by the least squared difference (LSD) multiple range test, the students t-test, and Spearman test was used to establish potential correlations between MRC complex IV activity, CS activity, and age. A p value < 0.05 was considered to be statistically significant.

3. Results

Recombinant β-interferon (4 and 16 million units) was not found to have an effect on MRC complex IV or CS activities in vitro. No correlation was found between age and BMNC MRC complex IV ($r = 0.688$; $n = 21$; $p = 0.7703$) or CS ($r = -0.276$; $n = 21$; $p = 0.742$) activities, respectively, in the control population. Gender was also not found to influence the activities of these enzymes in BMNCs, with no significant difference being found between male and female complex IV ($p = 0.675$) or CS ($p = 0.691$) activities.

BMNC MRC complex IV activity (expressed as a ratio to CS activity) was found to be significantly decreased ($p < 0.05$) in MS patients not on β-interferon (2.1 ± 0.8 k/nmol $\times 10^{-3}$; mean $\pm$ SD) when compared to the controls (7.2 ± 2.3 k/nmol $\times 10^{-3}$) (Figure 1). Complex IV activity in MS patients on β-interferon (4.9 ± 1.5 k/nmol $\times 10^{-3}$) was not found to be significantly different from that of the controls (Figure 1). No significant difference in BMNC CS activity was found between the control (45.24 ± 18.77 nmol/min/mg) and MS patients (33.65 ± 10.02 nmol/min/mg).

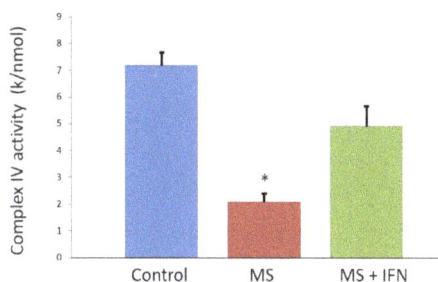

Figure 1. Blood mononuclear cell Complex IV activity, expressed as a ratio to citrate synthase, in control individuals, MS patients and MS patients receiving β-interferon (IFN). * Statistically different from both control and MS patients receiving β-interferon.

4. Discussion

The results of this study have indicated evidence of a deficiency in MRC complex IV activity in BMNCs of MS patients. The impairment of BMNC MRC complex IV activity may result in an altered immune response, which may contribute to disease pathophysiology. At present, the factors responsible for this MRC dysfunction in the MS patients are as yet uncertain. However, the absence of a significance decrease in BMNC MRC complex IV activity in MS patients receiving β-interferon suggests that the loss of enzyme activity may be the result of a disease process that is reversed by β-interferon. One of the mechanisms of action by which β-interferon elicits its beneficial effect in MS patients appears to be by its ability to inhibit astrocytic NO production [21], and thereby decreasing the availability of circulatory RNS that have the potential to induce MRC impairment, particularly at the level of complex IV. Whether such a mechanism occurs in the periphery requires further investigation. However, it is of note that serum levels of nitrite and nitrate (indices of RNS production) are reported to be elevated in MS patients [17]. Alternatively, the decrease in MRC complex IV activity detected in the MS patients may be the result of mitochondrial DNA deletions as reported in the neurons and choroid plexus of progressive MS patients [22]. Although, evidence of mitochondrial DNA mutations, and effects of β-interferon, in peripheral BMNCs has yet to be determined in MS patients [23]. Nonetheless, a study by Amorini el al. has reported a threefold elevation in serum lactate levels in MS patients [24]. Although this study supports evidence of mitochondrial dysfunction in MS, previous studies assessing both serum [25] and CSF (cerebral spinal fluid) [26] lactate levels in this disorder have failed to show any evidence of an increase in the level of this metabolite. Importantly, lactate levels may not necessarily be raised as a consequence of MRC dysfunction as evidenced in patients with primary mitochondrial disorders [27]. Furthermore, elevated serum lactate levels may not be a specific biomarker of MRC dysfunction, since this phenomenon has been reported to result from number of other clinical sequelae [27]. Therefore, the determination of MRC complex IV activity in BMNCs may serve as a more specific means of evaluating evidence of MRC dysfunction in MSA patients. In addition, in view of the association between oxidative and nitrosative stress and MRC dysfunction [11–14], BMNCs may also serve as a means of assessing the cellular antioxidant status of

MS patients. In view of its association with MRC dysfunction, the status of the cellular antioxidant, reduced glutathione may be judicious for this assessment [28]. The possibility arises that MRC complex IV dysfunction may also be a contributory factor to the pathophysiology of other diseases that are associated with nitrosative stress, such as diabetes, cancer, and stroke [29–31], and therefore, BMNCs may have utility is assessing evidence of mitochondrial impairment in these disorders.

In this study we have highlighted the feasibility of using BMNCs to assess evidence of MRC complex IV deficiency in MS patients. However, due to the limited amount of biological material available from BMNCs, it has not been possible to determine the activities of the other MRC enzymes (complexes I, II, and III) in the present study, and therefore, we cannot exclude the possibility that the MRC dysfunction in the MS patients is not solely restricted to complex IV. In spite of its limitations, this is the first study as far as the authors are aware to use BMNCs are a relatively low invasive surrogate to assess evidence of mitochondrial dysfunction in MS. This surrogate may also be of value in monitoring the therapeutic potential of pharmacotherapies on mitochondrial function in MS patients in view of the paucity of reliable biomarkers that are currently available.

Acknowledgments: We acknowledge the invaluable help of Dr. Peter Rudge, Consultant Neurologist, National Hospital, Queen Square London, for his help with patient recruitment. We also gratefully acknowledge financial support from the National Institute for Health Research, Biomedical Research Centre.

Author Contributions: I.H., J.L. and S.H. conceived and designed experiments; N.M. performed experiments and helped analyze and interpret the data along with S.H. and I.H.; S.H. and J.L. contributed reagents, materials and analysis tools; I.H. and S.H. wrote the paper.

Conflicts of Interest: The authors declare no conflict of interest.

References

1. Cannella, B.; Raine, C.S. The adhesion molecule and cytokine profile of multiple sclerosis lesions. *Ann. Neurol.* **1995**, *37*, 424–435. [CrossRef] [PubMed]
2. Keegan, B.M.; Noseworthy, J.H. Multiple sclerosis. *Annu. Rev. Med.* **2002**, *53*, 285–302. [CrossRef] [PubMed]
3. Stadelmann, C.; Wegner, C.; Bruck, W. Inflammation, demyelination, and degeneration—Recent insights from MS pathology. *Biochim. Biophys. Acta* **2011**, *1812*, 275–282. [CrossRef] [PubMed]
4. Popescu, B.F.; Pirko, I.; Lucchinetti, C.F. Pathology of multiple sclerosis: Where do we stand? *Continuum (Minneap. Minn.)* **2013**, *19*, 901–921. [CrossRef] [PubMed]
5. Mao, P.; Reddy, P.H. Is multiple sclerosis a mitochondrial disease? *Biochim. Biophys. Acta* **2010**, *1802*, 66–79. [CrossRef] [PubMed]
6. Lublin, F.D.; Reingold, S.C. Defining the clinical course of multiple sclerosis: Results of an international survey. National Multiple Sclerosis Society (USA) Advisory Committee on Clinical Trials of New Agents in Multiple Sclerosis. *Neurology* **1996**, *46*, 907–911. [CrossRef] [PubMed]
7. Rogers, J.M.; Panegyres, P.K. Cognitive impairment in multiple sclerosis: Evidence based analysis and recommendations. *J. Clin. Neurosci.* **2007**, *14*, 919–927. [CrossRef] [PubMed]
8. Lu, F.; Selak, M.; O'Connor, J.; Croul, S.; Lorenzana, C.; Butunoi, C.; Kalman, B. Oxidative damage to mitochondrial DNA and activity of mitochondrial enzymes in chronic active lesions of multiple sclerosis. *J. Neurol. Sci.* **2000**, *177*, 95–103. [CrossRef]
9. Dutta, R.; McDonough, J.; Yin, X.; Peterson, J.; Chang, A.; Torres, T.; Gudz, T.; Macklin, W.B.; Lewis, D.A.; Fox, R.J.; et al. Mitochondrial dysfunction as a cause of axonal degeneration in multiple sclerosis patients. *Ann. Neurol.* **2006**, *59*, 478–489. [CrossRef] [PubMed]
10. Sadeghian, M.; Mastrolia, V.; Rezaei Haddad, A.; Mosley, A.; Mullali, G.; Schiza, D.; Sajic, M.; Hargreaves, I.; Heales, S.; Duchen, M.R.; et al. Mitochondrial dysfunction is an important cause of neurological deficits in an inflammatory model of multiple sclerosis. *Sci Rep.* **2016**, *6*, 33249. [CrossRef] [PubMed]
11. Giovannoni, G.; Heales, S.J.; Land, J.M.; Thompson, E.J. The potential role of nitric oxide in multiple sclerosis. *Mult. Scler.* **1998**, *4*, 212–216. [CrossRef] [PubMed]
12. Bö, L.; Dawson, T.M.; Wesselingh, S.; Mörk, S.; Choi, S.; Kong, P.A.; Hanley, D.; Trapp, B.D. Induction of nitric oxide synthase in demyelinating regions of multiple sclerosis brains. *Ann. Neurol.* **1994**, *36*, 778–786. [CrossRef] [PubMed]

13. Jersild, C.; Fog, T.; Hansen, G.S.; Thomsen, M.; Svejgaard, A.; Dupont, B. Histocompatibility determinants in multiple sclerosis, with special reference to clinical course. *Lancet* **1993**, *2*, 1121–1125. [CrossRef]
14. Johnson, A.W.; Land, J.M.; Thompson, E.J.; Bolaños, J.P.; Clark, J.B.; Heales, S.J. Evidence for increased nitric oxide production in multiple sclerosis. *J. Neurol. Neurosurg. Psychiatry* **1995**, *58*, 107. [CrossRef] [PubMed]
15. Niedziela, N.; Adamczyk-Sowa, M.; Niedziela, J.T. Assessment of serum nitrogen species and inflammatory parameters in relapsing remitting multiple sclerosis patients treated with different therapeutic approaches. *Biomed. Res. Int.* **2016**, *2016*, 4570351. [CrossRef] [PubMed]
16. Yamamoto, T.; Maruyama, W.; Kato, Y.; Yi, H.; Shamoto-Nagai, M.; Tanaka, M.; Sato, Y.; Naoi, M. Selective nitration of mitochondrial complex I by peroxynitrite: Involvement in mitochondria dysfunction and cell death of dopaminergic SH-SY5Y cells. *J. Neural Transm.* **2002**, *109*, 1–13. [CrossRef] [PubMed]
17. Giovannoni, G.; Heales, S.J.; Silver, N.C.; O'Riordan, J.; Miller, R.F.; Land, J.M.; Clark, J.B.; Thompson, E.J. Raised serum nitrate and nitrite levels in patients with multiple sclerosis. *J. Neurol. Sci.* **1997**, *145*, 77–81. [CrossRef]
18. Kumleh, H.H.; Riazi, G.H.; Houshmand, M.; Sanati, M.H.; Gharagozli, K.; Shafa, M. Complex I deficiency in Persian multiple sclerosis patients. *J. Neurol. Sci.* **2006**, *243*, 65–69. [CrossRef] [PubMed]
19. Stewart, V.C.; Sharpe, M.A.; Clark, J.B.; Heales, S.J. Astrocyte-derived nitric oxide causes both reversible and irreversible damage to the neuronal mitochondrial respiratory chain. *J. Neurochem.* **2000**, *7592*, 694–700. [CrossRef]
20. Poser, C.M.; Paty, D.W.; Scheinberg, L.; McDonald, W.I.; Davis, F.A.; Ebers, G.C.; Johnson, K.P.; Sibley, W.A.; Silberberg, D.H.; Tourtellotte, W.W. New diagnostic criteria for multiple sclerosis: Guidelines for research protocols. *Ann. Neurol.* **1983**, *13*, 227–231. [CrossRef] [PubMed]
21. Stewart, V.C.; Land, J.M.; Clark, J.B.; Heales, S.J.R. Pretreatment of astrocytes with interferon α/β prevents mitochondrial respiratory chain damage. *J. Neurochem.* **1998**, *70*, 432–434. [CrossRef] [PubMed]
22. Campbell, G.R.; Ziabreva, I.; Reeve, A.K.; Krishnan, K.J.; Reynolds, R.; Howell, O.; Lassmann, H.; Turnbull, D.M.; Mahad, D.J. Mitochondrial DNA deletions and neurodegeneration in multiple sclerosis. *Ann. Neurol.* **2011**, *69*, 481–492. [CrossRef] [PubMed]
23. Campbell, G.R.; Reeve, A.K.; Ziabreva, I.; Reynolds, R.; Turnbull, D.M.; Mahad, D.J. No excess of mitochondrial DNA deletions within muscle of progressive multiple sclerosis. *Mult. Scler.* **2013**, *19*, 1858–1866. [CrossRef] [PubMed]
24. Amorini, A.M.; Nociti, V.; Petzold, A.; Gasperini, C.; Quartuccio, E.; Lazzarino, G.; Di Pietro, V.; Belli, A.; Signoretti, S.; Vagnozzi, R.; et al. Serum lactate as a novel biomarker in multiple sclerosis. *Biochim. Biophys. Acta* **2014**, *1842*, 1137–1143. [CrossRef] [PubMed]
25. Mähler, A.; Steiniger, J.; Bock, M.; Brandt, A.U.; Haas, V.; Boschmann, M.; Paul, F. Is metabolic flexibility altered in multiple sclerosis patients? *PLoS ONE* **2012**, *7*, e43675. [CrossRef] [PubMed]
26. Fonalledas Perelló, M.A.; Politi, J.V.; Dallo Lizarraga, M.A.; Cardona, R.S. The cerebrospinal fluid lactate is decreased in early stages of multiple sclerosis. *P. R. Health Sci. J.* **2008**, *27*, 171–174. [PubMed]
27. Koenig, M.K. Presentation and diagnosis of mitochondrial disorders in children. *Pediatr. Neurol.* **2008**, *38*, 305–313. [CrossRef] [PubMed]
28. Hargreaves, I.P.; Sheena, Y.; Land, J.M.; Heales, S.J. Glutathione deficiency in patients with mitochondrial disease: Implications for pathogenesis and treatment. *J. Inherit. Metab. Dis.* **2005**, *28*, 81–88. [CrossRef] [PubMed]
29. Bolanos, J.P.; Peuchen, S.; Heales, S.J. Nitric oxide-mediated inhibition of the mitochondrial respiratory chain in cultured astrocytes. *J. Neurochem.* **1994**, *63*, 910–916. [CrossRef] [PubMed]
30. Pandit, A.; Vadnal, J.; Houston, S.; Freeman, E.; McDonough, J. Impaired regulation of electron transport chain subunit genes by nuclear respiratory factor 2 in multiple sclerosis. *J. Neurol. Sci.* **2009**, *279*, 14–20. [CrossRef] [PubMed]
31. Pacher, P.; Beckman, J.S.; Liaudet, L. Nitric oxide and peroxynitrite in health and disease. *Physiol. Rev.* **2007**, *87*, 315–424. [CrossRef] [PubMed]

Journal of
Clinical Medicine

MDPI

Article

Measurement of Respiratory Chain Enzyme Activity in Human Renal Biopsy Specimens

Arun Ghose [1], Christopher M. Taylor [1], Alexander J. Howie [2], Anapurna Chalasani [3],
Iain Hargreaves [3] and David V. Milford [1,*]

[1] Department of Nephrology, Birmingham Children's Hospital, Birmingham B4 6NH, UK;
    arun.ghose@bch.nhs.uk (A.G.); cmarktaylor@hotmail.com (C.M.T.)
[2] Department of Histopathology, Birmingham Children's Hospital, Birmingham B4 6NH, UK;
    alexander.howie@nhs.net
[3] Neurometabolic Unit, National Hospital for Neurology and Neurosurgery, London WC1N 3BG, UK;
    annapurnachalasani@nhs.net (A.C.); i.hargreaves@ucl.ac.uk (I.H.)
*   Correspondence: david.milford@bch.nhs.uk; Tel.: +44-121-333-9228

Academic Editor: Mark S. Sands
Received: 6 July 2017; Accepted: 13 September 2017; Published: 19 September 2017

Abstract: *Background*: Mitochondrial disorders can present as kidney disease in children and be difficult to diagnose. Measurement of mitochondrial function in kidney tissue may help diagnosis. This study was to assess the feasibility of obtaining renal samples and analysing them for respiratory chain enzyme activity. *Methods*: The subjects were children undergoing a routine diagnostic renal biopsy, in whom a clinical condition of renal inflammation, scarring and primary metabolic disorder was unlikely. A fresh sample of kidney was snap frozen and later assayed for the activities of respiratory chain enzyme complexes I, II/III, and IV using spectrophotometric enzyme assay, and expressed as a ratio of citrate synthase activity. *Results*: The range of respiratory chain enzyme activity for complex I was 0.161 to 0.866 (mean 0.404, SD 0.2), for complex II/III was 0.021 to 0.318 (mean 0.177, SD 0.095) and for complex IV was 0.001 to 0.025 (mean 0.015, SD 0.006). There were correlations between the different activities but not between them and the age of the children or a measure of the amount of chronic damage in the kidneys. *Conclusion*: It is feasible to measure respiratory chain enzyme activity in routine renal biopsy specimens.

Keywords: mitochondrial disorders; renal biopsy; respiratory chain enzymes

---

1. Introduction

Mitochondrial disorders are difficult to diagnose clinically because of phenotypic variation. Renal presentations include the renal Fanconi syndrome (tubular wasting of bicarbonate, sodium and potassium, glucose and amino acids) and steroid unresponsive nephrotic syndrome as a result of glomerular damage; progression to chronic kidney disease is also described [1–3]. Mitochondrial cytopathy is important to diagnose, as there may be reversibility of those with coenzyme Q10 biosynthesis defects as well as the implication for treatment options, including renal transplantation. A clue to the diagnosis may come from clinical abnormalities of other organs that require a high rate of oxidative metabolism, such as the brain (encephalopathy, retinopathy or stroke), skeletal muscle (myopathy), heart (cardiomyopathy) and liver; a full review of the clinical features of mitochondrial diseases is beyond the scope of this paper, but is extensively reviewed by Gorman et al. [4]. Without these other features, diagnosis may be difficult in children who have a predominantly renal presentation. Also, the proportion of normal to abnormal mitochondria may vary in different organs (heteroplasmy). Consequently, a diagnostic procedure performed on skeletal muscle, for example, may not identify an abnormality in a child whose main organ involvement is kidney or brain. A logical approach is to investigate tissue from the affected organ if possible [5–8].

Renal biopsy is a routine diagnostic procedure in children, principally to provide material for histopathological examination, but this can also be a source of appropriate tissue for enzymatic studies. Because mitochondrial disorders are likely to affect respiratory chain enzyme activity (RCEA), assay of this activity in renal biopsies could facilitate the diagnosis of these disorders. Before the measurement of RCEA could be investigated for its value in the diagnosis of mitochondrial disorders, it was necessary to see whether RCEA could be assayed in human renal tissue obtained by routine renal biopsy. The aim of this study was to determine the feasibility of measuring RCEA in renal biopsy specimens from children. The results of this study will provide pilot data for a more thorough validation study to assess the reliability of renal RCEA assessment as a diagnostic investigation.

2. Experimental Section

2.1. Renal Biopsies

Research and ethical approval was gained from the North Staffordshire Local Research Ethics Committee prior to the study (REC reference number: 08/H1204/123). Information for parents and information appropriate for the age of children was provided. Written parental consent was obtained, as was written consent from children over 14 years old.

Children aged from three to 16 years of age undergoing routine renal biopsy were invited to participate in the study. Samples were obtained using a 14-gauge biopsy needle to give two cores, and these were inspected immediately with a dissecting microscope. If there were more than 18 mm of cortex, a 3-mm piece of cortex was snap frozen in liquid nitrogen. In biopsies without 18 mm of cortex, and if there were more than 15 mm of medulla, a 3 mm core of medulla was taken. Only one research sample was taken from each patient. The remaining tissue was processed routinely for diagnosis. The frozen sample was only released for analysis of RCEA once the pathologist's report confirmed no more than minor structural abnormalities. If the diagnosis required more tissue, the frozen sample was formalin fixed for routine processing.

Biopsies from 13 children were considered suitable for RCEA assay. A sample from another child was not used as there was insufficient material for routine diagnostic studies. All these children had two kidneys of normal size, normal estimated glomerular filtration rate, and normal blood pressure. The indication for biopsy was persistent haematuria in six, nephrotic syndrome in four, persistent proteinuria in one, and treatment of the nephrotic syndrome with a calcineurin inhibitor, to check if chronic renal damage had resulted, in two. No child had any abnormalities outside the kidney to suggest a mitochondrial disorder.

Diagnoses were made in orthodox ways, and are given in the Results. Because some specimens had microscopically visible chronic damage, seen as tubular atrophy and/or global sclerosis of glomeruli, a morphometric method was used to measure this. Areas of chronic damage in the cortex were outlined on computer images of sections stained by periodic acid-methenamine silver, expressed as a percentage of the whole cortical cross-sectional area, and called the index of chronic damage [9].

2.2. Tissue Homogenization

Samples were immediately put on dry ice and then stored at −70 °C. Before RCEA assay, samples were homogenized on ice, 1:9 (*w/v*), in 320 mmol/L sucrose, 1 mmol/L ethylenediamine tetra acetic acid dipotassium salt, and 10 mmol/L Trizma-base, pH 7.4, using a pre-chilled hand-held glass homogenizer. Approximately 10–15 mg of renal tissue was homogenized per sample, distributed into four Eppendorf tubes. One Eppendorf tube was used for each enzyme determination following three cycles of freeze/thawing [7].

2.3. RCEA Assays

Mitochondrial respiratory chain enzyme and citrate synthase (CS) activities were determined by spectrophotometric enzyme assay [8,10]. Complex I (NADH:ubiquinone reductase, EC 1.6.5.3) activity was measured by the rotenone sensitive oxidation of NADH at 340 nm. Complex II/III (succinate:cytochrome c reductase, EC 1.3.5.1 and EC 1.10.2.2) activity was measured by antimycin A-sensitive succinate dependent reduction of cytochrome c at 550 nm. Complex IV (cytochrome c oxidase, EC 1.9.3.1) activity was measured by the potassium cyanide sensitive oxidation of reduced cytochrome c at 550 nm. CS (EC 2.3.3.1) activity was determined by the formation of 5-thio-2-nitrobenzoic acid following the incubation of tissue homogenate with acetyl-CoA, oxaloacetate and 5,5'-dithiobis-(2-nitrobenzoic acid), at 412 nm. No reference wavelength was used in the enzyme assays.

All mitochondrial RCEAs were expressed as a ratio to the activity of CS, a mitochondrial marker enzyme, to standardise the mitochondrial enrichment of the sample [11]. Because activities of CS and complexes I and II/III were expressed as nmol/min/mL, but activity of complex IV was expressed as k/min/mL, the ratios of complex I and complex II/III activities to CS activity have no units, but the ratio of complex IV activity to CS activity has units of k/nmol. The range, mean and standard deviation were calculated for each RCEA. Correlations between age, index of chronic damage and each RCEA were determined by Spearman's rank correlation coefficient (r), and significant correlations were considered to be those with the conventional p value of under 0.05.

3. Results

Patient details, diagnoses, indexes of chronic damage, measurements of RCEA, and whether measurements were on cortex or medulla or on undetermined renal tissue are given in Table 1.

The range of RCEA for renal complex I was 0.161 to 0.866 (mean 0.404, SD 0.2), for complexes II/III was 0.021 to 0.318 (mean 0.177, SD 0.095) and for complex IV was 0.001 to 0.025 (mean 0.015, SD 0.006). There were no significant correlations between age and index of chronic damage, age and any RCEA, or index of chronic damage and any RCEA. RCEA of complex I showed a trend to correlate with RCEA of complexes II/III ($r = 0.539$, $p = 0.057$) and was significantly correlated with RCEA of complex IV ($r = 0.704$, $p = 0.007$). RCEA of complexes II/III was significantly correlated with RCEA of complex IV ($r = 0.660$, $p = 0.014$). The inter-assay coefficient of variation values for the RCEA and CS enzyme measurements calculated from the in-house QC human skeletal muscle homogenate data are as follows: Complex I: 9.11% ($n = 29$); Complex II/III: 10.70% ($n = 29$); Complex IV: 8.9% ($n = 29$); CS: 5.6% ($n = 29$). QC muscle samples were prepared from skeletal muscle tissue from patients with no biochemical evidence of a RCEA defect by the same method as the kidney homogenates were prepared.

Table 1. Clinical details, index of chronic damage, and assay of respiratory chain enzyme activities (RCEA) in cortex or medulla, if site of tissue known. RCEA are expressed as a ratio of citrate synthase activity. Normal RCEA for skeletal muscle and liver are included for comparison [8].

Sex, Age (year)	Diagnosis	Index of Chronic Damage	Complex I	Complex II/III	Complex IV (k/nmol)	Cortex or Medulla or Not Known
M3	minimal change	0%	0.176	0.021	0.003	Cortex
M7	minimal change	3%	0.161	0.166	0.015	Cortex
M9	CNI effects	4%	0.542	0.042	0.018	Cortex
M10	Alport syndrome	2%	0.522	0.253	0.025	Cortex
F11	IgA nephropathy	3%	0.220	0.120	0.016	Medulla
F13	IgA nephropathy	3%	0.229	0.105	0.001	cortex
M13	CNI effects	15%	0.388	0.238	0.017	Cortex
F14	thin gbm disease	3%	0.866	0.313	0.018	not known
F14	IgA nephropathy	5%	0.569	0.225	0.016	not known
M15	thin gbm disease	0%	0.516	0.224	0.018	Cortex
F15	seg. sclerosis	2%	0.333	0.105	0.014	Medulla
F16	no abnormality	0%	0.452	0.318	0.018	Cortex
M16	minimal change	0%	0.278	0.172	0.017	Cortex
	normal RCEA values for skeletal muscle and liver					
skeletal muscle			0.104–0.268	0.040–0.204	0.014–0.034	
liver			0.054–0.22	0.057–0.204	0.011–0.031 Reference ranges for RCEA in liver and muscle are not thought to be age dependent according to the study of Chretien et al. [12]	

Abbreviations: CNI, calcineurin inhibitor; gbm, glomerular basement membrane; seg, segmental.

13

4. Discussion

The DNA coding for mitochondrial respiratory chain enzymes NADH-Q oxidoreductase (complex I), succinate-Q reductase (complex II), Q-cytochrome c oxidoreductase (complex III) and complex IV (cytochrome c oxidase) is a mixture of nuclear genes and mitochondrial genes. Most disorders of RCEA in children are caused by mutations in nuclear genes, although mutations in mitochondrial DNA have also been found [13]. In some cases, genetic screening does not give a diagnosis, and so further investigations may be required. Another problem with mitochondrial respiratory chain disorders is that mutations may only be expressed in affected tissues. Consequently, assessment of the affected organ may be appropriate. Confirmation of abnormal RCEA may then be helpful in determining the most appropriate genetic tests. This study has indicated that it is possible to measure RCEA in kidney tissue obtained by the routine clinical procedure of renal biopsy.

Using CS activity as a reference controlled for variables such as the mitochondrial enrichment of the sample [11]. There was a wide range of RCEA for each of the complexes, with a 5-fold difference between the minimum value and maximum value for complex I, a 15-fold difference for complex II/III, and a 25-fold difference for complex IV (Table 1), although RCEA of the different complexes had significant or nearly significant correlations between each other. The mean RCEA of complex I was higher than the ranges for skeletal muscle and liver, the mean RCEA of complex II/II was towards the upper end of the ranges for skeletal muscle and liver, and the mean RCEA of complex IV was towards the lower end of the ranges for skeletal muscle and liver, although the RCEA in kidney were derived from a small sample. We cannot also exclude the possibility that poor sample handling of some of the renal biopsies may have contributed to loss of RCEA and therefore may account for the wide range of enzyme activities [14].

This study has a number of limitations. The number of patients sampled was small and most children had patchy chronic damage as assessed by the index of chronic damage measured in the histological specimen. However, as there was no way of knowing if the sample used for RCEA contained areas of chronic damage and the percentage of the biopsy specimen affected by chronic change was generally low (greater than 5% in only one child) it is not unreasonable to consider the values obtained as being representative of activity in normal renal tissue. Unfortunately, the small sample size precluded repeat analyses to confirm reproducibility, and so could explain the range of values obtained. Furthermore, no child with known mitochondrial disease presented during the time of this study to allow comparison with the values we obtained. Some workers have proposed validating the methodology by undertaking RCEA studies in nephrectomy samples, as these would provide a larger amount of renal tissue and allow reproducibility studies. However, we believe these samples would be unrepresentative; firstly, because the tissue would be exposed to significant ischaemia following clamping of the renal artery prior to the nephrectomy; and secondly, because the nephrectomy specimen would predominantly comprise diseased tissue. Importantly, given the heterogeneity of mitochondrial respiratory chain disorders, a reference range derived from true ethical controls may be required to accurately determine RCEA dysfunction in human renal tissue.

We have shown no relationship between the age of the children and RCEA in renal tissue when enzyme activities are expressed as a ratio to CS. Although this has previously been reported for RCEA in muscle and liver [12], this is the first time it has been reported for renal tissue. This is important since it indicates that age-specific reference intervals may not be necessary for renal RCEA if enzyme activities are expressed as a ratio to CS activity. The site of origin of the tissue analysed, whether cortex or medulla, also appeared unimportant. However, at present it cannot be excluded that there may be a difference in RCEA between these two regions of the kidney and future studies to confirm or refute this will be undertaken.

The lack of a relation between the index of chronic damage and RCEA is encouraging for future studies of RCEA, because the measured activities presumably reflect those in surviving tubules, suggesting that assays can still be applied to samples from children with a suspected mitochondrial disorder but with chronic kidney disease.

We have shown that it is feasible to obtain samples from routine renal biopsies and freeze them for later analysis of RCEA. However, as there was insufficient tissue to duplicate analysis, the effect of poor sample handling, together with uncertainty about whether tissue from the renal medulla or cortex have different RCEA, this study is unable to establish a RCEA reference interval for renal tissue. It will require a larger and more rigorous validation study using ethically obtained control tissue, as well as renal biopsies from patients with confirmed mitochondrial respiratory chain dysfunction, to establish a reference range and to confirm the diagnostic value of this test. It remains to be seen if such a study is possible, given the difficulty in obtaining sufficient tissue from human renal biopsy.

Acknowledgments: The authors are grateful to Susan Cavanagh for her invaluable help in freezing, storing and retrieving the samples for this study.

Author Contributions: C.M.T. and D.V.M. conceived the study and obtained specimens; A.J.H. undertook the pathology review; A.C. and I.H. performed the RCEA; A.G. wrote the initial draft and all authors subsequently contributed to and approved the final version.

Conflicts of Interest: The authors declare no conflict of interest.

References

1. Niaudet, P.; Rotiq, A. Renal involvement in mitochondrial cytopathies. *Pediatr. Nephrol.* **1996**, *10*, 368–373. [CrossRef] [PubMed]
2. Munnich, A.; Rustin, P. Clinical spectrum and diagnosis of mitochondrial disorders. *Am. J. Med. Genet. Part A* **2001**, *106*, 4–17. [CrossRef] [PubMed]
3. Rotiq, A. Renal disease and mitochondrial genetics. *J. Nephrol.* **2003**, *16*, 286–292.
4. Gorman, G.S.; Chinnery, P.F.; DiMauro, S.; Hirano, M.; Koga, Y.; McFarland, R.; Suomalainen, A.; Thorburn, D.R.; Zeviani, M.; Turnbull, D.M. Mitochondrial diseases. *Nat. Rev. Dis. Primers* **2016**, *2*, 1–22. [CrossRef] [PubMed]
5. Kirby, D.M.; Thorburn, D.R.; Turnbull, D.M.; Taylor, R.W. Biochemical assays of the respiratory chain complex activity. *Methods Cell Biol.* **2007**, *80*, 93–119. [PubMed]
6. Janssen, A.J.; Smeitink, J.A.; van den Heuval, L.P. Some practical aspects of providing a diagnostic service for respiratory chain defects. *Ann. Clin. Biochem.* **2003**, *40*, 3–8. [CrossRef] [PubMed]
7. Thorburn, D.R. Practical problems in detecting abnormal mitochondrial function and genomes. *Hum. Reprod.* **2015**, *2*, 57–67. [CrossRef]
8. Heales, S.J.R.; Hargreaves, I.P.; Olpin, S.E. Diagnosis of mitochondrial electron transport chain defects in small muscle biopsies. *J. Inherit. Metab. Dis.* **1996**, *19*, 76.
9. Howie, A.J.; Ferreira, M.A.S.; Adu, D. Prognostic value of simple measurement of chronic damage in renal biopsy specimens. *Nephrol. Dial. Transplant.* **2001**, *16*, 1163–1169. [CrossRef] [PubMed]
10. Duberley, K.E.C.; Abramov, A.Y.; Chalasani, A.; Heales, S.J.; Rahman, S.; Hargreaves, I.P. Human neuronal coenzyme Q10 deficiency results in global loss of mitochondrial respiratory chain activity, increased mitochondrial oxidative stress and reversal of ATP synthase activity: Implications for pathogenesis and treatment. *J. Inherit. Metab. Dis.* **2013**, *36*, 63–73. [CrossRef] [PubMed]
11. Selak, M.A.; de Chadarevian, J.P.; Melvin, J.J.; Grover, W.D.; Salganicoff, L.; Kaye, E.M. Mitochondrial activity in Pompe's disease. *Pediatr. Nephrol.* **2000**, *23*, 54–57. [CrossRef]
12. Chretien, D.; Rustin, P.; Bourgeron, T.; Rötig, A.; Saudubray, J.M.; Munnich, A. Reference charts for respiratory chain activities in human tissue. *Clin. Chim. Acta* **1994**, *228*, 53–70. [CrossRef]
13. Scaglia, F. Nuclear gene defects in mitochondrial disorders. *Biochem. Mol. Anal.* **2012**, *837*, 17–34.
14. Berger, A.; Bruschek, M.; Grethen, C.; Sperl, W.; Kofler, B. Poor storage and handling of tissue mimics mitochondrial DNA depletion. *Diagn. Mol. Pathol.* **2001**, *10*, 55–59. [CrossRef] [PubMed]

Journal of
Clinical Medicine

MDPI

Article

Use of FGF-21 as a Biomarker of Mitochondrial Disease in Clinical Practice

Alireza Morovat [1], Gayani Weerasinghe [1], Victoria Nesbitt [2], Monika Hofer [3], Thomas Agnew [4], Geralrine Quaghebeur [5], Kate Sergeant [6], Carl Fratter [6], Nishan Guha [1], Mehdi Mirzazadeh [1] and Joanna Poulton [7,*]

[1] Department of Clinical Biochemistry, Oxford University Hospitals, Oxford OX3 9DU, UK; reza.morovat@ouh.nhs.uk (A.M.); gayani.weerasinghe@ouh.nhs.uk (G.W.); nishan.guha@ouh.nhs.uk (N.G.); mehdi.mirzazadeh@nhs.net (M.M.)
[2] Department of Paediatrics, The Children's Hospital, Oxford OX3 9DU, UK; victoria.nesbitt@ouh.nhs.uk
[3] Department of Neuropathology and Ocular Pathology, West Wing, Oxford University Hospitals, Oxford OX3 9DU, UK; monika.hofer@ouh.nhs.uk
[4] Sir William Dunn School of Pathology, University of Oxford, Oxford OX1 3RE, UK; thomas.agnew@path.ox.ac.uk
[5] Department of Neuroradiology, West Wing, Oxford University Hospitals, Oxford OX3 9DU, UK; gerardine.quaghebeur@ouh.nhs.uk
[6] NHS Specialised Services for Rare Mitochondrial Disorders of Adults and Children UK, Oxford Medical Genetics Laboratories, Oxford University Hospitals, Oxford OX3 7LE, UK; kate.sergeant@ouh.nhs.uk (K.S.); carl.fratter@ouh.nhs.uk (C.F.)
[7] Nuffield Department of Obstetrics and Gynaecology, University of Oxford, Oxford OX3 9DU, UK
* Correspondence: joanna.poulton@obs-gyn.ox.ac.uk; Tel.: +44-(0)1865-221007

Academic Editor: Iain Hargreaves
Received: 19 June 2017; Accepted: 2 August 2017; Published: 21 August 2017

Abstract: Recent work has suggested that fibroblast growth factor-21 (FGF-21) is a useful biomarker of mitochondrial disease (MD). We routinely measured FGF-21 levels on patients who were investigated at our centre for MD and evaluated its diagnostic performance based on detailed genetic and other laboratory findings. Patients' FGF-21 results were assessed by the use of age-adjusted z-scores based on normalised FGF-21 values from a healthy population. One hundred and fifty five patients were investigated. One hundred and four of these patients had molecular evidence for MD, 27 were deemed to have disorders other than MD (non-MD), and 24 had possible MD. Patients with defects in mitochondrial DNA (mtDNA) maintenance ($n = 32$) and mtDNA rearrangements ($n = 17$) had the highest median FGF-21 among the MD group. Other MD patients harbouring mtDNA point mutations ($n = 40$) or mutations in other autosomal genes ($n = 7$) and those with partially characterised MD had lower FGF-21 levels. The area under the receiver operating characteristic curve for distinguishing MD from non-MD patients was 0.69. No correlation between FGF-21 and creatinine, creatine kinase, or cardio-skeletal myopathy score was found. FGF-21 was significantly associated with plasma lactate and ocular myopathy. Although FGF-21 was found to have a low sensitivity for detecting MD, at a z-score of 2.8, its specificity was above 90%. We suggest that a high serum concentration of FGF-21 would be clinically useful in MD, especially in adult patients with chronic progressive external ophthalmoplegia, and may enable bypassing muscle biopsy and directly opting for genetic analysis. Availability of its assay has thus modified our diagnostic pathway.

Keywords: fibroblast growth factor-21; FGF-21; mitochondrial disease; diagnosis

1. Introduction

Mitochondrial disease (MD) is common, with 1 in 400 individuals harbouring the m.3243A>G mutation that usually causes mild disease such as presbycusis [1]. The vast majority of these are never identified, hence the reported prevalence of around 1:20,000 in children and 1:8000 in adults [2,3]. Mitochondrial diseases are notoriously difficult to diagnose because of their heterogeneity, the presence of mitochondrial heteroplasmy, poor genotype-phenotype relationships, and the bluntness of the generally used laboratory tests for assessing mitochondrial dysfunction. The diagnosis of mitochondrial disease often involves an invasive muscle biopsy for studies of mitochondrial function. Respiratory chain enzyme studies are technically demanding and available only in a few specialist centres. Furthermore, they may generate false positive results if tissue samples are poorly preserved [4]. The diagnosis usually requires confirmatory genetic analysis. Conventional blood tests, such as plasma lactate, pyruvate, and creatine kinase (CK), are too insensitive and non-specific to be of significant help in most cases. Depending on the presentation, amino acids, acylcarnitines, and urine organic acids results may help modify the index of suspicion, but their value is very limited.

Recent studies on serum fibroblast growth factor-21 (FGF-21) have suggested that it may offer a much better diagnostic power than conventional serum tests [5–7]. FGF-21 is a hormone-like cytokine that is involved in intermediary metabolism of carbohydrates and lipids [8]. Unlike the mouse, in which hepatic FGF-21 is induced by peroxisomal proliferator-activated receptor-α (PPARα) and in response to fasting [9,10], the function of FGF-21 in humans is still not completely understood. Circulating FGF-21 is mainly hepatic in origin in humans, but the protein is also expressed in adipocytes, myocytes, and the pancreas [11–13]. Some investigators hold that myocytic FGF-21 is induced as part of a protective mechanism against metabolic stress [14,15], although this is controversial [16]. Patients with respiratory chain deficiency have an increased myocytic expression of FGF-21, and treatment of myoblasts with FGF-21 increases PPARδ coactivator-1α (PGC-1α), thereby increasing ATP synthesis through the mammalian target of rapamycin—Yin Yang 1—PPAR gamma coactivator-1α (PGC-1α) pathway [17]. The induction of PGC-1α has lipolytic effects, and systemic FGF-21 improves glucose tolerance by stimulating insulin independent glucose-uptake in human myocytes and adipocytes [18,19]. FGF-21 was shown to increase in mitochondrial dysfunction and myopathy secondary to iron-sulphur cluster deficiency [20]. In trying to elucidate the mechanism linking FGF-21 and mitochondrial stress, other studies have found myocytic FGF-21 expression to be controlled by myogenic factor MyoD and to be driven by mitochondrial reactive oxygen species [21].

In the light of previous data showing that FGF-21 may be useful in a subgroup of patients with mitochondrial disease [5], in 2013 we adopted measurement of FGF-21 as an adjunct to first-line routine laboratory tests. This was aimed at improving diagnosis of patients in whom there was a high index of suspicion for the presence of mitochondrial disease, with a view to reducing the need for muscle biopsy in the assessment of patients. This report describes our findings.

2. Methods

In 2013, the Department of Clinical Biochemistry at the John Radcliffe Hospital in Oxford started offering FGF-21 assay to specialist clinicians for assessing patients who were being investigated for mitochondrial disease. The service was offered based on previous publications suggesting that FGF-21 is a useful marker for mitochondrial disease compared with other routine biochemical tests. We have retrospectively reviewed the FGF-21 data in the light of genetic findings and final diagnoses.

2.1. Individual Populations and Investigations

We reviewed all adult and paediatric patients referred to the Oxford centre for rare mitochondrial disorders at Oxford University Hospitals (OUH) between June 2013 and February 2017. Referrals were either from OUH consultants or from other counties in the United Kingdom. Patients were referred largely from clinical geneticists, neurologists, paediatricians, endocrinologists, and ophthalmologists because of suspected mitochondrial disease. In many cases, the DNA investigations were carried out prior to each individual being seen in the clinic. The referred patients were reviewed by a mitochondrial genetics consultant and her team and were then investigated further, usually while attending an outpatient appointment. Patients' demographics were recorded, and clinical details were reviewed extensively with special attention to the presence of ocular myopathy (ptosis or progressive external ophthalmoplegia (PEO)), cardio-skeletal myopathy, hearing loss, neuropathy, enteropathy, epilepsy, brain-stem signs and symptoms, diabetes, and liver disease.

Following clinical assessments, investigations carried out were routine biochemical tests that included plasma CK, lactate, and FGF-21. A small number of patients had their FGF-21 repeated during follow-up. Depending on clinical presentation, many patients were investigated further with a muscle biopsy that frequently included respiratory chain enzyme studies. Furthermore, genetic testing was offered to all patients, often being undertaken before clinical review in the specialist clinic.

Patients with a confirmed genetic diagnosis of mitochondrial disease (MD) were grouped according to mitochondrial DNA point mutation, single DNA rearrangement, defects in mitochondrial maintenance, partly-characterised mitochondrial myopathy, and other autosomal MD. All (i) patients in whom alternative non-mitochondrial diagnoses (such as inclusion body myositis, optic neuritis and tyrosine hydroxylase deficiency) had been made, (ii) patients who lacked biochemical, molecular, and clinical features of mitochondrial disease and were hence discharged from the clinic, and (iii) unaffected relatives were all classified as non-mitochondrial disease (non-MD) group. Patients with no identifiable genetic mutation but with either (i) muscle biopsies suggestive of mitochondrial disease (respiratory chain function and/or muscle histochemistry) or (ii) where biochemical, molecular, and/or clinic features of mitochondrial disease were sufficient for inclusion as likely mitochondrial disease in the NHS England's 100,000 genome study were classed as "possible-MD".

Data from clinical case notes and electronic records, the laboratory information system, and mitochondrial genetic reports were collected by clinicians directly involved in the care of patients.

Ethics approval and consent were obtained for collecting specimens for FGF-21 measurement from a population of 28 children who had been diagnosed with congenital hypothyroidism but had been on successful treatment and were shown to be euthyroid, and separately from 50 healthy adult volunteers.

2.2. Laboratory Analyses

Routine biochemical tests were performed by the use of verified automated methods. Respiratory chain enzyme studies were performed by the University College London Neurometabolic Unit at the National Hospital for Neurology and Neurosurgery.

Genetic analyses were undertaken under the Highly Specialised NHS Service for Rare Mitochondrial Disorders of Adults and Children, largely by the Molecular Genetics Laboratory at the Churchill Hospital, Oxford, but some specimens were referred to the arms of the service operating in Newcastle or London. This included exome sequencing, which was carried out as a research project funded by Lily Foundation, based at Guy's and St Thomas' NHS Foundation Trust, London, UK. Histopathology on muscle biopsies was performed at the Neuropathology Department at the John Radcliffe Hospital.

Specimens for FGF-21 were centrifuged, and serum was separated and stored at $-80\ °C$ until analysis. FGF-21 was measured by ELISA (BioVendor, Brno, Czech Republic) according to the manufacturer's instructions, with the exception of a reduced incubation time with the substrate (the final step) in order to prevent top calibrants' absorbances reaching values above 2.5 units. We have unpublished data to indicate that FGF-21 is stable for up to three days in unseparated and separated serum, and its assay is unaffected by haemolysis. In our hands, the assay had an inter-batch coefficient of variation (CV) of $\leq 8.0\%$ at concentrations of 108–2642 ng/L.

2.3. Statistics

The Biovendor's FGF-21 kit insert describes FGF-21 mean and SD values from a healthy population. Since these data indicated a skewed distribution, we requested and were kindly supplied with the raw data, from which we were able to normalise them, derive SD, and assess patients' FGF-21 values based on z-scores. Analyse-it software (Analyse-it Software Ltd, Leeds, UK) was used for all statistical analyses. Anderson-Darling A2 normality test was used to assess the healthy population's descriptive data. Logarithmic transformation was found to be the best mode for normalising the data, and was also employed to assess the distribution of values separated into various age bins. Mean and SD of transformed data were calculated. To assess patients' FGF-21 values, they were similarly log-transformed, and their z-scores calculated from the number of SDs above or below the mean for each age bin. In practice, values with z-scores of ≥ 2 were considered clinically significant. Receiver operating characteristic (ROC) curves were used to assess the performance of FGF-21 z-scores for diagnosing MD. Comparisons of values between different groups of patients, and also between kit manufacturer' and Oxford's healthy adult and paediatric populations were made by the Mann-Whitney test. Association between variables was assessed by linear regression.

3. Results

3.1. FGF-21 Reference Values

Serum FGF-21 raw data from a healthy population (76 men and 108 women; age range 2–85 years) were obtained from the manufacturer of FGF-21 kits. The data showed serum concentrations to range from undetectable to 1210 ng/L, with a distribution that had a skewness of 2.01 and a kurtosis of 5.48 ($p < 0.0001$). There was no difference between FGF-21 values in men and women ($p = 0.907$), but FGF-21 showed an increase with age (mean 3.0 ng/L per year (95% CI: 1.4–4.6 ng/L per year); $p < 0.0001$). Dividing values into age groups 21–60 and >60 years gave log-transformed distributions that were not significantly different from normal ($n = 97$, skewness -0.49, $p = 0.056$, and $n = 68$, skewness -0.07, $p = 0.628$, respectively). Log-transformed mean and SD values were 2.21 and 0.383 for the age group 21–60 years, and 2.41 and 0.296 for the age group >60 years. The number of healthy individuals aged ≤ 20 years was only 16, and the addition of these to the 21–60 years age group resulted in a significant negative skew in the log-transformed distribution.

FGF-21 values obtained on Oxford's healthy adults and children were significantly lower than those supplied by the kit manufacturer ($p < 0.0001$ and $p = 0.0013$, respectively). Oxford's healthy adults (age range 19–68 years) and children (age range 1–15 years) had log-transformed mean (SD) values of 1.93 (0.499) and 1.68 (0.422), respectively. In view of this difference and despite fewer numbers of individuals in the Oxford's group, these data were used to establish FGF-21 z-scores for patients' populations.

3.2. Patients' Demographics and Diagnoses

One hundred and eighty four patients were investigated. Of these, 29 patients were omitted from our data analyses as insufficient clinical information related to diagnosis was available. The remaining 155 patients were aged between one day and 87 years at the time of investigations, with mean and median age of 38.0 and 37.5 years, respectively.

Diagnosis of mitochondrial disease was made in 104 patients (MD group) based on DNA data (Table 1). The category of patients with mitochondrial DNA (mtDNA) point mutations was the largest (40 cases), with m.3243A>G mutation comprising the majority (29 patients). The category of mtDNA maintenance defect had 32 cases, including 12 DNA polymerase subunit gamma (*POLG*), 6 ribonucleoside-diphosphate reductase subunit M2 B (*RRM2B*) and 5 *TWNK* gene defects. Single mtDNA rearrangements constituted 17 cases. There were 27 patients, who did not have mitochondrial defect (non-MD group). In nine of these 27, a clearly defined diagnosis was made, and this was mostly a metabolic enzyme defect. In a further 14 of the non-MD patients, a diagnosis of MD was deemed clinically unlikely, and there were a further four individuals in the non-MD group who were unaffected relatives of patients. Finally, there were 24 patients in whom a definitive diagnosis could not be made, but these patients were deemed likely to have MD, often on the basis of muscle biopsy histochemistry (possible-MD group). Patients in the possible-MD group were much younger than those in either the MD or the non-MD groups (median ages 11.0, 39.5, and 45.0 years, respectively; Table 1).

Table 1. Demographics and FGF-21 data in patients diagnosed with mitochondrial disease (MD), those without MD (non-MD), and patients with a possibility of having MD. FGF-21 z-scores are based on log-transformed values. Data on categories of MD, as well as those on patients with and without (cardio) myopathy and ophthalmoplegia (progressive external ophthalmoplegia or ptosis), have been given.

Diagnostic Category	n	Median Age (Range)	Median FGF-21 z-Score (95% C.I.)	FGF-21 z-Score IQR
Non-MD	27	45 (0.5–78)	0.86 (−0.39–1.80)	2.00
Possible MD	24	11 (0–63)	1.87 (1.81–2.78)	2.72
MD	104	39.5 (2–87)	1.72 (1.46–1.99)	1.65
-mtDMA maintenance defects	32	50 (7–87)	1.99 (1.46–2.73)	1.69
-mtDNA rearrangements	17	35 (2–82)	1.99 (1.56–2.76)	1.30
-mt DNA point mutations	40	40 (14–72)	1.40 (0.99–1.94)	1.26
-Partially characterised mitochondrial myopathy	8	22.5 (2–71)	1.64 (−0.47–4.15)	1.88
-Other autosomal MD	7	36 (3–56)	0.90 (0.56–3.56)	1.88
Patients without (cardio)myopathy	*117*	*37 (0–84)*	*1.13 (1.30–1.94)*	*1.85*
Patients with (cardio)myopathy	*39*	*40 (3–87)*	*1.74 (1.52–2.11)*	*1.35*
Patients without ophthalmoplegia	*93*	*35*	*1.33 (0.99–1.88)*	*1.88*
Patients with ophthalmoplegia	*63*	*43*	*1.74 (1.49–2.11)*	*1.51*

3.3. Biochemical Data

The laboratory reported serum FGF-21 concentrations up to a value 5700 ng/L and the exact concentrations above this upper limit were not quantified. Thus, the highest FGF-21 z-scores of 3.67 for adults and 4.91 for children, relate to FGF-21 values of 5700 ng/L. Patients with mitochondrial rearrangements or defects in mitochondrial maintenance had the highest FGF-21 (median z-score of 1.99 for both groups) (Figure 1 and Table 1). On the other hand, patients with defects in mitochondrial dynamics (mostly *OPA1* and *MFN2* gene defects) had the lowest FGF-21 z-scores ($n = 7$; median 0.90; "other autosomal MD" category in Figure 1). Many of the patients with m.3243A>G point mutation had FGF-21 values that were below those of non-MD patients. FGF-21 was neither predictable by any single clinical feature nor by the degree of heteroplasmy in blood.

As a group, patients with MD had a median FGF-21 z-score of 1.72, compared with 0.86 for non-MD patients ($p = 0.0037$). The median FGF-21 z-score for patients with possible MD was 1.87, comparable with the MD group ($p = 0.852$), but these patients also had a wide range of values (Table 1 and Figure 2). Using MD and non-MD patients' FGF-21 z-scores for ROC analysis gave an area under the curve that was 0.68 (Figure 3). A z-score of 2.8 was associated with a specificity of 0.93 but a sensitivity of only 0.20 (Table 2). The assay had positive predictive values that were around 0.90, but poor negative predictive values (Table 2). Given the high median FGF-21 in patients with mtDNA maintenance and rearrangement defects, we assessed the power of the test for discriminating between these patients and the non-MD group.

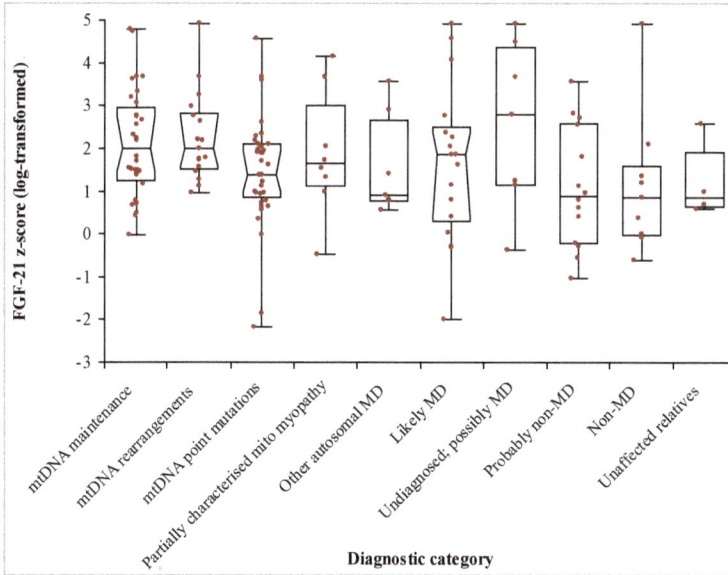

Figure 1. FGF-21 z-scores obtained on specimens from 155 patients investigated for mitochondrial disease. FGF-21 values were normalised by log-transformation, and z-scores were calculated based on age-dependent distributions. Patients were divided into a mitochondrial disease (MD) group consisting of those with mitochondrial DNA or maintenance defects, 30 patients without MD, and 24 patients who were deemed likely to have MD.

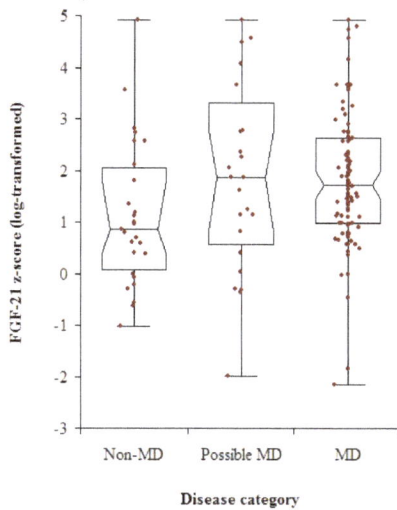

Figure 2. Distribution of FGF-21 z-scores in patients investigated for mitochondrial disease (MD). Patients were grouped according to the final diagnoses, with those who were likely to have MD but whose further investigations were still pending placed in the possible MD group.

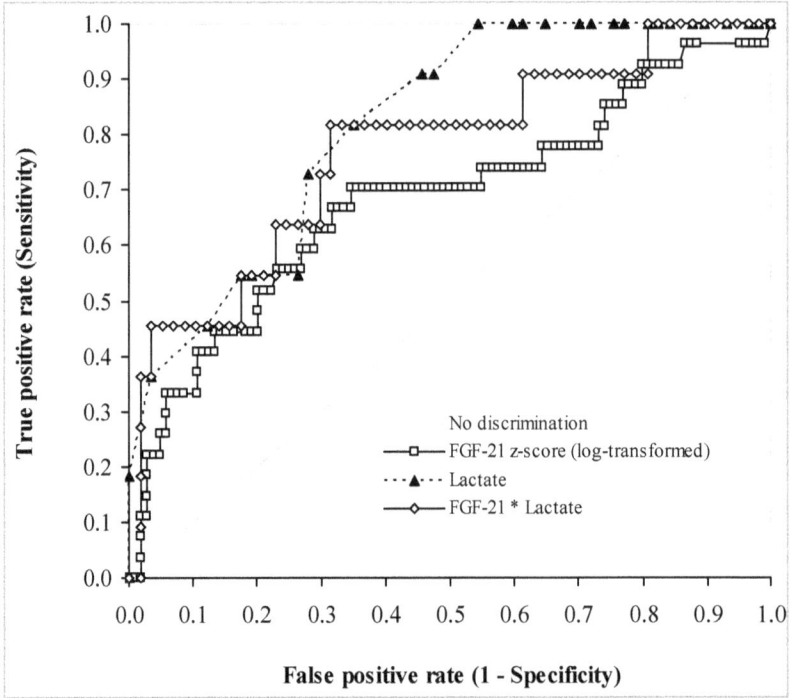

Figure 3. Receiver operating characteristic curves comparing FGF-21 z-scores, lactate concentration, and the product of FGF-21 and lactate concentrations for the diagnosis of mitochondrial disease. The curves were established using data from 104 MD and 27 non-MD patients. See Table 2 for further details on the diagnostic performance of FGF-21 z-scores.

Table 2. Performance of log-transformed FGF-21 z-scores for diagnosing mitochondrial disease. FGF-21 z-scores refer to thresholds. Figures in brackets are 95% confidence intervals. PV is predictive value.

FGF-21 z-Score	Sensitivity	Specificity	Positive PV	Negative PV
1.35	0.65 (0.55–0.74)	0.70 (0.50–0.86)	0.90	0.35
2.82	0.20 (0.13–0.29)	0.93 (0.76–0.99)	0.91	0.23

There were 39 patients in whom myopathy was a prominent feature (assessed clinically as we do not routinely measure respiratory chain activities in adult patients), but there was no difference between the FGF-21 z-scores in these patients and in the ones without myopathy ($p = 0.320$; Table 1). Also, patients who had ptosis and/or PEO also had median FGF-21 z-score that were comparable with those in patients without ocular myopathy ($p = 108$; Table 1). Repeated measurements of FGF-21 in 10 out of 11 patients suggested that FGF-21 concentrations were fairly stable, with a good correlation between repeat measures ($p < 0.001$). For these, repeat FGF-21 measures had CVs that ranged between 3.1% and 68.1% (median 18.0%), with the highest CV relating to duplicate measures that were 90 and 257 ng/L. However, in the case of a child with tyrosine hydroxylase deficiency in whom FGF-21 was >5700 ng/L, a repeat measurement a year later showed FGF-21 to be 70 ng/L. One other patient from whom five specimens were collected had serum FGF-21 concentrations that ranged from 227 to 421 ng/L (CV of 29.0%). In a further patient whose clinical conditions fluctuated, three FGF-21 measurements showed changes with results that reflected the clinical status of the patient (FGF-21 range of 800 to 1595 ng/L).

There was no correlation between FGF-21 and plasma creatinine ($p = 0.835$). CK was measured in 66 patients, and there was no relationship between it and FGF-21 z-scores ($p = 0.239$). There was no difference between CK values in patients with and without (cardio) myopathy ($p = 0.084$) or patients with and without ophthalmoplegia ($p = 0.275$). Lactate was measured in 80 patients, 57 of whom had MD with median (IQR) lactate concentrations of 1.50 (1.23) mmol/L (range 0.7–6.5 mmol/L), compared with only 11 non-MD lactates that had a median (IQR) of 0.90 (0.48) mmol/L (range 0.5–1.5 mmol/L) (Figure 4). There was a significant positive association between FGF-21 z-scores and lactate ($p < 0.0001$; Figure 5). Patients without MD all had plasma lactate concentrations below 1.5 mmol/L, well within lactate reference interval, with a plasma lactate of 1.5 mmol/L displaying a sensitivity of 46% for MD. Figure 3 compares the ROC curves for FGF-21 z-score, lactate and the product of FGF-21 and lactate concentrations. There was no significant difference between the areas under the ROC curves, although both lactate and the arithmetic product of FGF-21 and lactate concentrations gave higher areas of 0.81 and 0.77, respectively. Amongst all patients investigated, the presence of (cardio) myopathy was associated with a higher plasma lactate (median (IQR) of 1.9 (1.5) mmol/L; $p = 0.021$), and so was ophthalmoplegia (median (IQR) of 1.5 (1.1) mmol/L; $p = 0.047$) compared with lactate in other patients (median (IQR) of 1.2 (0.6) mmol/L for both groups of patients without (cardio) myopathy and those without ophthalmoplegia).

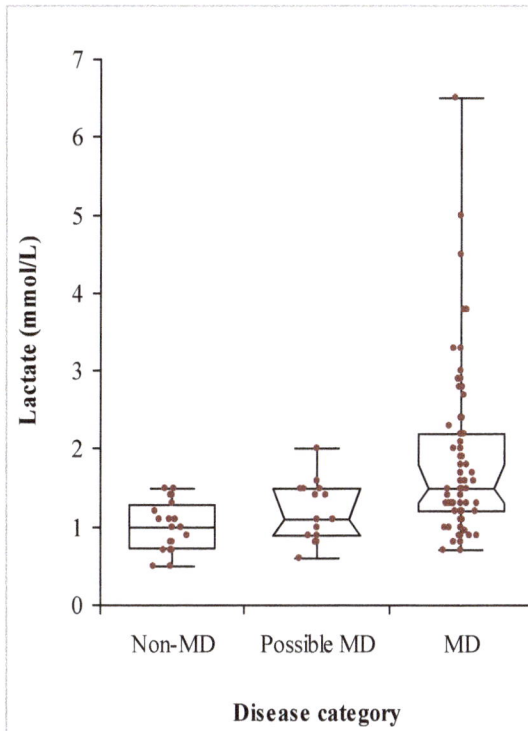

Figure 4. Distribution of plasma lactate concentrations in patients investigated for mitochondrial disease (MD). Patients were grouped according to the final diagnoses, with those who were likely to have MD by further investigations were still pending placed in the possible MD group.

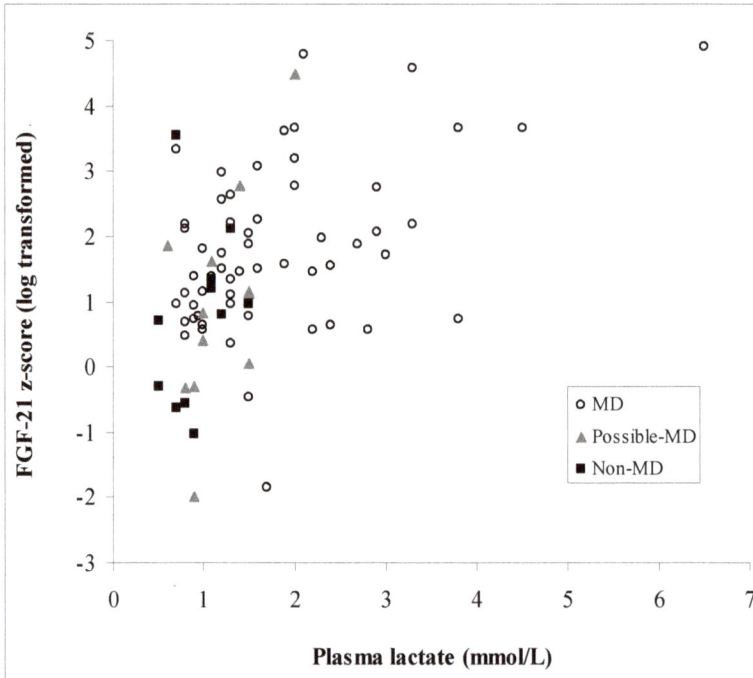

Figure 5. Relationship between log-transformed FGF-21 z-scores and plasma lactate in 80 patients (57 MD, 12 possible MD, and 11 non-MD), on whose specimens lactate values were available. A formula of FGF-21 z-score = 0.53x lactate + 0.84 describes the relationship between the two measures (R^2 = 0.205).

4. Discussion

In search of better biochemical markers for MD, a few previous studies have suggested serum FGF-21 to have adequate sensitivity and specificity for it to be employed in MD diagnostic pathway, and for its values to correlate with disease severity [5–7,22]. In order to measure FGF-21 for assessing referred patients, we employed the same ELISA kit used by previous studies. We studied two sets of healthy populations' FGF-21 data: one largely adult from the kit manufacturer, and a smaller one from local populations of adults and children. Differences between the results from the two sets of population made us use only the locally derived mean and SD values from normalised data, but this had limitations. Thus, although based on the external data, FGF-21 is affected by age, and the local population was too small to allow us to derive age-related values in adults. Nevertheless, based on separate log-normalised distributions, we used z-scores to assess the magnitude of increase in serum FGF-21 of patients.

We found serum FGF-21 concentrations to be higher in patients with MD compared with others. Those with mtDNA maintenance defects had the highest FGF-21, followed by patients with mtDNA rearrangements. Overall, however, there was a significant degree of overlap in FGF-21 values between MD and non-MD patients, such that FGF-21 z-score had an ROC curve area of 0.68. Patients with either mtDNA maintenance defects or mtDNA rearrangement had higher FGF-21. In this respect, our findings are consistent with those of the Finnish group [5], who found that serum FGF-21 discriminated most effectively from controls those patients with defects in mtDNA maintenance and single rearrangements. In particular, we did not find increased FGF-21 in the majority of our m.3243A>G patients, with only 6/29 (21%) having a z-score of over 2.0. This may reflect the contrast between our m.3243A>G patient population, which had mainly maternally inherited diabetes and deafness, and the previously published populations that were based on neurological disorders [23]. Nevertheless, although our data show that FGF-21 improves the biochemical investigation of patients suspected of having MD, they do not display the diagnostic powers that have been reported previously. Furthermore, FGF-21 was less useful in patients with non-myopathic disease affecting a single system, such as mitochondrial optic neuropathies.

We also found that plasma lactate displayed an ROC curve that was at least as good as that of FGF-21 z-scores. Patients with myopathy also had higher lactate concentrations than others. However, the highest lactate concentration in the non-MD group was 1.5 mmol/L. Although this conferred a better diagnostic performance in our cohort, giving a specificity of 100% and a sensitivity of 46% for detecting MD, this was based on a small population of only 11 non-MD patients. Furthermore, unlike FGF-21 z-score of 2.8, which is well above the upper limit of normal values, a lactate of 1.5 mmol/L is well within the reference interval (up to 2.2 mmol/L), potentially posing problems in clinical practice with a true specificity that would be lower than seen here. Nevertheless, given the diagnostic use of plasma lactate, we assessed the power of the product of FGF-21 and lactate concentrations for detecting MD. However, this did not show any superiority over either analyte.

We measured FGF-21 routinely and in a non-discriminatory fashion in our patients and found several non-MD patients to have high FGF-21 values. FGF-21 is not specific to MD and has been shown to increase in many other conditions, including obesity, diabetes, fatty liver disease, and metabolic syndrome [24–26]. In contrast to mice [27], starvation and ketogenic states do not appear to increase FGF-21 significantly in humans [9], and we did not observe any association between serum FGF-21 and renal function, as has been described previously [28]. To what extent our data in non-MD patients has been influenced by such physiological factors is unknown. Furthermore, since FGF-21 increases significantly after fructose consumption [29], we propose that a more standardised approach to sample collection that records relevant clinical parameters and pre-analytical variables, such as feeding state, would be needed.

Other investigators have been able to demonstrate a positive correlation between FGF-21 levels and the Newcastle mitochondrial disease scale for adults (NMDAS) clinical assessment scale [30]. We do not routinely use NMDAS score and hence could not assess this. However, given that the majority of the individuals with both proven MD and high FGF-21 levels have ocular myopathies (100% and 87% of patients who were seen in the clinic with single rearrangements and defects of mtDNA maintenance, respectively), the test may be less sensitive than skilled clinical examinations.

Although FGF-21 has been described as a "mitokine", its physiological role is uncertain. Under most physiological conditions the liver is the main contributor to circulating levels, followed by adipose tissue [31]. However, cardiac stresses may increase serum FGF-21 both in human cardiovascular disease [14] and in animal models of progressive MD [32,33]. The increase in circulating FGF-21 associated with human MD is believed to be a response to stress, originating from liver, adipose tissues, and skeletal muscle [34,35]. Although we did not measure respiratory chain activities in all patients, our data did not show any significant difference between FGF-21 values in patients with and without clinical evidence of significant myopathy. For example, we found an FGF-21 z-score of only 0.56 in a patient with complex-I deficiency due to mutations in the gene encoding an assembly factor, known as acyl-CoA dehydrogenase family member 9 (ACAD9). However, we found patients with ophthalmoplegia to have higher FGF-21 concentrations. Studies in mice suggest that mitochondrial dysfunction results in an increase in serum FGF-21 that is distinct from the PPARα-dependent response to starvation. Therefore, based on previous studies in mice, we suggest that FGF-21 increase may depend upon activating transcription factor 4 and the integrated stress response (ISR) [32,36], which itself is influenced by many factors including nutritional status and potentially the type of MD [22]. It is increasingly clear that several different mitochondrial stresses, including the mitochondrial unfolded protein response [37], mtDNA depletion, and inhibition of mitochondrial translation [38], can activate ISR. The type of mitochondrial stress caused by specific molecular defects that ultimately activate ISR-dependent FGF-21 expression may therefore underlie the variability of plasma FGF-21 levels observed in the MD group. Furthermore, its rather indirect relationship to MD limits its usefulness as a biomarker for the disease.

In general, we tended to find relatively consistent FGF-21 values in duplicate samples. Anecdotally, in two patients undergoing intensive care, the highest values were associated with respiratory deterioration (but without lactic acidosis), consistent with the high serum FGF-21 concentrations in patients undergoing intensive care [15]. Given the known relationship between FGF-21 and body weight [39], this type of response may play an important role both in progressive weight loss that is frequently observed in MD [40] and in the poor response of patients to critical care. Further investigation is needed in order to determine whether any apparent longitudinal change in serum FGF-21 may be related to the clinical course of the disease. If so, measuring FGF-21 may prove to be more useful for assessing disease progression and for the monitoring of the effects of therapies, rather than as a diagnostic biomarker. If having high FGF-21 levels benefit mitochondrial patients, the FGF-21 analogue that has been trialled as a drug treatment for hyperglycaemia and obesity [41,42] would also have potential as a therapy for MD.

In conclusion, our experience has shown that in clinical practice, FGF-21 measurements would be helpful in only around one-third of patients who are investigated in a specialist MD clinic. In our hands, a z-score of above 2.8 has given a specificity and a sensitivity of 93% and 21%, respectively. This indicates that although the test may not be used for the diagnosis of MD, its availability would affect the diagnostic pathway. In many patients, finding a high FGF-21 would suggest MD, and a targeted diagnostic approach in such patients would be preferable to next-generation sequencing. In the past, we used muscle biopsy to triage those patients needing detailed genetic investigation. Now, based on high serum FGF-21 values, we are able to identify those adult patients that would benefit from genetic investigations without prior muscle biopsy. The availability of this assay has thus modified our diagnostic pathway in adults, whereas muscle biopsy remains more important in children [43].

FGF-21 is now a useful adjunct to our clinical examination. Furthermore, because it is stable in samples at room temperature, we can analyse specimens sent in from peripheral clinics. However, further work is needed to determine whether it is useful in children, as currently there is insufficient reference data, and we saw a high proportion of false positives in this age group. Our clinic cohort was limited by the low number of disease controls having either non-mitochondrial myopathies or conditions that secondarily affect mitochondria. Further data are also needed to determine whether the apparent longitudinal changes in FGF-21 are related to the clinical course. Finally, growth and differentiation factor 15 (GDF-15) was recently identified in transcriptosome profiling and has been found to be a biomarker for MD. GDF-15 is held to have a superior sensitivity but a lower specificity compared with FGF-21 [22,44]. It remains to be seen whether combining FGF-21 with GDF-15 and conventional lactate measurements offers any advantage in stratifying patients investigated for MD.

Acknowledgments: We would like to thank the patients and their carers for participating in this study and the Neurometabolic Unit at the National Hospital for Neurology and Neurosurgery in London for respiratory chain analyses. J.P., C.F., K.S., and N.G. receive salary support from the UK Rare Mitochondrial Disorders of Adults and Children (NHS Highly Specialized Services). J.P. acknowledges grant support from the Lily Foundation, NewLife (SG/14-15/11), the Medical Research Council (MR/J010448/1) and the Wellcome Trust (0948685/Z/10/Z). The NHS and sponsors had no role in the design of study or interpretation of results.

Author Contributions: J.P., M.M. and A.M. conceived and designed the study; G.W., A.M., C.F. and K.S. performed the laboratory analyses; A.M., J.P. and G.W. analyzed the data; M.M., N.G., V.N., M.H., G.Q. contributed to patient care; A.M., J.P., T.A., M.M. and G.W. wrote the paper.

Conflicts of Interest: The authors declare no conflict of interest.

References

1. Manwaring, N.; Jones, M.M.; Wang, J.J.; Rochtchina, E.; Howard, C.; Mitchell, P.; Sue, C.M. Population prevalence of the MELAS A3243G mutation. *Mitochondrion* **2007**, *7*, 230–233. [CrossRef] [PubMed]
2. Skladal, D.; Halliday, J.; Thorburn, D.R. Minimum birth prevalence of mitochondrial respiratory chain disorders in children. *Brain* **2003**, *126*, 1905–1912. [CrossRef] [PubMed]

3. Gorman, G.S.; Schaefer, A.M.; Ng, Y.; Gomez, N.; Blakely, E.L.; Alston, C.L.; Feeney, C.; Horvath, R.; Yu-Wai-Man, P.; Chinnery, P.F.; et al. Prevalence of nuclear and mitochondrial DNA mutations related to adult mitochondrial disease. *Ann. Neurol.* **2015**, *77*, 753–759. [CrossRef] [PubMed]
4. Oglesbee, D.; Freedenberg, D.; Kramer, K.A.; Anderson, B.D.; Hahn, S.H. Normal muscle respiratory chain enzymes can complicate mitochondrial disease diagnosis. *Pediatr. Neurol.* **2006**, *35*, 289–292. [CrossRef] [PubMed]
5. Suomalainen, A.; Elo, J.M.; Pietilainen, K.H.; Hakonen, A.H.; Sevastianova, K.; Korpela, M.; Isohanni, P.; Marjavaara, S.K.; Tyni, T.; Kiuru-Enari, S.; et al. FGF-21 as a biomarker for muscle-manifesting mitochondrial respiratory chain deficiencies: A diagnostic study. *Lancet Neurol.* **2011**, *10*, 806–818. [CrossRef]
6. Suomalainen, A. Fibroblast growth factor 21: A novel biomarker for human muscle-manifesting mitochondrial disorders. *Expert Opin. Med. Diagn.* **2013**, *7*, 313–317. [CrossRef] [PubMed]
7. Davis, R.L.; Liang, C.; Edema-Hildebrand, F.; Riley, C.; Needham, M.; Sue, C.M. Fibroblast growth factor 21 is a sensitive biomarker of mitochondrial disease. *Neurology* **2013**, *81*, 1819–1826. [CrossRef] [PubMed]
8. Woo, Y.C.; Xu, A.; Wang, Y.; Lam, K.S. Fibroblast growth factor 21 as an emerging metabolic regulator: Clinical perspectives. *Clin. Endocrinol.* **2013**, *78*, 489–496. [CrossRef] [PubMed]
9. Christodoulides, C.; Dyson, P.; Sprecher, D.; Tsintzas, K.; Karpe, F. Circulating fibroblast growth factor 21 is induced by peroxisome proliferator-activated receptor agonists but not ketosis in man. *J. Clin. Endocrinol. Metab.* **2009**, *94*, 3594–3601. [CrossRef] [PubMed]
10. Lundåsen, T.; Hunt, M.C.; Nilsson, L.M.; Sanyal, S.; Angelin, B.; Alexson, S.E.; Rudling, M. PPARalpha is a key regulator of hepatic FGF21. *Biochem. Biophys. Res. Commun.* **2007**, *360*, 437–440. [CrossRef] [PubMed]
11. Nishimura, T.; Nakatake, Y.; Konishi, M.; Itoh, N. Identification of a novel FGF, FGF-21, preferentially expressed in the liver. *Biochim. Biophys. Acta* **2000**, *1492*, 203–206. [CrossRef]
12. Fisher, F.M.; Maratos-Flier, E. Understanding the physiology of FGF21. *Annu. Rev. Physiol.* **2016**, *78*, 223–241. [CrossRef] [PubMed]
13. Hojman, P.; Pedersen, M.; Nielsen, A.R.; Krogh-Madsen, R.; Yfanti, C.; Akerstrom, T.; Nielsen, S.; Pedersen, B.K. Fibroblast growth factor-21 is induced in human skeletal muscles by hyperinsulinemia. *Diabetes* **2009**, *58*, 2797–2801. [CrossRef] [PubMed]
14. Planavila, A.; Redondo-Angulo, I.; Ribas, F.; Garrabou, G.; Casademont, J.; Giralt, M.; Villarroya, F. Fibroblast growth factor 21 protects the heart from oxidative stress. *Cardiovasc. Res.* **2015**, *106*, 19–31. [CrossRef] [PubMed]
15. Thiessen, S.E.; Vanhorebeek, I.; Derese, I.; Gunst, J.; Van den Berghe, G. FGF21 Response to critical illness: Effect of blood glucose control and relation with cellular stress and survival. *J. Clin. Endocrinol. Metab.* **2015**, *100*, E1319–E1327. [CrossRef] [PubMed]
16. Ost, M.; Coleman, V.; Voigt, A.; van Schothorst, E.M.; Keipert, S.; van der Stelt, I.; Ringel, S.; Graja, A.; Ambrosi, T.; Kipp, A.P.; et al. Muscle mitochondrial stress adaptation operates independently of endogenous FGF21 action. *Mol. Metab.* **2015**, *5*, 79–90. [CrossRef] [PubMed]
17. Ji, K.; Zheng, J.; Lv, J.; Xu, J.; Ji, X.; Luo, Y.B.; Li, W.; Zhao, Y.; Yan, C. Skeletal muscle increases FGF21 expression in mitochondrial disorders to compensate for energy metabolic insufficiency by activating the mTOR-YY1-PGC1α pathway. *Free Radic. Biol. Med.* **2015**, *84*, 161–170. [CrossRef] [PubMed]
18. Mashili, F.L.; Austin, R.L.; Deshmukh, A.S.; Fritz, T.; Caidahl, K.; Bergdahl, K.; Zierath, J.R.; Chibalin, A.V.; Moller, D.E.; Kharitonenkov, A.; et al. Direct effects of FGF21 on glucose uptake in human skeletal muscle: Implications for type 2 diabetes and obesity. *Diabetes Metab. Res. Rev.* **2011**, *27*, 286–297. [CrossRef] [PubMed]
19. Lee, D.V.; Li, D.; Yan, Q.; Zhu, Y.; Goodwin, B.; Calle, R.; Brenner, M.B.; Talukdar, S. Fibroblast growth factor 21 improves insulin sensitivity and synergizes with insulin in human adipose stem cell-derived (hASC) adipocytes. *PLoS ONE* **2014**, *9*, e111767. [CrossRef] [PubMed]
20. Crooks, D.R.; Natarajan, T.G.; Jeong, S.Y.; Chen, C.; Park, S.Y.; Huang, H.; Ghosh, M.C.; Tong, W.H.; Haller, R.G.; Wu, C.; et al. Elevated FGF21 secretion, PGC-1α and ketogenic enzyme expression are hallmarks of iron-sulfur cluster depletion in human skeletal muscle. *Hum. Mol. Genet.* **2014**, *23*, 24–39. [CrossRef] [PubMed]
21. Ribas, F.; Villarroya, J.; Hondares, E.; Giralt, M.; Villarroya, F. FGF21 expression and release in muscle cells: Involvement of MyoD and regulation by mitochondria-driven signalling. *Biochem. J.* **2014**, *463*, 191–199. [CrossRef] [PubMed]

22. Lehtonen, J.M.; Forsström, S.; Bottani, E.; Viscomi, C.; Baris, O.R.; Isoniemi, H.; Höckerstedt, K.; Österlund, P.; Hurme, M.; Jylhävä, J.; et al. FGF21 is a biomarker for mitochondrial translation and mtDNA maintenance disorders. *Neurology* **2016**, *87*, 2290–2299. [CrossRef] [PubMed]

23. Koene, S.; de Laat, P.; van Tienoven, D.H.; Vriens, D.; Brandt, A.M.; Sweep, F.C.; Rodenburg, R.J.; Donders, A.R.; Janssen, M.C.; Smeitink, J.A. Serum FGF21 levels in adult m.3243A>G carriers: Clinical implications. *Neurology* **2014**, *83*, 125–133. [CrossRef] [PubMed]

24. Li, H.; Dong, K.; Fang, Q.; Hou, X.; Zhou, M.; Bao, Y.; Xiang, K.; Xu, A.; Jia, W. High serum level of fibroblast growth factor 21 is an independent predictor of non-alcoholic fatty liver disease: A 3-year prospective study in China. *J. Hepatol.* **2013**, *58*, 557–563. [CrossRef] [PubMed]

25. Dushay, J.; Chui, P.C.; Gopalakrishnan, G.S.; Varela-Rey, M.; Crawley, M.; Fisher, F.M.; Badman, M.K.; Martinez-Chantar, M.L.; Maratos-Flier, E. Increased fibroblast growth factor 21 in obesity and nonalcoholic fatty liver disease. *Gastroenterology* **2010**, *139*, 456–463. [CrossRef] [PubMed]

26. Tyynismaa, H.; Raivio, T.; Hakkarainen, A.; Ortega-Alonso, A.; Lundbom, N.; Kaprio, J.; Rissanen, A.; Suomalainen, A.; Pietiläinen, K.H. Liver fat but not other adiposity measures influence circulating FGF21 levels in healthy young adult twins. *J. Clin. Endocrinol. Metab.* **2011**, *96*, E351–E355. [CrossRef] [PubMed]

27. Badman, M.K.; Pissios, P.; Kennedy, A.R.; Koukos, G.; Flier, J.S.; Maratos-Flier, E. Hepatic fibroblast growth factor 21 is regulated by PPARalpha and is a key mediator of hepatic lipid metabolism in ketotic states. *Cell Metab.* **2007**, *5*, 426–437. [CrossRef] [PubMed]

28. Lin, Z.; Zhou, Z.; Liu, Y.; Gong, Q.; Yan, X.; Xiao, J.; Wang, X.; Lin, S.; Feng, W.; Li, X. Circulating FGF21 levels are progressively increased from the early to end stages of chronic kidney diseases and are associated with renal function in Chinese. *PLoS ONE* **2011**, *6*, e18398. [CrossRef] [PubMed]

29. Dushay, J.R.; Toschi, E.; Mitten, E.K.; Fisher, F.M.; Herman, M.A.; Maratos-Flier, E. Fructose ingestion acutely stimulates circulating FGF21 levels in humans. *Mol. Metab.* **2014**, *4*, 51–57. [CrossRef] [PubMed]

30. Davis, R.L.; Liang, C.; Sue, C.M. A comparison of current serum biomarkers as diagnostic indicators of mitochondrial diseases. *Neurology* **2016**, *86*, 2010–2015. [CrossRef] [PubMed]

31. Planavila, A.; Redondo-Angulo, I.; Villarroya, F. FGF21 and Cardiac Physiopathology. *Front. Endocrinol.* **2015**, *6*, 133–139. [CrossRef] [PubMed]

32. Dogan, S.A.; Pujol, C.; Maiti, P.; Kukat, A.; Wang, S.; Hermans, S.; Senft, K.; Wibom, R.; Rugarli, E.I.; Trifunovic, A. Tissue-specific loss of DARS2 activates stress responses independently of respiratory chain deficiency in the heart. *Cell Metab.* **2014**, *19*, 458–469. [CrossRef] [PubMed]

33. Tyynismaa, H.; Carroll, C.J.; Raimundo, N.; Ahola-Erkkilä, S.; Wenz, T.; Ruhanen, H.; Guse, K.; Hemminki, A.; Peltola-Mjøsund, K.E.; Tulkki, V.; et al. Mitochondrial myopathy induces a starvation-like response. *Hum. Mol. Genet.* **2010**, *19*, 3948–3958. [CrossRef] [PubMed]

34. Luo, Y.; McKeehan, W.L. Stressed liver and muscle call on adipocytes with FGF21. *Front. Endocrinol.* **2013**, *4*, 194. [CrossRef] [PubMed]

35. Jeanson, Y.; Ribas, F.; Galinier, A.; Arnaud, E.; Ducos, M.; André, M.; Chenouard, V.; Villarroya, F.; Casteilla, L.; Carrière, A. Lactate induces FGF21 expression in adipocytes through a p38-MAPK pathway. *Biochem. J.* **2016**, *473*, 685–692. [CrossRef] [PubMed]

36. Kim, K.H.; Jeong, Y.T.; Oh, H.; Kim, S.H.; Cho, J.M.; Kim, Y.N.; Kim, S.S.; Kim, D.H.; Hur, K.Y.; Kim, H.K.; et al. Autophagy deficiency leads to protection from obesity and insulin resistance by inducing FGF21 as a mitokine. *Nat. Med.* **2013**, *19*, 83–92. [CrossRef] [PubMed]

37. Rath, E.; Berger, E.; Messlik, A.; Nunes, T.; Liu, B.; Kim, S.C.; Hoogenraad, N.; Sans, M.; Sartor, R.B.; Haller, D. Induction of dsRNA-activated protein kinase links mitochondrial unfolded protein response to the pathogenesis of intestinal inflammation. *Gut* **2012**, *61*, 1269–1278. [CrossRef] [PubMed]

38. Michel, S.; Canonne, M.; Arnould, T.; Renard, P. Inhibition of mitochondrial genome expression triggers the activation of CHOP-10 by a cell signaling dependent on the integrated stress response but not the mitochondrial unfolded protein response. *Mitochondrion* **2015**, *21*, 58–68. [CrossRef] [PubMed]

39. Gómez-Ambrosi, J.; Gallego-Escuredo, J.M.; Catalán, V.; Rodríguez, A.; Domingo, P.; Moncada, R.; Valentí, V.; Salvador, J.; Giralt, M.; Villarroya, F.; et al. FGF19 and FGF21 serum concentrations in human obesity and type 2 diabetes behave differently after diet- or surgically-induced weight loss. *Clin. Nutr.* **2017**, *36*, 861–868. [CrossRef] [PubMed]

40. Wolny, S.; McFarland, R.; Chinnery, P.; Cheetham, T. Abnormal growth in mitochondrial disease. *Acta Paediatr.* **2009**, *98*, 553–554. [CrossRef] [PubMed]

41. Sonoda, J.; Chen, M.Z.; Baruch, A. FGF21-receptor agonists: An emerging therapeutic class for obesity-related diseases. *Horm. Mol. Biol. Clin. Investig.* 2017. [CrossRef] [PubMed]
42. Gaich, G.; Chien, J.Y.; Fu, H.; Glass, L.C.; Deeg, M.A.; Holland, W.L.; Kharitonenkov, A.; Bumol, T.; Schilske, H.K.; Moller, D.E. The effects of LY2405319, an FGF21 analog, in obese human subjects with type 2 diabetes. *Cell Metab.* **2013**, *18*, 333–340. [CrossRef] [PubMed]
43. Phadke, R. Myopathology of adult and paediatric mitochondrial diseases. *J. Clin. Med.* **2017**, *6*, 64. [CrossRef] [PubMed]
44. Kalko, S.G.; Paco, S.; Jou, C.; Rodriguez, M.A.; Meznaric, M.; Rogac, M.; Jekovec-Vrhovsek, M.; Sciacco, M.; Moggio, M.; Fagiolari, G.; et al. Transcriptomic profiling of TK2 deficient human skeletal muscle suggests a role for the p53 signalling pathway and identifies growth and differentiation factor-15 as a potential novel biomarker for mitochondrial myopathies. *BMC Genomics* **2014**, *15*, 91. [CrossRef] [PubMed]

Journal of
Clinical Medicine

MDPI

Article

An Effective, Versatile, and Inexpensive Device for Oxygen Uptake Measurement

Paule Bénit [1,2], Dominique Chrétien [1,2], Mathieu Porceddu [3], Constantin Yanicostas [1,2], Malgorzata Rak [1,2] and Pierre Rustin [1,2,*]

[1] INSERM UMR 1141, Hôpital Robert Debré, 75019 Paris, France; paule.benit@inserm.fr (P.B.); dominique.chretien@inserm.fr (D.C.); constantin.yanicostas@inserm.fr (C.Y.); malgorzata.rak@inserm.fr (M.R.)
[2] Faculté de Médecine Denis Diderot, Université Paris Diderot—Paris 7, Site Robert Debré, 75013 Paris, France
[3] MITOLOGICS S.A.S. Hôpital Robert Debré, 48 Bd Sérurier, 75019 Paris, France; mporceddu@mitologics.com
* Correspondence: pierre.rustin@inserm.fr; Tel.: +33-1-4003-1989

Academic Editor: Iain P. Hargreaves
Received: 3 April 2017; Accepted: 6 June 2017; Published: 8 June 2017

Abstract: In the last ten years, the use of fluorescent probes developed to measure oxygen has resulted in several marketed devices, some unreasonably expensive and with little flexibility. We have explored the use of the effective, versatile, and inexpensive Redflash technology to determine oxygen uptake by a number of different biological samples using various layouts. This technology relies on the use of an optic fiber equipped at its tip with a membrane coated with a fluorescent dye (www.pyro-science.com). This oxygen-sensitive dye uses red light excitation and lifetime detection in the near infrared. So far, the use of this technology has mostly been used to determine oxygen concentration in open spaces for environmental studies, especially in aquatic media. The oxygen uptake determined by the device can be easily assessed in small volumes of respiration medium and combined with the measurement of additional parameters, such as lactate excretion by intact cells or the membrane potential of purified mitochondria. We conclude that the performance of by this technology should make it a first choice in the context of both fundamental studies and investigations for respiratory chain deficiencies in human samples.

Keywords: respiration assay; oxygen uptake; glycolysis; mitochondriopathy

---

1. Introduction

A number of cell functions directly rely on the capacity of mitochondria to utilize oxygen through the mitochondrial respiratory chain (RC) [1]. Accordingly, in a number of clinical conditions, measuring the ability of mitochondria to use oxygen can shed light on the disease mechanism and/or help in establishing a diagnosis [2]. An impaired capacity for oxygen uptake is in particular observed in most primary mitochondriopathies of genetic origin. These are relatively rare disorders, but encompass numerous medical specialties [3]. In addition, both primary and secondary impairments of mitochondrial function are now regarded as instrumental in the course of a set of common diseases, including different cancers [4] and age-related neurodegenerative diseases [5]. This comes as no surprise given the role of mitochondria as a crucial turntable for the overall cell metabolism, acting as determining actor for cell differentiation, proliferation, and death. Finally, mitochondria represent a cellular sink for numerous toxins [6] to which organisms are exposed, potentially affecting their own function [7].

Significant defects of the RC generally result in most tissues in an elevation of the redox status of the matrix pyridine nucleotides and a reduced capacity of mitochondria to oxidize pyruvate [8]. This unused pyruvate is instead reduced to lactate by cytosolic lactate dehydrogenase and is excreted from the cells. Accordingly, the suspicion of an RC defect can be reinforced by the demonstration of abnormal acetoacetate/hydroxybutyrate (tracing the redox status of the mitochondrial pyridine nucleotide pool) and lactate/pyruvate ratios in the body fluids [2].

An impairment of the RC activity might also reduce mitochondrial ATP production. Under these circumstances, as to match a cellular unsatisfied demand for ATP, an activation of glycolysis, an alternative way to produce ATP (yet less efficient than the RC), will take place producing both ATP and pyruvate, thus again favoring lactate production and excretion [9].

Starting with the pioneer work of Otto Heinrich Warburg during the last century [10], a number of devices have been developed to quantify the capacity of biological samples to consume oxygen. Successively using gas pressure in a closed chamber (Warburg apparatus [10]), oxygen-dependent current flow at the surface of an electrode (Clark oxygen electrode [11]), or oxygen-sensing fluorophore (oxygen extracellular fluxes; Seahorse technology [12]), methods have substantially increased in sensitivity, reducing volumes to be used from several milliliters to a few tens of microliters; however, price varied inversely, from a few to now more than €150,000. As a sensitive, versatile, and cheap alternative, we describe here the use of the Redflash technology (FireSting O2; PyroScience; Aachen, Germany) to measure oxygen uptake by various biological systems in an aqueous medium. The method measures the luminescence of an oxygen-sensitive sensor molecule covalently attached to a polymer membrane, which covers the tip of an optic fiber connected to a PC-controlled meter (Figure 1). The luminescence measurement uses red light excitation and lifetime detection in the near infrared. This represents a quite sensitive, very low-cost alternative in terms of quantifying oxygen uptake by intact cells or isolated mitochondria.

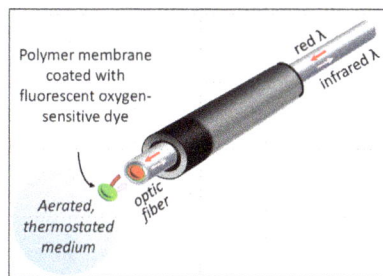

Figure 1. The optode device. The tip of the optic fiber (in red) is covered by a polymer membrane coated with a fluorescent oxygen-sensitive dye (in green) fixed to the optic fiber with silicone glue. The optic fiber receives red light (excitation) from, and re-emits infrared light (emission) to, an analyzer box that can be connected to a personal computer. The fluorescence of the dye is proportional to its oxygen-dependent oxidation state, which is fully reversible. Time-dependent variation of the infrared emission reflects variation of the oxygen at the membrane surface. By inserting the tip of the optic fiber into any aerated medium, it appears possible to determine at any time the oxygen tension in the medium and so to estimate oxygen consumption in the medium.

In addition, thanks to the convenient flexibility offered by the optic fiber, this device was fitted to the cuvette of a spectrophoto- or spectrofluorometer, allowing for concurrent measurement of oxygen uptake plus an additional optical signal. Using such a configuration, it was possible to concomitantly and continuously measure mitochondrial substrate oxidation and membrane potential, or cell respiration and glycolysis (specifically through lactate excreted).

2. Material and Methods

2.1. Zebra Fish Embryos

Zebrafish (*Danio rerio*) stocks of the wild-type AB strain were maintained at 28 °C in a standard zebrafish facility (Aquatic Habitat, Pentair, Minneapolis, MN, USA). Embryos were collected by natural spawning and raised under a standard 14:10 h light/dark photoperiod [13]. Developmental stages were determined as days post-fertilization (dpf), as described [14]. Three-day-old embryos were used throughout this study.

2.2. Rat Liver Mitochondria

Liver mitochondria from 6-week-old Wistar Han IGS female rats (Charles River, Saint-Germain-sur-l'Arbresle, France) were isolated and purified by isopycnic density-gradient centrifugation in Percoll, as previously described [15–17].

2.3. Cell Culture

Primary skin fibroblasts were derived from healthy individuals and grown under standard condition at 37 °C in a 5% CO_2, in DMEM with 4.5 g/L glucose, 4 mM glutamine as Glutamax, 10 mM pyruvate, 10% FCS, 200 µM uridine, and penicillin/streptomycin (100 U/mL). Upon confluence, cells were trypsinized, pelleted at $1500\times g$, 5 min, and used immediately for analysis.

2.4. Mouse Astrocytes

Astrocytes were prepared from meninges-free cerebellum of 6–7-day-old control and *Harlequin* mice with a mixed genetic background (B6CBACaAw-J/A-Pdcd8/J). The *Harlequin* mouse has been previously shown defective for complex I due to a mutation in the *Aif* gene [18]. Mice were housed with a 12 h light/dark cycle with free access to food and water. Astrocytes were plated into culture flasks in DMEM containing glucose (1 g/L) and 10% fetal calf serum at 37 °C in a 5% CO_2. Upon confluence, flasks were shaken (180 rpm × 30 min; RockingOrbital shaker, VWR, Fontenay sous Bois, France) to remove contaminated microglia cells. Astrocytes are then detached from the culture flash by trypsin and pelleted at $1500\times g$, 5 min [19].

2.5. Ethics Statement

Details of the mouse study were approved by the Robert Debré-Bichat Ethics Committee on Animal Experimentation (http://www.bichat.inserm.fr/comite_ethique.htm; Protocol Number 2010-13/676-003) in accordance with the French and European Laws on animal protection.

2.6. Organism, Organs, or Cells Respiration

Different layouts were selected to fit the conditions imposed by the biological material selected. A first set-up (Figure 2A) allowed us to set tissue or organ samples on a nylon net (1 mm² mesh) about halfway-up of the measuring chamber equipped with a handmade cap (HMC No. 1) allowing for substrate or inhibitor additions. The respiration was simultaneously measured with a macro-optode (3 mm tip diameter) inserted on the top of the chamber and with the oxygen Clark electrode at the bottom. Respiration medium A (400 µL) consisting in 0.25 mM sucrose, 10 mM KH_2PO_4 (pH 7.2), 5 mM $MgCl_2$, 5 mM KCl, and 1 mg/mL bovine serum albumin was thermostated (37.5 °C) and magnetically stirred (high speed, stirring bar 2.0 × 5.0 mm). A second layout (Figure 2B) was used to measure oxygen uptake by cells or mitochondria with the macro-optode in a smaller volume (200 µL medium A; HMC No. 1). Using a third layout, the respiration of one to five zebrafish embryos (3 dpf) was recorded with a micro-optode (50 µm tip diameter) in a minimal volume of PBS (30 µL) (PyroScience, Aachen, Germany) (Figure 3A). The assay was carried out in a flat-bottom glass tube (6 mm diameter) positioned on a magnetic stirrer, maintained at room temperature, and equipped with a handmade cape (HMC No. 2), allowing for the micro-optode insertion and the addition of chemicals with a 5 µL syringe. A 2 mm × 2 mm ball-shaped magnetic flea slowly rotating, harmlessly aside from the embryos, was placed in the glass tube.

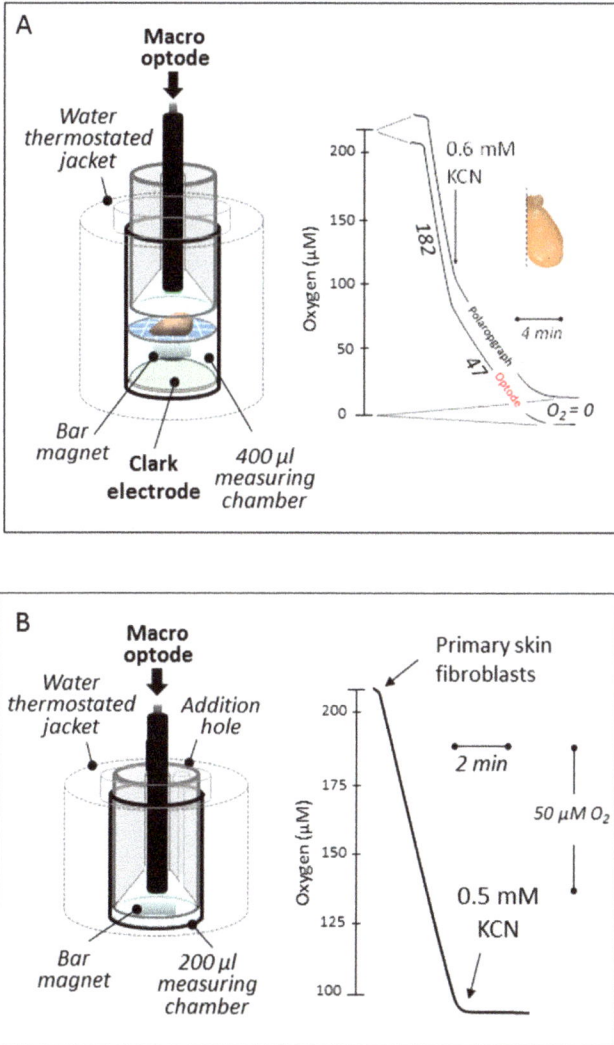

Figure 2. In vitro oxygen uptake by tissue sample and cells in suspension. (**A**) Both the Hansatech polarographic device (bottom) and the FireSting optode work simultaneously allowing to record strictly similar rates of oxygen uptake (about 25% resistant to 0.6 mM cyanide) by a mouse brain hemisphere place on a nylon net at mid-height in 400 μL of respiratory medium A (see Material and Methods). (**B**) Fully cyanide-sensitive human primary fibroblast respiration (about 1×10^6 cells for 50 μM O_2/min) recorded in 200 μL of respiratory medium A. Numbers along the traces are nmol/min/mg protein.

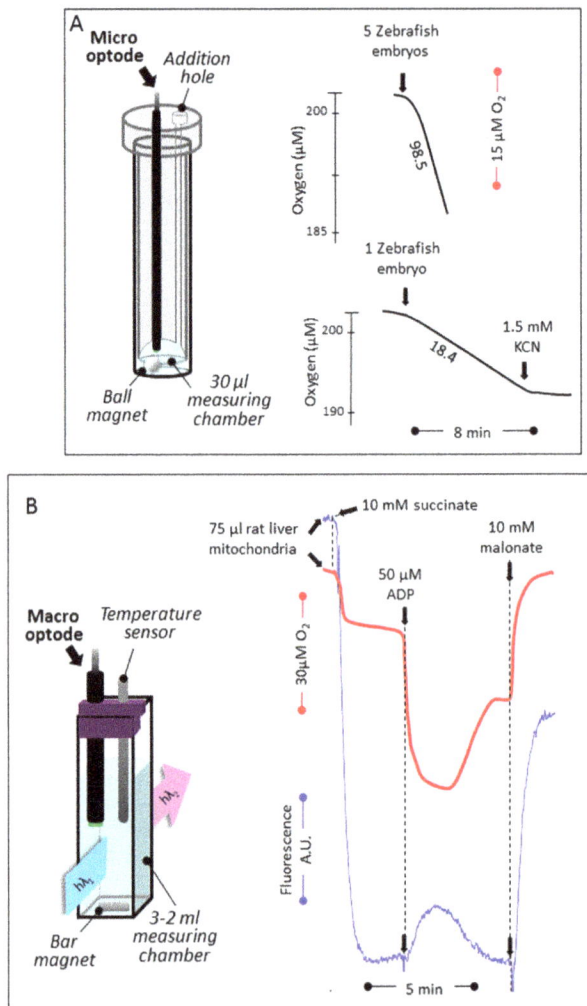

Figure 3. Respiration of Zebrafish embryos, and oxygen tension coupled to membrane potential determination by rat liver mitochondria. (**A**) Cyanide-sensitive respiration of Zebrafish embryos (1 and 5) measured at ambient temperature (20 °C) using a micro-optode in 30 μL of PBS (linear rates observed for 10 min). Notice the spherical magnetic stirrer avoiding to hurt the embryos. (**B**) Using an open layout, oxygen level changes linked to mitochondrial substrate oxidation was measured concomitantly to the membrane potential. Oxygen was measured using the macro-optode placed in a 3 mL quartz cell thermostated at 38 °C and magnetically stirred. Oxygen uptake (red trace) was started by the addition of succinate followed by the addition of a limiting amount of ADP (decreased level of oxygen, due to high rate of consumption; oxidation state 3 [20]), the exhaustion of which exhaustion in a higher oxygen level (oxidation state 4). The addition of malonate (a long established inhibitor of the succinate dehydrogenase [21]) fully inhibited oxygen uptake, and the level of oxygenation of the medium came back to initial value by re-equilibration with air. The membrane potential measured simultaneously (blue trace) rose upon succinate addition (quenching of rhodamine fluorescence) to drop down upon the ADP addition. After ADP exhaustion, the membrane potential rose again, while adding malonate worked to abolish most of it. Numbers along the traces are nmol/min/mg protein.

2.7. Mitochondrial Substrate Oxidation and Membrane Potential

Oxygen uptake and membrane potential were simultaneously measured in a 37.5 °C-water-jacket-thermostated, 1.5 mL quartz-cuvette using the Flx-Xenius XC spectrofluorometer (SAFAS, Monaco, France) with a modified optical path fitted to the magnetically stirred cuvette (Figure 3B). Measurements were made using rat liver mitochondria in 750 µL of respiratory medium A with a macro-optode fitted to a hand-made open cap (HMC No. 3). Mitochondria were successively given 100 nM rhodamine, 10 mM succinate, 50 µM ADP (to ensure state 3, phosphorylating condition), and, after ADP exhaustion (state 4), 10 mM malonate, a specific inhibitor of the succinate dehydrogenase. Membrane potential variation was determined by the fluorescence change of rhodamine (503 nm λ excitation; 527 nm λ emission).

2.8. Respiration and Lactate Excretion

Cell oxygen consumption and lactate excretion were measured using a similar device in 750 µL of respiratory medium A except for the handmade cap (HMC n°3), closing the cuvette yet allowing for micro-syringe (5 and 10 µL) insertion. Purified rabbit muscle lactate dehydrogenase (5 IU; EC 1.1.1.27) and 2 mM NAD$^+$ were added to the cuvette to measure the lactate excreted by the cells, plus 17 mM glutamate and pig heart glutamate–pyruvate transaminase (6 IU; EC 2.6.1.2) to avoid any accumulation of pyruvate that might decrease LDH activity [22,23]. A final addition of known amounts of an NADH solution (4 µM) enabled the calibration of NADH fluorescence. Under these conditions, the rate of lactate excretion by the cell can be calculated from the rate of NAD$^+$ reduction (365 nm λ excitation; 460 nm λ emission).

2.9. Protein Determination and Chemicals

Protein was determined using the Bradford method [24], and all chemicals were of the highest purity grade from Sigma-Aldrich.

2.10. Free 3D Printable Model Accessories

STS files corresponding to several of the layouts described in this paper (HMC No. 1, 3 and 4) are available free on demand.

3. Results

3.1. Reducing the Volume for Oxygen Consumption

We initially tested the macro- (extremity diameter, 3 mm) and micro-optode (extremity diameter, 50 µm) devices under the standard conditions of polarographic analysis used for more than 30 years in our laboratory to measure oxygen consumption by tissues, intact cells, or isolated mitochondria [2] (Figure 2). The macro-optode was first inserted in the top compartment of a closed, magnetically stirred, thermostated chamber equipped with an oxygen-recording Clark-electrode at the bottom compartment. The compartments were separated by a nylon grid (1 mm holes) holding a piece of tissue, yet allowing a free magnetic stirring of the 400 µL of respiratory medium. Identical responses were obtained from the optode and polarographic device (Figure 2A). With a quite similar configuration but without the electrode disk, the assay medium could be reduced to 200 µL without affecting oxygen detection by the macro-optode device (Figure 2B). Noticeably, this latter does not significantly consume oxygen (at variance with a Clark electrode).

We next manufactured a device to use a micro-optode in a much smaller volume of respiratory medium (30–50 µL). Using a magnetically stirred (ball stirrer) 1 mL glass tube, containing 30 µL of PBS and the micro-optode device fitted to a handmade cap, it was possible to quantify the respiration of as few as 1 to 5 Zebrafish embryos, the respiration of which being proportional to the number of Zebrafish embryos studied.

3.2. Simultaneous Determination of Mitochondrial Oxygen Consumption and Membrane Potential

In order to simultaneously measure oxygen tension and membrane potential, we next placed the macro-optode in a magnetically stirred, thermostated quartz cell (750 μL) using an open handmade cap (Figure 3B). We then measured the changes in oxygen tension by the optode signal (red trace) and the mitochondrial membrane potential (blue trace) inversely proportional to the quenching of rhodamine-123 fluorescence (Figure 3B).

3.3. Simultaneous Determination of Cell Respiration and Lactate Excretion

The macro-optode was finally inserted into a handmade cap closing a 37 °C-thermostated, magnetically stirred, 1.5 mL quartz-cell containing 750 μL of respiration medium (Figure 4A). By supplementing the medium with NAD^+, lactate dehydrogenase (LDH), glutamate, and glutamate–pyruvate transaminase (GPT), it was possible to spectrofluorimetrically estimate the NADH accumulation due to the LDH-catalyzed oxidation of any excreted lactate to pyruvate (Figure 4B). Noticeably, in the presence of an excess of added glutamate and GPT, the pyruvate is readily transaminated to alanine and α-ketoglutarate, avoiding LDH substrate inhibition by pyruvate. To quantify the fluorescent signal, a known amount of NADH was added at the end of the assay. This allowed us in a few minutes to accurately measure the rates of oxygen consumption by respiring intact cells together with lactate production indicative of glycolytic flux. This is exemplified in the case of astrocytes prepared from control or *Harlequin* mice, the latter being defective for respiratory chain complex I [18,22].

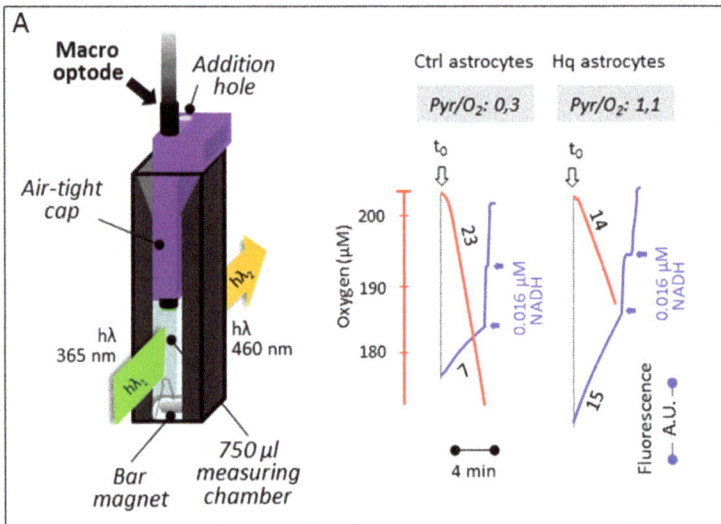

Figure 4. *Cont.*

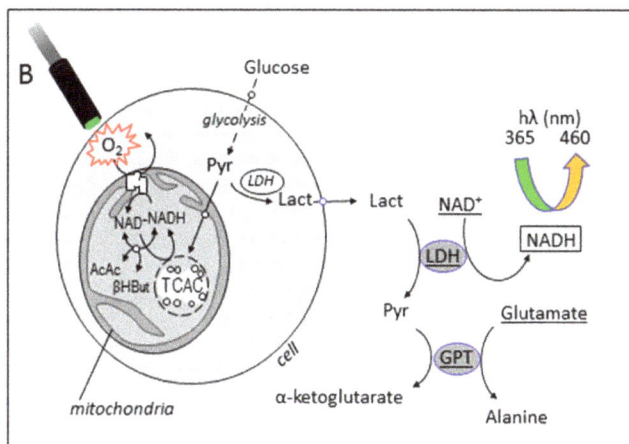

Figure 4. Respiration and lactate excretion by mouse-cultured astrocytes. (**A**) The macro-optode fitted to a magnetically stirred, 37.5 °C-thermostated quartz-cell by a closed cap (yet allowing for the addition of substrates and inhibitors) measures oxygen uptake due to cyanide-sensitive respiration (red traces) by control astrocytes or astrocytes prepared from the CI-defective *Harlequin* mice. The concomitant fluorometric determination of NADH (blue traces) allows for a determination of the rate of the excretion of lactate, thanks to its conversion to pyruvate brought about by added lactate dehydrogenase in the presence of added NAD$^+$. Numbers along the traces are nmol/min/mg protein. (**B**) The additional presence of glutamate and glutamate transaminase avoided inhibition of the LDH reaction by accumulated pyruvate.

4. Discussion

The comprehensive diagnostic of suspected oxidative phosphorylation (OXPHOS) defect requires the complementary assays of RC complex activity and of mitochondrial oxygen consumption [2]. Similarly, the significance of numerous base changes in the several hundred genes encoding OXPHOS components revealed by systematic sequencing can only be established by an extensive characterization of OXPHOS activities [3]. In addition to the determination of the activity of OXPHOS complexes, when possible, this includes the study of the cell respiration, the mitochondrial oxidation of various respiratory substrates, the determination of the ADP/O and respiratory control values. A complete investigation of oxidative properties supposes the use of an adaptable device allowing for the addition of multiple substrates and inhibitors in the assay medium and to register the oxygen consumption in real time. To this end, the Clark oxygen electrode that replaced the previous Warburg apparatus represented major progress, allowing for the use of much less precious material.

Here we have shown that it is possible to use RedFlash technology to reduce (by at least two-thirds) the amount of biological sample to be studied, as compared with previous devices. This represents similar progress in terms of the biological material required and the ease of use. The device is stable for months/years, as long as the probe is kept dry and not exposed to strong light. Various optodes have been used for several years in other fields of biology [25–27] and a careful comparison between these devices and the Clark electrode already been reported [28].

In the context of screening for OXPHOS defects, an immediate benefit of using this technology is smaller muscle biopsies or blood samples needed to be taken from patients and a reduction of the amount of cultured cells to be used, i.e., fewer traumas for patients and a lower cost in terms of cell cultures. The flexibility of the optic fiber allows one to adapt the device to various specific environments, such as spectrophoto- or spectrofluorometer cuvettes. As such, it is suitable for the simultaneous determination of cell respiration and lactate cell excretion. More specific than suspending medium acidification [29], an increased rate of lactate excretion can be taken as an indication of a reduced rate of mitochondrial pyruvate oxidation or increased pyruvate production by glycolysis [22].

Acknowledgments: This work was supported by French (ANR MITOXDRUGS-16-CE18-0015 to PB, DC, MP, PR, MR) and European (E-rare Genomit (16-CE18-0010-02 to PB, DC, PR, MR) institutions, and patient associations: Association Française contre les Myopathies (AFM; Project No. 11639 to PB, PR), Association d'Aide aux Jeunes Infirmes (AAJI to PR), Association contre les Maladies Mitochondriales (AMMi to PB, DC, PR), Association Française contre l'Ataxie de Friedreich (AFAF to PB, PR), and Ouvrir Les Yeux (OLY to PB, PR, MG).

Author Contributions: Work was done on mouse astrocytes (P.B., M.R.), Zebrafish embryos (C.Y., P.R.), mouse brain (P.B., P.R.) human skin fibroblasts (D.C.), and rat liver mitochondria (M.P., M.R.). P.B. and P.R. wrote the manuscript.

Conflicts of Interest: The authors declare no conflict of interest.

References

1. Galluzzi, L.; Kepp, O.; Trojel-Hansen, C.; Kroemer, G. Mitochondrial control of cellular life, stress, and death. *Circ. Res.* **2012**, *111*, 1198–1207. [CrossRef] [PubMed]
2. Rustin, P.; Chretien, D.; Bourgeron, T.; Gerard, B.; Rotig, A.; Saudubray, J.M.; Munnich, A. Biochemical and molecular investigations in respiratory chain deficiencies. *Clin. Chim. Acta* **1994**, *228*, 35–51. [CrossRef]
3. Turnbull, D.M.; Rustin, P. Genetic and biochemical intricacy shapes mitochondrial cytopathies. *Neurobiol. Dis.* **2016**, *92*, 55–63. [CrossRef] [PubMed]
4. Benit, P.; Letouze, E.; Rak, M.; Aubry, L.; Burnichon, N.; Favier, J.; Gimenez-Roqueplo, A.P.; Rustin, P. Unsuspected task for an old team: Succinate, fumarate and other krebs cycle acids in metabolic remodeling. *Biochim. Biophys. Acta* **2014**, *1837*, 1330–1337. [CrossRef] [PubMed]
5. Golpich, M.; Amini, E.; Mohamed, Z.; Azman Ali, R.; Mohamed Ibrahim, N.; Ahmadiani, A. Mitochondrial dysfunction and biogenesis in neurodegenerative diseases: Pathogenesis and treatment. *CNS Neurosci. Ther.* **2017**, *23*, 5–22. [CrossRef] [PubMed]
6. Horobin, R.W.; Trapp, S.; Weissig, V. Mitochondriotropics: A review of their mode of action, and their applications for drug and DNA delivery to mammalian mitochondria. *J. Control. Release* **2007**, *121*, 125–136. [CrossRef] [PubMed]
7. Pearson, B.L.; Ehninger, D. Environmental chemicals and aging. *Curr. Environ. Health Rep.* **2017**, *4*, 38–43. [CrossRef] [PubMed]
8. Munnich, A.; Rustin, P. Clinical spectrum and diagnosis of mitochondrial disorders. *Am. J. Med. Genet.* **2001**, *106*, 4–17. [CrossRef] [PubMed]
9. Potter, M.; Newport, E.; Morten, K.J. The warburg effect: 80 years on. *Biochem. Soc. Trans.* **2016**, *44*, 1499–1505. [CrossRef] [PubMed]
10. Warburg, O.; Krippahl, G. Further development of manometric methods. *J. Natl. Cancer Inst.* **1960**, *24*, 51–55. [PubMed]
11. Severinghaus, J.W. The invention and development of blood gas analysis apparatus. *Anesthesiology* **2002**, *97*, 253–256. [CrossRef] [PubMed]
12. Ferrick, D.A.; Neilson, A.; Beeson, C. Advances in measuring cellular bioenergetics using extracellular flux. *Drug Discov. Today* **2008**, *13*, 268–274. [CrossRef] [PubMed]
13. Westerfield, M. *The Zebrafish Book. A Guide for the Laboratory Use of Zebrafish (Danio Rerio)*; University of Oregon Press: Eugene, OR, USA, 2000.
14. Kimmel, C.B.; Ballard, W.W.; Kimmel, S.R.; Ullmann, B.; Schilling, T.F. Stages of embryonic development of the zebrafish. *Dev. Dyn.* **1995**, *203*, 253–310. [CrossRef] [PubMed]
15. Porceddu, M.; Buron, N.; Roussel, C.; Labbe, G.; Fromenty, B.; Borgne-Sanchez, A. Prediction of liver injury induced by chemicals in human with a multiparametric assay on isolated mouse liver mitochondria. *Toxicol. Sci.* **2012**, *129*, 332–345. [CrossRef] [PubMed]
16. Buron, N.; Porceddu, M.; Brabant, M.; Desgue, D.; Racoeur, C.; Lassalle, M.; Pechoux, C.; Rustin, P.; Jacotot, E.; Borgne-Sanchez, A. Use of human cancer cell lines mitochondria to explore the mechanisms of bh3 peptides and abt-737-induced mitochondrial membrane permeabilization. *PLoS ONE* **2010**, *5*, e9924. [CrossRef] [PubMed]

17. Lecoeur, H.; Langonne, A.; Baux, L.; Rebouillat, D.; Rustin, P.; Prevost, M.C.; Brenner, C.; Edelman, L.; Jacotot, E. Real-time flow cytometry analysis of permeability transition in isolated mitochondria. *Exp. Cell Res.* **2004**, *294*, 106–117. [CrossRef] [PubMed]
18. Vahsen, N.; Cande, C.; Briere, J.J.; Benit, P.; Joza, N.; Larochette, N.; Mastroberardino, P.G.; Pequignot, M.O.; Casares, N.; Lazar, V.; et al. Aif deficiency compromises oxidative phosphorylation. *EMBO J.* **2004**, *23*, 4679–4689. [CrossRef] [PubMed]
19. Schildge, S.; Bohrer, C.; Beck, K.; Schachtrup, C. Isolation and culture of mouse cortical astrocytes. *J. Vis. Exp.* **2013**. [CrossRef] [PubMed]
20. Chance, B.; Williams, G.R. The respiratory chain and oxidative phosphorylation. *Adv. Enzymol. Relat. Subj. Biochem.* **1956**, *17*, 65–134. [PubMed]
21. Thorn, M.B. Inhibition by malonate of succinic dehydrogenase in heart-muscle preparations. *Biochem. J.* **1953**, *54*, 540–547. [CrossRef] [PubMed]
22. Benit, P.; Pelhaitre, A.; Saunier, E.; Bortoli, S.; Coulibaly, A.; Rak, M.; Schiff, M.; Kroemer, G.; Zeviani, M.; Rustin, P. Paradoxical inhibition of glycolysis by pioglitazone opposes the mitochondriopathy caused by aif deficiency. *EBioMedicine* **2017**, *17*, 75–87. [CrossRef] [PubMed]
23. Fernandez-Mosquera, L.; Diogo, C.V.; Yambire, K.F.; Santos, G.L.; Luna Sanchez, M.; Benit, P.; Rustin, P.; Lopez, L.C.; Milosevic, I.; Raimundo, N. Acute and chronic mitochondrial respiratory chain deficiency differentially regulate lysosomal biogenesis. *Sci. Rep.* **2017**, *7*, 45076. [CrossRef] [PubMed]
24. Bradford, M.M. A rapid and sensitive method for the quantitation of microgram quantities of protein utilizing the principle of protein-dye binding. *Anal. Biochem.* **1976**, *72*, 248–254. [CrossRef]
25. Oellermann, M.; Portner, H.O.; Mark, F.C. Simultaneous high-resolution pH and spectrophotometric recordings of oxygen binding in blood microvolumes. *J. Exp. Biol.* **2014**, *217*, 1430–1436. [CrossRef] [PubMed]
26. Svendsen, M.B.; Bushnell, P.G.; Steffensen, J.F. Design and setup of intermittent-flow respirometry system for aquatic organisms. *J. Fish Biol.* **2016**, *88*, 26–50. [CrossRef] [PubMed]
27. Steffensen, J.F. Some errors in respirometry of aquatic breathers: How to avoid and correct for them. *Fish Physiol. Biochem.* **1989**, *6*, 49–59. [CrossRef] [PubMed]
28. Shaw, A.D.; Li, Z.; Thomas, Z.; Stevens, C.W. Assessment of tissue oxygen tension: Comparison of dynamic fluorescence quenching and polarographic electrode technique. *Crit. Care* **2002**, *6*, 76–80. [CrossRef] [PubMed]
29. Sica, V.; Bravo-San Pedro, J.M.; Pietrocola, F.; Izzo, V.; Maiuri, M.C.; Kroemer, G.; Galluzzi, L. Assessment of glycolytic flux and mitochondrial respiration in the course of autophagic responses. *Methods Enzymol.* **2017**, *588*, 155–170. [PubMed]

Journal of
Clinical Medicine

MDPI

Article

The Relationship between Mitochondrial Respiratory Chain Activities in Muscle and Metabolites in Plasma and Urine: A Retrospective Study

Corinne Alban [1,2], Elena Fatale [2], Abed Joulani [2], Polina Ilin [2] and Ann Saada [1,2,*]

[1] Monique and Jacques Roboh Department of Genetic Research, Hadassah-Hebrew University Medical Center, P.O. Box 12000, 91120 Jerusalem, Israel; korin@hadassah.org.il
[2] Metabolic Laboratory, Department of Genetics and Metabolic Diseases, Hadassah-Hebrew University Medical Center, P.O. Box 12000, 91120 Jerusalem, Israel; fatalelena@yahoo.com (E.F.); abedg@hadassah.org.il (A.J.); ilin@hadassah.org.il (P.I.)
* Correspondence: annsr@hadassah.org.il; Tel.: +972-02-6776844

Academic Editor: Iain P. Hargreaves
Received: 20 February 2017; Accepted: 9 March 2017; Published: 10 March 2017

Abstract: The relationship between 114 cases with decreased enzymatic activities of mitochondrial respiratory chain (MRC) complexes I-V (C I-V) in muscle and metabolites in urine and plasma was retrospectively examined. Less than 35% disclosed abnormal plasma amino acids and acylcarnitines, with elevated alanine and low free carnitine or elevated C4-OH-carnitine as the most common findings, respectively. Abnormal urine organic acids (OA) were detected in 82% of all cases. In CI and CII defects, lactic acid (LA) in combination with other metabolites was the most common finding. 3-Methylglutaconic (3MGA) acid was more frequent in CIV and CV, while Tyrosine metabolites, mainly 4-hydroxyphenyllactate, were common in CI and IV defects. Ketones were present in all groups but more prominent in combined deficiencies. There was a significant strong correlation between elevated urinary LA and plasma lactate but none between urine Tyrosine metabolites and plasma Tyrosine or urinary LA and plasma Alanine. All except one of 14 cases showed elevated FGF21, but correlation with urine OA was weak. Although this study is limited, we conclude that urine organic acid test in combination with plasma FGF21 determination are valuable tools in the diagnosis of mitochondrial diseases.

Keywords: mitochondrial disease; mitochondrial respiratory chain; plasma amino acids; plasma carnitines; urine organic acid; FGF21

1. Introduction

Mitochondria are essential organelles present in eukaryotic cells. They perform vital roles in many cellular pathways; however, their main function is to supply cellular energy in the form of adenosine triphosphate (ATP) via oxidative phosphorylation (OXPHOS) performed by the mitochondrial respiratory chain (MRC). The MRC located in mitochondrial inner membrane is comprised of ninety proteins organized into five multi-subunit enzymatic protein complexes. These proteins and many other auxiliary proteins are essential for maintaining MRC, and OXPHOS is encoded by two genomes, the nuclear genome and the mitochondrial genome (mtDNA). Consequences of OXPHOS dysfunction include not only energy (ATP) depletion but also elevated oxidative stress, disturbed mitochondrial membrane potential, subsequently leading to imbalanced calcium homeostasis, autophagy/mycophagy, apoptosis, and ultimately to cell death. Mitochondrial diseases affecting one or more MRC complexes are common (prevalence 5 to 15 cases per 100,000 individuals), disabling, progressive, or fatal disorders affecting several vital organs including the brain, optic nerves, the liver, skeletal muscles, and the heart, manifesting at birth, during infancy, and in adulthood [1–3]. They are mostly caused by mutations in genes leading to dysfunction of one or multiple MRCs or auxiliary proteins crucial for OXPHOS but could also be secondary to other conditions [4]. The extreme heterogeneity of mitochondrial disorders makes diagnosis complex and sometimes requires the investigation of MRC function in the affected tissues. Before an invasive procedure such as muscle biopsy is performed, a metabolic workup is usually performed in blood and/or urine. Exome sequencing has recently become a valuable diagnostic tool; however, the interpretation is complex when numerous pathogenic and/or new variants are detected [1–3,5]. In this retrospective study, we investigated the relationship between common metabolic tests and MRC dysfunction detected in muscle in order to facilitate the diagnostic workup of mitochondrial diseases.

2. Materials and Methods

2.1. Mitochondrial Respiratory Chain Enzymatic Analysis

The enzymatic activities of MRC complexes I-V and citrate synthase CS (a mitochondrial control enzyme) were determined by spectrophotometric methods in isolated muscle mitochondria, as we have previously described [6].

2.2. Metabolic Tests

Acylcarnitines were determined by electrospray–tandem mass spectrometry in plasma or dry bloodspots (Micromass, Waters, Milford, MA, USA) [7]. Organic acids in urine were determined qualitatively by gas chromatography–mass spectrometry (GC-MS) (Agilent, Santa Clara, CA, USA) [8]. Plasma amino acid analysis was performed on a Biochrom 30 amino acid analyzer according to the manufacturer's instructions (Biochrom, Holliston, MA, USA).

2.3. Fibroblast Growth Factor 21

Fibroblast growth 21 (FGF 21) was determined in plasma, by a solid phase sandwich enzyme-linked immunosorbant assay using the Human FGF-21 Quantikine ELISA kit (R&D systems, Minneapolis, MN, USA) according to the manufacturer's instructions.

2.4. Statistical Analysis

Where indicated, the relationship between two parameters was assessed by the Spearman's rank-order correlation test using IBM SPSS statistics for Windows, version 24.0 (IBM Co., Armonk, NY, USA).

3. Results

3.1. Distribution of MRC Deficiencies

Enzymatic analysis of MRC complexes I-IV was performed in 1163 muscle samples, referred to our laboratory for diagnostic purposes over a ten-year period (2006–2016). Mean age of the patients was 4.9 years ranging between 1 day and 67 years. Of these, 193 were found to be defective (17% diagnostic yield) in one or several MRC complexes (<50% residual activity of control mean, normalized to CS) (Figure 1A). The most common defect found (63 samples) were combined deficiencies including two or more MRC complexes but with normal or elevated CII activity. Other frequent defects were isolated CI (40 samples) and CIV (44 samples). Less frequent were deficiencies in CV (28 samples) and CII (6 samples). The rarest isolated defect was CIII with only one sample. Two cases which were designated "CoQ level" as the combined activity of CI + III and CII + III were decreased, while each measured separately disclosed normal activity. We also included a group designated as a general decrease where all respiratory chain complexes activities were decreased relative to CS. Metabolic workup data obtained from urine and/or plasma for diagnostic purposes was available for 114 of the 193 (59%)

patients with decreased MRC activities in muscle (Figure 1B). This distribution reflected the total deficiencies (Figure 1C), with the exception of CIII, where no data was available and the proportion of CV defects somewhat increased. For comparison, we also obtained one or more metabolic parameters from 150 patients with clinical suspicion of mitochondrial diseases but with normal activities of MRC complexes I-V.

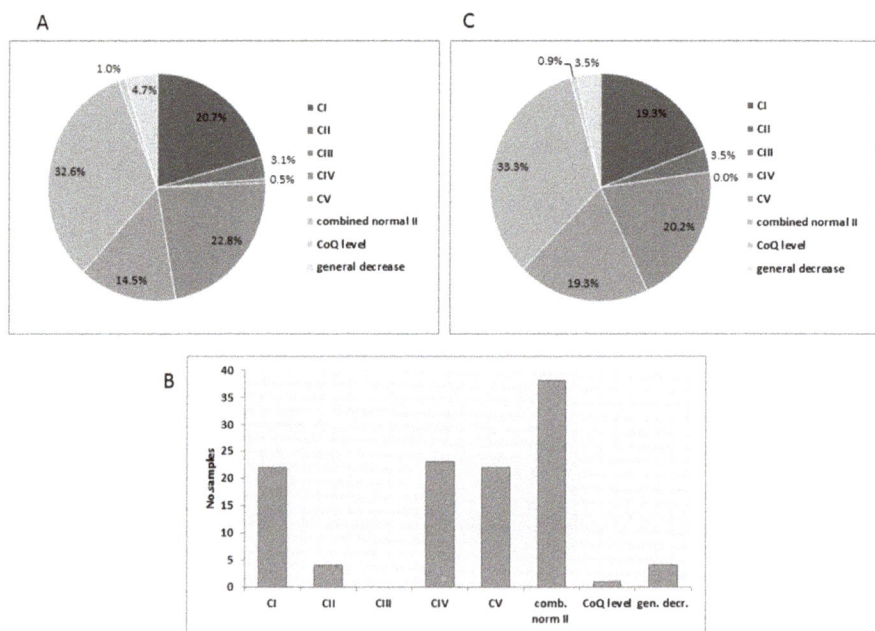

Figure 1. Distribution of MRC deficiencies: (**A**) The relative proportion of MRC defects according to groups. (**B**) The number of samples from each group with available metabolic test data. (**C**) The relative proportion of MRC defects with available metabolic test data.

3.2. Plasma Amino Acids

Plasma amino acids were measured in 49 samples but of these only 17 (34%) were abnormal, mostly disclosing elevated Alanine in alone or in combination with elevated Glutamine and/or tyrosine. Thus 50% or more in each group tested normal with the exception of two generally decreased cases that were both abnormal (Figure 2A). The most frequent abnormality was Alanine followed by Glutamine and Tyrosine (Figure 2B). Although quantitative data are available for many samples (analysis performed in our laboratory) we opted not to include these measurements as some samples were reported as abnormal without quantification (data reported from other sources). 10 tests disclosed elevated Alanine only, 4 with elevated Alanine and Glutamine and 4 elevated Alanine Glutamine and tyrosine. Other amino acids were inconsistent, for example Citrulline was elevated in two samples while decreased in one and branched chain amino acids, Proline, Phenylalanine and Methionine were elevated in in one case each (results not shown). As Alanine is also derived from lactate via transamination of pyruvate we assessed the correlation to plasma lactate in 19 samples with data available for both parameters; however, the association was weak at best and statistically not significant. Only 9 of 90 (10%) of cases with normal MRC disclosed abnormal plasma amino acids of these, 3 had elevated Alanine. Taken together, the amino acid analysis was not very informative with respect to predicting MRC dysfunction but is vital for the differential diagnosis of other conditions such as urea cycle disorders.

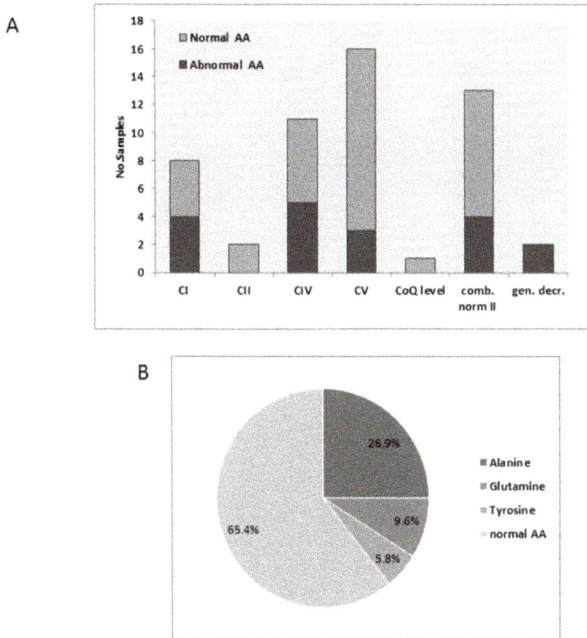

Figure 2. Plasma amino acids: (**A**) The relative proportion abnormal to abnormal plasma amino acid (AA) tests results according to groups; (**B**) The relative proportion of the most frequent abnormalities.

3.3. Acylcarnitines

Acylcarnitine analysis test results were available from either plasma or dry bloodspots from 61 cases, of which 21 (34%) were abnormal. e most affected groups are CI defects, and approximately half were abnormal, as were the two cases of general defects. In the other groups, the test was mostly normal (Figure 3A). The most common abnormalities were low free C0-carnitine and elevated C4OH-(3-hyroxybutyryl)-carnitine. Elevation of medium- or long chain- (MC/LC) carnitines was only detected in two CV defects and one combined deficiency case (Figure 3B). Since elevated C4OH– carnitine is associated with ketosis, we evaluated the association with urinary ketones as detected by the organic acid analysis (see next section) from 17 samples and did not find any statistically significant correlation between the two parameters. Of 101 tests from patients with normal MRC, only nine (9%) had abnormal acylcarnitines, mostly with decreased in free carnitine and only one showing elevated C4OH-carnitine. Accordingly, as was the case with amino acid analysis, the acylcarnitine test was not significantly informative with respect to most MRC defects, but abnormality was more frequent than in cases with normal MRC.

3.4. Urinary Organic Acids

Urinary organic acids were qualitatively evaluated in 75 cases, of which 66 (82%) disclosed elevated levels of one or more metabolites. Notably, all samples in the groups CI and CII, and the general decrease, were abnormal, as was the majority of CIV and CV, and the combined defects. The single CoQ level sample tested normal (Figure 4A). The most frequent abnormalities were lactic acid (LA) concomitantly with ketones and/or TCA (Krebs cycle) metabolites and/or Tyrosine metabolites and/or dicarboxylic acids (DCA) and/or 3-methylglutaconic acid (3MGA). In all groups together, LA and ketones (mainly 3-hydroxybutyrate) were equally common followed by TCA and tyrosine metabolites (mainly 4-hydroxyphenyllactate), whereas 3MGA and DCA were less common

(Figure 4B). Other occasionally occurring metabolites were methylmalonic acid, branched chain ketoacids, glutaric acid, and acylglycines but without any specific pattern (not shown). Interestingly, the distribution of the metabolites varied according the MRC defects (Figure 5).

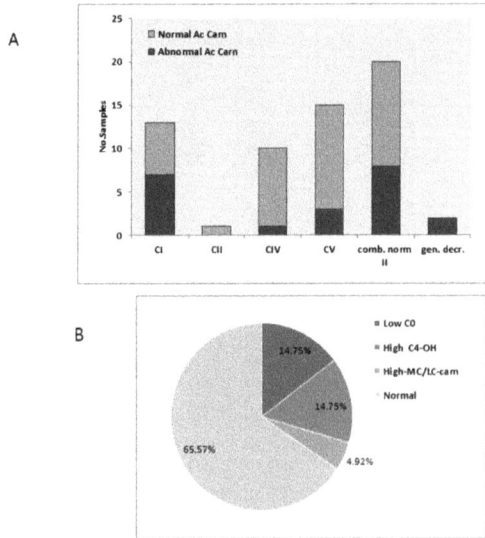

Figure 3. Plasma acylcarnitines: (**A**) The relative proportion abnormal to abnormal acylcarnitines (Ac. Carn) tests results according to groups; (**B**) The relative proportion of the most common abnormalities.

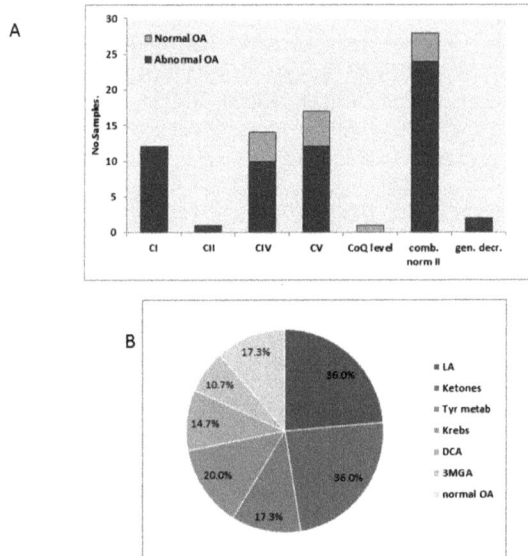

Figure 4. Urinary organic acids: (**A**) The relative proportion abnormal to abnormal urinary organic acids (OA) tests results according to groups; (**B**) The relative proportion of the most common abnormalities.

In CI defects, 75% excreted elevated levels of LA, while this proportion was less than 36% or less in the other groups (Figure 5A). The combined defects group is characterized by a higher proportion of ketones (50%) and DCA (25%) (Figure 5D). Both CIV and CV defects disclosed a higher percentage of 3MGA than the other groups. However, almost a third of the samples in this group were normal (Figure 5B,D). Tyrosine metabolites (mainly 4-hydroxyphenyllactate) were common in groups CI and CIV. For comparison, 32 out of 88 (36%) cases with normal MRC disclosed an abnormal organic acid test; however, no specific pattern resembling the findings in MCR defects was recognized. For example, four cases had elevated ketones, six elevated LA, and four elevated Tyrosine metabolites, but mostly separately, LA was never observed in combination with another metabolite. On the contrary, in the MRC defective group, LA was elevated concomitantly with ketones and/or Tyrosine metabolites in 22 (91%) of 24 samples with elevated lactate. Interestingly, the "normal" group contained six samples with low/moderately elevation of methylmalonic acid in combination with other metabolites, while only one case was detected in the MRC defect group. This finding could possibly be related to a nutritional deficiency (vitamin B12) rather than an inborn metabolic disease. Consequently, the organic acid test is quite informative, and the results could be helpful in guiding the investigation towards a specific group of MRC defects. The number of cases with CII and III, and the general defects, was too small to be evaluated. Still, one out of two CII cases disclosed elevated succinate and fumarate. As expected, the correlation between urinary and plasma LA examined in 23 samples was statistically significant (R 0.843 P0.0). Additionally, we examined the relationship between Tyrosine metabolites in urine and plasma Tyrosine in 10 samples and did not find any statistically significant correlation, as only two samples disclosed elevated Tyrosine.

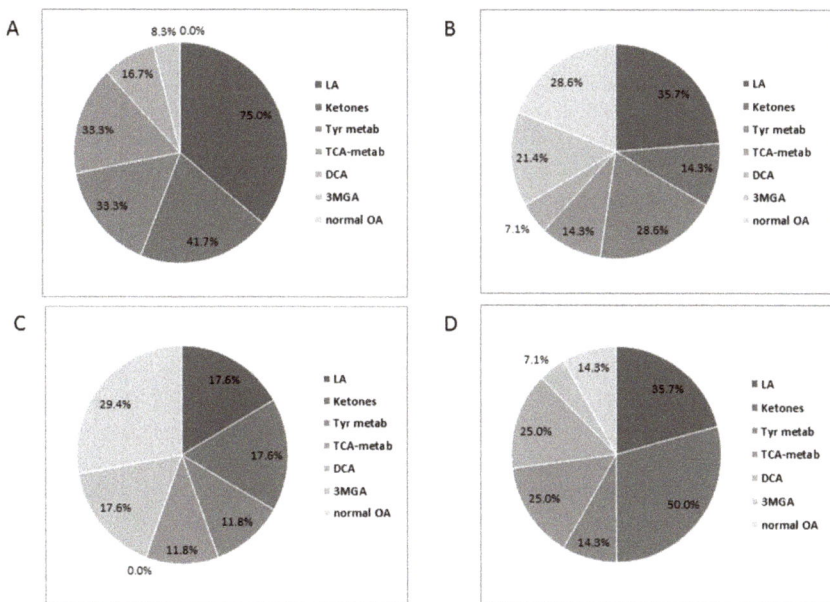

Figure 5. Urinary organic acids in (**A**) CI defects; (**B**) CIV defects; (**C**) CV defects; (**D**) combined defects with normal CII.

3.5. Plasma FGF21

The level of plasma FGF21 was measured in 14 samples, and all except one disclosed high levels (331–10,700 pg/mL (Normal < 250 pg/mL). Notably, eight samples with high FGF21 were found in the combined group with normal complex II. Nevertheless, we did not find any significant relationship to the urinary organic acid test. The correlation with urinary OA was at best weak (R 0.377, P0.356), as four samples with high FGF21 showed normal urinary OD, while one sample with normal FGF21 had an abnormal urinary organic acid profile. We also measured FGF 21 in 25 cases with normal MRC of which four disclosed elevated FGF21. Still, taken together, the correlation between abnormal MRC activities and plasma FGF21 is significant (R0.690, P0.0).

4. Discussion

The purpose of this study was to retrospectively examine the correlation between biochemical findings and MRC enzymatic activity in an unbiased manner. Obviously, this study is limited as it includes neither clinical (not reported in a systematic manner) nor molecular data (not available for most cases; however, from our limited knowledge, we estimate that at least 20% of the cases are molecularly defined). We made an effort, but were unable to establish correlations between the clinical symptoms and the different groups. We kindly refer to the literature for this complex issue [1–5]. Moreover, only data obtained from muscle are presented since we did not have pertinent data for liver tissue; therefore, conditions such as certain mtDNA depletion syndromes with variable tissue expression could have been overlooked in this study [9]. The limit of less than 50% residual activity normalized to citrate synthase was determined to potentially include cases of mtDNA heteroplasmy, and, from our experience, this cutoff is relevant [10]. On the other hand, a partial deficiency could also be secondary [4] to other conditions linked or apparently not linked to MRC function, as exemplified by decreased cytochrome c oxidase activity in Troyer syndrome [11], thus the inclusion criteria is indeed a complex, unresolved issue. Additionally, we did not include cases with pyruvate dehydrogenase E1 or E3 deficiency. Nevertheless, this study provides some pertinent information that could facilitate the decision of how to proceed with the workup of a patient clinically suspected to harbor a mitochondrial disease, and provides a complement to other studies [1]. According to the presented data, plasma amino acids did not disclose a significant correlation with MRC dysfunction as most showed a normal profile while a small proportion disclosed mainly elevated Alanine without significant correlation with plasma LA. We also did not detect any specific correlation with Citrulline and/or Arginine, as has been previously reported in MELAS [12], the reason for this, could be that our cohort did not include this specific condition. According to our data, acylcarnine test is also not very informative. Theoretically, impaired MRC should affect mitochondrial fatty acid oxidation with subsequent accumulation of acylcarnitine; however, we mostly detected signs of ketosis or decreased free carnitine. These findings are in accord with the consensus reported by the Mitochondrial Medicine Society [5]. Obviously, if indicated, both amino acid and acylcarnitine tests are important to rule out urea cycle disorders, amino acid degradation defects, fatty acid oxidation defects, primary/secondary carnitine deficiency, etc. and should therefore not be neglected. It is also anticipated that patients exhibiting abnormal acylcarnitine or amino acids, patterns consistent with a known inborn error of metabolism, would not be referred for a muscle biopsy. Still in the context of mitochondrial diseases urinary organic acid analysis seems to be considerably more informative, as the four out of five groups of MRC defects disclosed abnormally elevated excretion of one of several metabolites. Moreover, it was possibly to link certain patterns with certain MRC defects. LA and ketones, mostly concurrently, were prevalent in CI and combined defects. Notably, the combined defects encompass combinations of CI,III,IV,V defects, while complex II, which is solely encoded by the nuclear genome, is normal, so this group includes both mtDNA depletion and translation defects [1,9,13]. 3MGA aciduria is relatively prevalent in CIV and CV defect and is among several other disorders, been linked TMEM70 mutations in CV [14]. Recently elevated urinary 3MGA levels were reported in mitochondrial membrane defects with or without enzymatic MRC dysfunction, thus 3MGA remains an important biomarker for mitochondrial disease [15,16]. Although quantitative measurement in urine was not performed, elevated urine and LA were, as expected, closely correlated, serving as an additional marker of hyper lactic acidemia >6–10 mM [17]. Elevated 4-hydroxyphenyllactate, a Tyrosine metabolite in urine, and Tyrosine in plasma are also associated with liver dysfunction [18]. In this respect, the organic acid analysis is more sensitive as a marker since many cases showed increased 4-hydroxyphenyllactate while only a few had elevated Tyrosine. The more prevalent occurrence of urinary tyrosine metabolites in CI and CIV indicate liver involvement in the pathogenesis of these conditions. The significant correlation between abnormal MRC activities and plasma FGF21 confirms that this protein is a good marker for MRC dysfunction in accord with previously reported findings [19]. Moreover, the finding that elevated FGF21 was prevalent in the combined group and suspected of mitochondrial translation or maintenance defects is certainly consistent with the recent report by Lehtonen et al. [20]. Nevertheless, as some discrepancies were observed between plasma FGF21 and urine organic acids, it seems that the combination of these two tests would be more informative than each one alone. As we have previously pointed out, this study is limited, and there are several other biochemical parameters that were not included in this because of a lack of sufficient data. Among these are growth differentiation factor 15, amino acids, liver function tests, pyruvate, muscle pathology, etc. [5,20,21]. Nevertheless, according to our findings, testing patients suspected of a mitochondrial diseases, for organic acids in urine and FGF21 in plasma is informative and the results facilitate the decision whether to perform a biopsy or not. Alternatively, or in addition, these tests could also be useful in guiding the filtering of variants in exome analysis.

We conclude that urine organic acid test in combination with plasma FGF21 determination are valuable tools in the diagnosis of mitochondrial diseases.

Acknowledgments: AS is supported by the Pakula family via AFHU. Orly Elpeleg and Stanley Korman are acknowledged for interpretations.

Author Contributions: C.A. performed enzymatic workup, E.F., and P.I., performed metabolic tests, A.G. tabulated data, A.S. interpreted, analyzed data and wrote the manuscript.

Conflicts of Interest: The authors declare no conflict of interest.

References

1. Gorman, G.S.; Chinnery, P.F.; DiMauro, S.; Hirano, M.; Koga, Y.; McFarland, R.; Suomalainen, A.; Thorburn, D.R.; Zeviani, M.; Turnbull, D.M. Mitochondrial diseases. *Nat. Rev. Dis. Prim.* **2016**, *20*, 16080. [CrossRef] [PubMed]
2. DiMauro, S.; Schon, E.A.; Carelli, V.; Hirano, M. The clinical maze of mitochondrial neurology. *Nat. Rev. Neurol.* **2013**, *9*, 429–444. [CrossRef] [PubMed]
3. Chinnery, P.F. Mitochondrial disorders overview. In *GeneReviews*; Pagon, R.A., Adam, M.P., Ardinger, H.H., Wallace, S.E., Amemiya, A., Bean, L.J.H., Bird, T.D., Ledbetter, N., Mefford, H.C., Stephens, K., et al., Eds.; University of Washington: Seattle, WA, USA, 2014.
4. Niyazov, D.; Kahler, S.; Frye, R. Primary Mitochondrial Disease and Secondary Mitochondrial Dysfunction: Importance of Distinction for Diagnosis and Treatment. *Mol. Syndromol.* **2016**, *7*, 122–137. [CrossRef] [PubMed]
5. Parikh, S.; Goldstein, A.; Koenig, M.K.; Scaglia, F.; Enns, G.M.; Saneto, R.; Anselm, I.; Cohen, B.H.; Falk, M.J.; Greene, C.; et al. Diagnosis and management of mitochondrial disease: A consensus statement from the Mitochondrial Medicine Society. *Genet. Med.* **2015**, *17*, 689–701. [CrossRef] [PubMed]
6. Saada, A.; Bar-Meir, M.; Belaiche, C.; Miller, C.; Elpeleg, O. Evaluation of enzymatic assays and compounds affecting ATP production in mitochondrial respiratory chain complex I deficiency. *Anal. Biochem.* **2004**, *335*, 66–72. [CrossRef] [PubMed]
7. Korman, S.H.; Andresen, B.S.; Zeharia, A.; Gutman, A.; Boneh, A.; Pitt, J.J. 2-ethylhydracrylic aciduria in short/branched-chain acyl-CoA dehydrogenase deficiency: Application to diagnosis and implications for the R-pathway of isoleucine oxidation. *Clin. Chem.* **2005**, *51*, 610–617. [CrossRef] [PubMed]
8. Elpeleg, O.N.; Amir, N.; Christensen, E. Variability of clinical presentation in fumarate hydratase deficiency. *J. Pediatr.* **1992**, *121*, 752–754. [CrossRef]
9. Elpeleg, O.; Mandel, H.; Saada, A. Depletion of the other genome-mitochondrial DNA depletion syndromes in humans. *J. Mol. Med.* **2002**, *80*, 389–396. [CrossRef] [PubMed]

10. Schon, E.A.; Hirano, M.; DiMauro, S. Mitochondrial encephalomyopathies: Clinical and molecular analysis. *J. Bioenerg. Biomembr.* **1994**, *26*, 291–299. [CrossRef] [PubMed]

11. Spiegel, R.; Soiferman, D.; Shaag, A.; Shalev, S.; Elpeleg, O.; Saada, A. Novel Homozygous Missense Mutation in SPG20 Gene Results in Troyer Syndrome Associated with Mitochondrial Cytochrome c Oxidase Deficiency. *JIMD Rep.* **2016**. [CrossRef]

12. Naini, A.; Kaufmann, P.; Shanske, S.; Engelstad, K.; de Vivo, D.C.; Schon, E.A. Hypocitrullinemia in patients with MELAS: An insight into the "MELAS paradox". *J. Neurol. Sci.* **2005**, *229–230*, 187–193. [CrossRef] [PubMed]

13. Boczonadi, V.; Horvath, R. Mitochondria: Impaired mitochondrial translation in human disease. *Int. J. Biochem. Cell. Biol.* **2014**, *48*, 77–84. [CrossRef] [PubMed]

14. Magner, M.; Dvorakova, V.; Tesarova, M.; Mazurova, S.; Hansikova, H.; Zahorec, M.; Brennerova, K.; Bzduch, V.; Spiegel, R.; Horovitz, Y.; et al. TMEM70 deficiency: Long-term outcome of 48 patients. *J. Inherit. Metab. Dis.* **2015**, *38*, 417–426. [CrossRef] [PubMed]

15. Zeharia, A.; Friedman, J.R.; Tobar, A.; Saada, A.; Konen, O.; Fellig, Y.; Shaag, A.; Nunnari, J.; Elpeleg, O. Mitochondrial hepato-encephalopathy due to deficiency of QIL1/MIC13 (C19 or f70), a MICOS complex subunit. *Eur. J. Hum. Genet.* **2016**, *24*, 1778–1782. [CrossRef] [PubMed]

16. Mandel, H.; Saita, S.; Edvardson, S.; Jalas, C.; Shaag, A.; Goldsher, D.; Vlodavsky, E.; Langer, T.; Elpeleg, O. Deficiency of HTRA2/Omi is associated with infantile neurodegeneration and 3-methylglutaconic aciduria. *J. Med. Genet.* **2016**, *53*, 690–696. [CrossRef] [PubMed]

17. Phypers, B.; Pierce, T. Lactate physiology in health and disease. *Contin. Educ. Anaesth. Crit. Care Pain* **2006**, *6*, 128–132. [CrossRef]

18. Kumps, A.; Duez, P.; Mardens, Y. Metabolic, nutritional, iatrogenic, and artifactual sources of urinary organic acids: A comprehensive table. *Clin. Chem.* **2002**, *48*, 708–717. [PubMed]

19. Suomalainen, A. Fibroblast growth factor 21: A novel biomarker for human muscle-manifesting mitochondrial disorders. *Expert Opin. Med. Diagn.* **2013**, *4*, 313–317. [CrossRef] [PubMed]

20. Lehtonen, J.M.; Forsström, S.; Bottani, E.; Viscomi, C.; Baris, O.R.; Isoniemi, H.; Höckerstedt, K.; Österlund, P.; Hurme, M.; Jylhävä, J.; et al. FGF21 is a biomarker for mitochondrial translation and mtDNA maintenance disorders. *Neurology* **2016**, *87*, 2290–2299. [CrossRef] [PubMed]

21. Yatsuga, S.; Fujita, Y.; Ishii, A.; Fukumoto, Y.; Arahata, H.; Kakuma, T.; Kojima, T.; Ito, M.; Tanaka, M.; Saiki, R.; et al. Growth differentiation factor 15 as a useful biomarker for mitochondrial disorders. *Ann. Neurol.* **2015**, *78*, 814–823. [CrossRef] [PubMed]

Journal of
Clinical Medicine

MDPI

Article

Mitochondrial Modification Techniques and Ethical Issues

Lucía Gómez-Tatay [1,2,3], José M. Hernández-Andreu [2,3] and Justo Aznar [3,*]

1   Escuela de Doctorado Universidad Católica de Valencia San Vicente Mártir, Valencia 46001, Spain;
    lucia.gomez@ucv.es
2   Facultad de Medicina y Odontología, Universidad Católica de Valencia San Vicente Mártir,
    Departamento de Ciencias Médicas Básicas, Grupo de Medicina Molecular y Mitocondrial, Valencia 46001,
    Spain; jmiguel.hernandez@ucv.es
3   Institute of Life Sciences, Universidad Católica de Valencia San Vicente Mártir, Valencia 46001, Spain
*   Correspondence: justo.aznar@ucv.com; Tel.: +34-605-845-544

Academic Editor: Iain P. Hargreaves
Received: 27 December 2016; Accepted: 20 February 2017; Published: 24 February 2017

Abstract: Current strategies for preventing the transmission of mitochondrial disease to offspring include techniques known as mitochondrial replacement and mitochondrial gene editing. This technology has already been applied in humans on several occasions, and the first baby with donor mitochondria has already been born. However, these techniques raise several ethical concerns, among which is the fact that they entail genetic modification of the germline, as well as presenting safety problems in relation to a possible mismatch between the nuclear and mitochondrial DNA, maternal mitochondrial DNA carryover, and the "reversion" phenomenon. In this essay, we discuss these questions, highlighting the advantages of some techniques over others from an ethical point of view, and we conclude that none of these are ready to be safely applied in humans.

Keywords: mitochondrial disease; mitochondrial replacement; gene editing; ethics; pronuclear transfer; maternal spindle transfer; polar body transfer; CRISPR; TALENs

---

1. Introduction

Mitochondria are organelles present in the cytoplasm of most eukaryotic cells. Although their main function is the production of cellular energy, they also play an important role in other cell processes, such as calcium signalling, regulation of cell metabolism, embryonic development and programmed cell death [1]. In addition, they are implicated in the pathogenesis of numerous diseases, in particular neurodegenerative disorders [2].

These organelles contain their own DNA, known as mitochondrial DNA (mtDNA) [3], which is a circular double-helix DNA molecule, which in humans contains 37 genes: 13 of these code for a polypeptide involved in the respiratory chain, 22 for transfer RNAs (tRNA) and two for ribosomal RNAs (rRNA), all responsible for the translation of these 13 peptides [4]. The mitochondrial electron transport chain is composed of freely moving respiratory complexes and mobile electron carriers that coexist with larger structures called respiratory supercomplexes [5].

Mitochondrial biogenesis and function is dual, depending on both the nuclear and mitochondrial genome. Thus, the replication of mtDNA, its packaging in nucleoids (DNA-protein complexes), and its transcription and translation are processes that depend, to a large extent, on nuclear-encoded proteins [6]. Furthermore, the processes of mitochondrial fusion and fission, which enable intermitochondrial cooperation and compartmentalisation of organelles, respectively, are controlled by products that come completely from the expression of the nDNA [7]. With respect to mitochondrial function, 79 of the 92 subunits that comprise the oxidative phosphorylation (OXPHOS) system are encoded by the nDNA [6]. Thus, primary respiratory chain defects may be due to hereditary alterations in the mtDNA (deletions, rearrangements or point mutations) or nDNA genes that encode subunits for this system, as well as somatic mutations resulting from the action of free radicals, which either directly damage the mtDNA or prevent correct repair of the damage [8].

Mitochondrial alterations cause a decrease in cellular energy that can affect different organs, expressing various clinical phenotypes [6], and can cause significant morbidity and mortality [9,10]. The prevalence of diseases due to mutations in mtDNA is approximately 1 per 5000 individuals, although it may be much higher in certain regions due to genetic founder mutations and high consanguinity [2]. Furthermore, one in 200 healthy individuals is a carrier of a pathogenic mitochondrial mutation that can affect the offspring of female carriers [11].

The proportion of mutant mtDNA can vary between tissues and over time. In the case of the most common point mutations, the disease manifests at cellular level if a threshold of 80%–90% mutated mitochondria is exceeded [12,13]. The proportion necessary for the disease to manifest varies depending on the mutation, the tissue and even on the individual, since environmental factors, physical exercise or the nuclear genetic load itself can also have an effect [14].

mtDNA is inherited exclusively from the mother, however, the level of heteroplasmy varies between individuals descended from the same mutant mtDNA mutation carrier mother. This is due to the "bottleneck" effect that occurs in mitochondrial transmission. Only a fraction of the mother's mitochondria pass to the offspring, which explains the variation in the level of heteroplasmy between different generations and between siblings. This genetic drift has been thought to be random, but recent studies point towards differences in the behaviour of the mtDNA bottleneck, depending on the specific mtDNA mutation [15]. Knowing the expected probability for the heteroplasmy values in the offspring is important for genetic counselling of the future parents [16–18].

2. Treatment of Mitochondrial Diseases

Although vitamin supplements, drugs and physical exercise have been used as treatment in isolated cases and small clinical trials, there is currently no evidence on the effectiveness of these interventions on mitochondrial disorders [19], so new treatment approaches are being developed [2]. However, the highest expectations have been placed on two types of novel techniques that seem to have great potential for application, so that both women affected by a disease due to an alteration in their mtDNA and asymptomatic carriers can have children free from the mutation.

The first group is based on the use of healthy donor mitochondria. These are known as mitochondrial replacement techniques: maternal spindle transfer (MST), pronuclear transfer (PNT), and the most recent, polar body transfer (PBT). Although the first two were authorised for clinical use in the United Kingdom in October 2015 [20], the Human Fertilisation and Embryology Authority (HFEA) announced in June 2016 that the safety and efficacy of these techniques had to be confirmed before any medical centre could request a license to offer mitochondrial donation. To that end, a group of scientists was convened to review the latest advances in this respect [21]. The review by the panel of experts was published in November 2016, and recommends that "in specific circumstances, MST and PNT are cautiously adopted in clinical practice where inheritance of the disease is likely to cause death or serious disease and where there are no acceptable alternatives" [22]. After that, on 15 December 2016, the HFEA approved the use of mitochondrial donation in certain specific cases. Clinics wishing to offer these techniques to patients can now apply to the HFEA for permission to do so and then two committees will assess the suitability of the clinic and each particular clinic case [23].

These techniques are aimed at eradicating the maternal mtDNA in the individual's cells. Nevertheless, there is always some carryover, which can mean that mutant mtDNA levels increase during subsequent development—a phenomenon known as "genetic instability" [24], "genetic drift" [25] or "reversion" [26]—and the disease reappears in later generations. A recent study calculated that, for a clinical threshold of 60%, reducing the mutant mtDNA transferred to below 5% would eradicate the disease forever in that lineage, while if this figure is exceeded, the likelihood that the disease will reappear in subsequent generations is high, so it is important to limit mutant mtDNA transmission to levels below 3% [27]. These low levels have already been achieved with the MST technique in primates [28] and human oocytes [29], with PNT in preimplantation human embryos [30], and with PBT in mouse oocytes and embryos [31]. However, these are probabilistic calculations, so no claims can be made with complete certainty in this respect. Moreover, there is no data of this type on the "reversion" phenomenon.

These techniques may be relatively easy to carry out and feasible for many clinics which can perform intracytoplasmic sperm injection (ICSI), if they can use donor eggs, but realizing them with the precision that is required for optimal results is not so simple. Thus, in the last scientific review of the safety and efficacy of mitochondrial donation to the HFEA, the panel of experts state that "key recommendations are conditional on a number of considerations, including a requirement for appropriate levels of skill being demonstrated by named practitioners within a named clinic, and relevant key performance indicators being met" [22]. In addition, in the report of the the Institute of Medicine (IOM) of the National Academies of Sciences, Engineering, and Medicine to the Food and Drug Administration (FDA), there is a section devoted to the "Expertise of Investigators and Centers" where they point out that "Most MRT approaches contemplated at present would involve highly intricate micro-manipulations of human gametes and/or embryos. Use of the techniques would therefore require operator skill, which evolves over time, varies from one individual to another, and resists specification in a protocol" ([32], p. 138). Finally, in the announcement of the HFEA on 15 December, it states that "HFEA's Licence Committee will first assess a clinic's suitability, looking at existing staff expertise, skill and experience at the clinic, as well as its equipment and general environment" [23].

The second group includes two gene editing techniques: CRISPR-Cas 9 (clustered regularly interspaced short palindromic repeats) and TALENs (transcription activator-like effector nucleases). These techniques, unlike the previous, have not been specifically designed to act on the mitochondria, but they can also be used to correct the mtDNA. Gene editing has been applied in some studies to reduce the levels of mutant mtDNA in heteroplasmic cells [33–36]. However, in order for its action to prevent the transgenerational transmission of mitochondrial diseases, it needs to act on the germline.

2.1. Mitochondrial Replacement Techniques

2.1.1. Pronuclear Transfer

PNT consists of performing in vitro fertilisation using the eggs of the affected woman—whose mitochondria contain mutant mtDNA—and the sperm of the future father, and subsequent extraction of the pronuclei on day 1 of development, leaving behind most of the mutated mitochondria. These pronuclei are transferred to an enucleated zygote with healthy mitochondria (Figure 1); they are transferred to an enucleated zygote, not an egg, since the developmental state must be the same. The hybrid zygote is then developed in vitro until it reaches an appropriate state for transfer to the uterus. Thus, this technique is not strictly preventive, since the gene transfer takes place once the zygote is produced.

Craven et al. applied this technique in embryos with an abnormal number of pronuclei, and succeeded in eliminating more than 98% of the maternal mitochondria [30], which, in principle, is sufficient to prevent clinical manifestation of the disease [37], and its transmission to subsequent generations [27]. PNT has also been applied in normally-fertilised human embryos, with a percentage of mtDNA carryover that did not exceed 5% in any case, and which in most embryos was less than 2% [25]. However, it was observed that, in a stem cell line derived from a blastocyst with 4% mtDNA carryover, the mother's mtDNA gradually increased its proportion with respect to that of the donor. The causes of this reversion are unknown, although it is speculated (among other reasons) that one haplotype may have a replicative advantage over another in specific combinations [22]. The case of an infertile woman who had become pregnant with triplets using this technique was later published, although none of the foetuses reached full term [38].

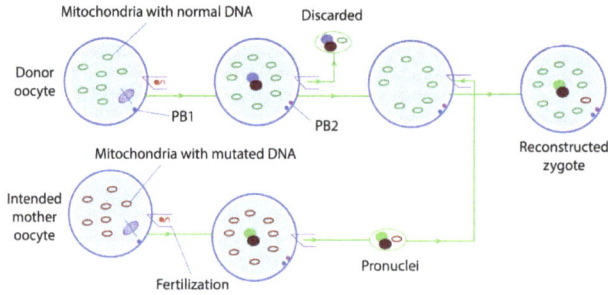

Figure 1. Pronuclear transfer.

2.1.2. Maternal Spindle Transfer

MST consists in extracting the chromosomes in metaphase II from the mother's egg—whose mtDNA has some mutation—to then transfer them to a healthy donor egg, in which the chromosomes have been removed. The hybrid egg is fertilised in vitro and then transferred to the mother's uterus (Figure 2). This technique, however, is strictly preventive, as the individual created will be free from mitochondrial disease from the moment of conception.

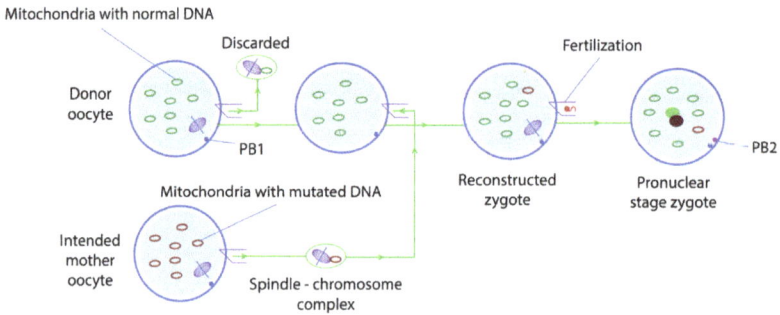

Figure 2. Maternal spindle transfer.

MST was performed in primates (*Macaca mulatta*) in 2009, resulting in the birth of four healthy monkeys, in which the presence of maternal mitochondria was not detected, with a sensitivity of 3% [28]. These were the first animals born following an MST procedure. The technique was then tested in human eggs, and although 52% of the zygotes were abnormally fertilised, the rest were able to develop to blastocysts and produce stem cells in a manner similar to the controls [29]. So far, the efficiency of the technique has improved to reach a carryover less than 1% [26]. However, Yamada et al. and Kang et al. observed that, despite the low levels of mtDNA carryover, there was sometimes gradual loss of the donor mtDNA and re-establishment of the maternal haplotype [24,26].

In April 2016, the first child resulting from this technique was born in Mexico [39]. The mother was an asymptomatic carrier of a mitochondrial mutation that caused Leigh syndrome, a fatal neurological disorder. Despite the fact that she did not have the syndrome, the disease could be transmitted to her children and, in fact, she had suffered four miscarriages and had two children with the disease, who died at the ages of six years and eight months, respectively. The child, who has 1% of its mother's mtDNA, was healthy at three months, although it is not known if any abnormality might appear in the future.

2.1.3. Polar Body Transfer

A potential new technique for mitochondrial replacement, PBT, was described in a 2014 publication [31].

The first polar body (PB1) is formed during egg maturation. In this process, the DNA duplicates, so that the egg contains four chromosome sets. Of these, two will remain within the egg, while the other two will package, forming the PB1, which is extruded and will not be present in the resulting embryo. The second polar body (PB2) is formed during fertilisation. One set of the remaining chromosomes is packaged, forming the PB2, while the other set will form the nuclear DNA of the embryo together with the sperm DNA.

Polar bodies contain very few mitochondria, which is an advantage for avoiding mitochondrial carryover. PBT consists in transferring the PB1 to an unfertilised enucleated donor egg (PB1T) or the PB2 to a half enucleated zygote (PB2T) (Figure 3). Thus, the first strategy is strictly preventive, while the second is not.

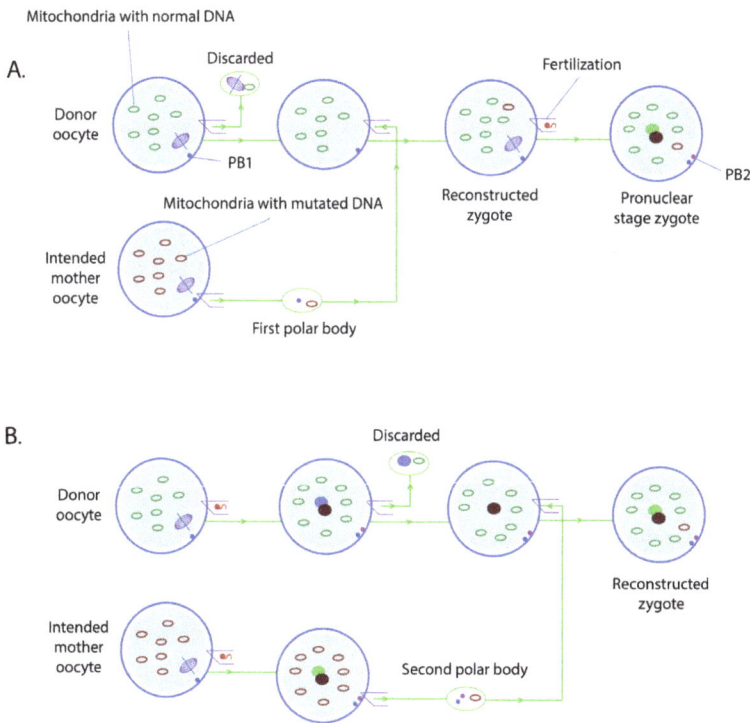

Figure 3. (**A**) first polar body transfer; (**B**) second polar body transfer.

Wang et al. compared PB1T with MST and PB2T with PNT [31]. With regard to the developmental outcomes, they found that the rate of embryos developing to the blastocyst stage was the same for PB1T and MST (87.5% and 85.7%, respectively), while PB2T was less efficient than PNT (55.5% and 81.3%, respectively). The number of live births was also similar to those they obtained for unmanipulated embryos (control). Regarding donor mtDNA carryovers in the F1 and F2 offspring, they found that mtDNA carryover with PNT was much higher than with PB1T, MST and PB2T, which achieved low or undetectable mtDNA carryover. This could be due to the mtDNA amplification around pronuclei that occurs in zygotic activation [31], or to the inexperience of Wang et al. with this technique [40].

2.2. Mitochondrial Gene Editing

2.2.1. CRISPR/Cas9

CRISPR/Cas9 was identified by Mojica as a natural system that provides bacteria with an adaptive response against viruses [41]. In 2012, Doudna and Charpentier published a study in which they detailed how this system could be used to perform programmed gene editing in different cell types [42]. The method consists in using a custom single guide RNA (sgRNA) fragment that has a dual function. On one hand, it acts as a guide to find the piece of DNA to be modified and binds to it. On the other, it recruits the enzyme Cas9, whose function is to cut the DNA. This enables the desired pieces of DNA to be cut, allowing the modification or removal of specific sequences. Unlike other gene editing methods, CRISPR-Cas9 is cheap, easy to use and very efficient, with the result that its use has become widespread in laboratories throughout the world within a very short period of time.

We are only beginning to glimpse the enormous possibilities offered by this new biotechnology tool, which as well as a multitude of applications in the medical field, has environmental, agricultural and livestock applications [43]. The healthcare applications arouse most interest due to their direct impact on people's lives, and at the same time the most controversy, mainly in relation to germline genetic modification (gametes and embryos).

Although there is considerable reluctance among the scientific community in the use of this technique in the germ line [44,45], studies using non-viable embryos have already been carried out [46,47]. These studies show that the technique worked with a low efficiency of on-target gene modification and that it generated off-target mutations and mosaicism in the embryos.

CRISPR/Cas9 has already been successfully employed to produce mitochondrial sequence-specific cleavage, as a proof of concept of the potential of this technique to target specific mitochondrial genes [48]. In the same work, researchers engineered a new version of the enzyme Cas9, mitoCas9, whose localization is restricted to mitochondria matrix. This is highly relevant, since it would reduce the risk of off-target mutations in the embryos and prospective children.

2.2.2. TALENs

TALENs are engineered nucleases composed of a transcription activator-like effector DNA-binding domain from Xanthomonas fused to a FokI nuclease domain [49]. When the action of the TALENs is directed specifically at the mtDNA, they are called mito-TALENs [33], which can be used to cleave the mutated mtDNA. In this respect, TALENs has already been used to selectively eliminate defective mtDNA in both unfertilised mouse eggs and in murine embryos; moreover, these genetically modified mice also gave birth to two successive generations of healthy mice [50]. When an mRNA encoding mito-TALENs was injected in an oocyte from a heteroplasmic mouse model carrying two different mtDNA haplotypes (NZB and BALB), mtDNA heteroplasmy shift was achieved. Furthermore, mito-TALENs successfully targeted and reduced the levels of human mtDNA mutation when injected in human patient cells fused to mouse oocytes.

3. Bioethical Issues

Techniques involving mitochondrial transfer raise a series of bioethical issues. Firstly, a significant increase in the number of egg donors would be required to conduct the necessary research and for its clinical application. It been proposed that in order to manage this increase, a regulation would have to be implemented that would guarantee the donor's well-being, through proper recruitment and support and a limitation on the number of donations per donor that protects them against the negative effects of repeated ovarian hyperstimulation [14]. In this regard, Baylis argues that there is a risk of coercion and exploitation in disadvantaged women [51]. Although we consider this as something to be taken very much into account, it does not constitute an ethical problem intrinsic to mitochondrial replacement techniques, but rather a consequence dependent on the human individuals involved, which must be strictly regulated.

We believe that the ethical issues really intrinsic to the clinical use of these techniques stems mainly from the fact that they involve a genetic modification of the germline, and that children born following their use would have a genetic link to three people: their parents and the donor. We must also consider the fact that it is currently planned to extend the application of these techniques to the field of infertility treatment. Thus, what it was first proposed as an exceptional application in the case of mitochondrial diseases, once it has gained acceptance, the intention would be to extend it to other fields, giving weight to the "slippery slope" argument used by those opposed to these techniques: that once the door has been opened to germline genetic modification, it is merely a matter of time until its use is extended to various applications [52].

3.1. Germline Genetic Modification

Genetic modification of the germline occurs when foreign DNA is introduced into the gametes or early embryo, which will pass to any children and, therefore, future generations. While somatic gene modification is generally accepted, since it does not alter the genome overall and is not transmissible to offspring, germline gene modification is more controversial. In this case, the risks of the genetic modification are very difficult to predict, exacerbated by the fact that the modification will be transmitted to the offspring. Furthermore, persons born following the application of these techniques cannot give their informed consent, and germline genetic manipulation could be used for human enhancement instead of therapeutic purposes.

Nevertheless, there is no general consensus among investigators as regards including these techniques in the field of germline gene modification [53]. The main argument against this inclusion is that the nucleus remains intact, and that no genome is modified, but whole mitochondria are replaced [14]. It is the donor mtDNA that, in different proportions according to the technique, will appear in the individual and in subsequent generations. Therefore, in principle, current ethical restrictions for modification of nDNA would not apply. However, the fact is that it is the nuclear genome that is transferred and replaced, so some authors have claimed that the term "mitochondrial replacement" or "mitochondrial transfer" is used to intentionally misguide the public debate [54–56]. Intentionally or not, the truth is that these terms do not faithfully reflect what is carried out in these techniques, not only because it is the nDNA that is actually transferred, but also because that nucleus is introduced into an egg or zygote which, as well as other mitochondria, has many other cytoplasmic factors of which very little is currently known, apart from the fact that they are critical for early embryonic development [57].

Another argument is that mtDNA in humans contains only 37 genes, approximately 0.1% of the genome [14]. However, oocytes can contain around 200,000 copies of the mitochondrial genome, which is 50% of the total amount of DNA [58]. In addition, if the number of genes are taken into account, more are modified with these techniques, since in the therapy directed to the nucleus a single gene or a few are modified, surely less than 37. In addition, the Y chromosome contains only 86 genes, but its modification would be considered ethically relevant, which leads to think that what is important is the function of these genes [14], which, in the case of mitochondrial DNA, is not known in depth, as explained later.

Finally, mtDNA does not follow Mendelian inheritance, but is transmitted exclusively via maternal means, the reason why the males would not transmit the modification. For this reason and for the aforementioned, Newson and Wrigley propose the term "conditionally inheritable genomic modification" (CIGM) to classify these techniques [53].

In our view, establishing this new category is not correct insofar as it seeks a conceptual dissociation from germline gene modification, which, in our opinion, is any genetic modification that, being carried out in gametes or embryos, will affect all the cells of the resultant organism, which is fulfilled in the case at hand. The method, inheritance mechanism, and type and degree of change are only associated factors that do not affect this definition. Likewise, the fact that it is transmitted to later generations is only a consequence that until now was fulfilled in all cases. In the same way, the

Nuffield Council on Bioethics states that it will "refer to the techniques […] as 'germline therapies' because they introduce a change that is incorporated into the (mitochondrial) genes of the resulting people, and so will be incorporated into the germline that they will go on to develop [14].

However, the really important question is whether the modification of mtDNA can be considered an event of sufficient genetic importance. Certainly, there is an objection to the establishment of a strict dichotomy between nDNA and mtDNA. The view that mitochondria are responsible only for the production of cellular energy, so that its modification would not pose the same ethical disadvantages as the modification of the nDNA, which could alter essential characteristics of the individual [59], is challenged by evidence that suggests that mtDNA may have a relevant effect on the phenotype, influencing our personal identity. This is due to the close interconnection between nDNA and mtDNA, which has become highly specific over evolutionary time [60].

Thus, one study reported the involvement of mtDNA in cognitive functioning in mice [61], while another detected associations between the mtDNA variant and susceptibility to alcoholism [62]. Another study showed that mice bred so that their nDNA and mtDNA came from different strains tended to age with better health than mice whose nDNA and mtDNA corresponded ancestrally [63]. This study suggest that the mtDNA variant in the individual has a great effect on its interaction with the nDNA, which, in turn, affects the synthesis, functionality and half-life of the mitochondrial proteins, oxidative stress, insulin signalling, obesity and the parameters of aging, including telomere shortening and mitochondrial dysfunction, resulting in profound differences in longevity. In other studies with both vertebrate and invertebrate models, these novel combinations have led to negative health effects [60]. In mice, altered respiration and growth in hybrid cell lines [64], reduced male exercise ability and growth [65] and reduced learning and explorative behavior in males [61] have been reported. In fruit flies, the negative health outcomes of mitochondrial replacement include altered male nuclear gene expression [66], altered male aging [67,68], altered female aging [69,70], male infertility [66,71], altered male fertility [72], altered male and female juvenile viability [73] and impaired mitochondrial function [74]. In seed beetles, different studies show altered egg-to-adult development time [75], altered metabolic rates [76], altered male fertility [77] and altered survival in females [78]. Lastly, reduced juvenile viability and reduced mitochondrial function and ATP production has been reported in copepods [79]. Therefore, novel combinations of nDNA and mtDNA occurring in MRT may be mismatched, that is, may not be fully compatible, which can lead to health complications. In our opinion, each specific case should be examined, weighing the possible positive and negative consequences of the application and/or omission of treatment. Thus, the severity of the disease, the probability of transmitting it and the risks derived from the technique must be considered. It is therefore essential to have as much information as possible about the symptoms of the disease, mechanism of transmission of the mtDNA to offspring and the interaction between the mtDNA and nDNA to determine—in addition to the safety problems that could arise with mismatch of these two genomes—if the mtDNA replacement would have any effect on our identity and, if so, to what extent. Thus, further research is necessary in order to minimize risk to children that would be born after the application of these techniques.

The report of the IOM points out that before clinical studies are started, preclinical research must be conducted, not only on animals but also on human gametes and embryos, since it "might be necessary to learn about and optimize the physical manipulations of oocytes and embryos required for MRT, establish optimal timing for applying the techniques in gamete provider and intended mother gametes, and provide a better understanding of the appropriate application of reagents to achieve desired effects" [32]. The last report to the Human Fertilisation and Embryology Authority (HFEA) carried out by an independent expert panel to undertake a review of mitochondrial donation techniques indicates some research areas of interest, such as whole genome sequencing approaches to investigating mitochondrial disease genetics, studies in ES cells to investigate mtDNA bottlenecks and the effects of specific variants on mtDNA dynamics in combination with specific nuclear alleles, or to study methods to reduce mitochondrial carryover when performing the techniques [22]. With regard to PBT, which is technically less developed, the panel recommends studies "carried out using normal human oocytes subjected to PBT and the embryos and embryonic stem (ES) cell lines derived from them, to explore whether they develop normally and have minimal carryover of mtDNA [...] and an examination of methods to prevent premature activation of oocytes or detect abnormally fertilised oocytes" [40].

In addition, we must examine whether there are other possibilities that are either free from, or have less severe negative effects, in both the medical and moral plane. In this respect, it is important to consider that MST and PB1T act on the egg, while PNT and PB2T require the destruction of one embryo for each embryo produced, which is a considerable and unacceptable ethical difficulty for many. The report of the IOM to the FDA states that "In addition to manipulation, MRT would involve the creation and possible destruction of embryos, both in the research phase and in clinical use. The ethical, social, and policy concerns surrounding the creation and destruction of embryos are long-standing [...]. The manipulation, creation, and destruction of embryos are opposed by a range of groups, and federal funding for research involving these processes is severely restricted [...]. Religious, ethical, social, and policy issues are associated with the creation, manipulation, and destruction of human embryos." ([32], pp. 102–106).

Taking this into account and given that the terms of safety and efficacy available data do not indicate whether one technique is preferable to the other [22], we think that, on ethical grounds, there is a strong reason to prefer MST and PB1T over PNT and PB2T. In fact, the couple who had the first child born of these techniques [39] chose MST for this reason, since they are Muslims [80].

3.2. Application of Mitochondrial Transfer in Cases of Infertility

The factors contained in the cytoplasm of the oocyte are crucial for embryonic development, such that they may be involved in certain fertility problems [57]. Specifically, it has been found that mitochondrial dysfunction is related with various fertility problems [81]. Hence, almost 20 years ago, a technique was developed to make it possible for women who had poor embryonic development and repeated failures of the embryos implantation to have children: so-called ooplasmic transfer [82].

Although the exact mechanism by which this technique contributes to the correct development of the pregnancy is unknown, it is thought that it acts by "rejuvenating" the eggs of infertile women, as it provides better quality cytoplasmic factors, such as mtDNA, mRNA, proteins and other molecules [83]. In 1997, the first baby resulting from this technique was born [84], and by 2001, around 30 had been born [14]. However, some security issues appeared. Two foetuses conceived after the application of the technique were affected by Turner's syndrome and were aborted, one spontaneously and the other induced [85], and a born child was diagnosed with pervasive developmental disorder (PDD) at 18 months [86].

After the publication of these cases, in 2001, the United States Food and Drug Administration banned the practice of this technique. However, it is offered in other countries, such as India, Turkish Republic of Northern Cyprus, Ukraine, Armenia, Georgia, Israel, Turkey, Thailand, Singapore, Germany and Austria [14].

The advent of the new mitochondrial replacement techniques has led to renewed interest in this approach to infertility, giving rise to a lively debate in the scientific community [87].

In relation to this, the company Ovascience (Waltham, MA, United States) offers its Augment treatment, which is designed to improve a patient's egg health, when it is compromised due to poor egg quality, age or other reasons. In this treatment, mitochondria from a patient's own immature egg cells are obtained from an ovarian tissue sample and added to the patient's mature eggs along with sperm during in vitro fertilisation (IVF). In 2013, the first child resulting from this technique was born [88].

Another possibility is to use donor mitochondria, as in the mitochondrial transfer techniques. Two Ukrainian women with fertility problems but with healthy mitochondria have already become pregnant using this method [89]. Similarly, Zhang et al. used PNT to achieve a pregnancy in a woman who had had two failed IVF cycles, achieving a triplet pregnancy, none of which reached full term [38].

One central question to be resolved is whether mitochondrial replacement really improves the health of the egg, thus improving fertility. As regards Augment, no animal studies with a control group have been done to determine the efficacy of this technique in improving fertility, or its safety for offspring [87]. With respect to mitochondrial transfer, Cohen, one of the doctors who participated in the development of ooplasmic transfer points out that this cannot be emphatically affirmed. Since only a small number of women participated in his studies, there was no control group and the ooplasm contains many other factors in addition to the mitochondria, which could explain the successful cases [87]. In fact, in the UK, the use of mitochondrial replacement techniques has been approved in cases of mitochondrial disease only and not to treat infertility.

When the ethical and safety problems of these techniques have not yet been resolved, they continue to take steps forward in its clinical application [90], strengthening the slippery slope argument. The fear is that boundaries for use of these techniques would continue to be eroded. Françoise Baylis states: "It provides scientists with 'a quiet way station' in which to refine the micromanipulations techniques essential for other human germline interventions (including nDNA germline modification) and human cloning" ([56], p. 12). Thus, research in this field approaches us to the possible modification of the nDNA in the germinal line not only conceptually, but also technically, which could eventually culminate in the production of "designer babies". In an earlier paper, the author suggests that these techniques may be used for creating genetic ties in lesbian couples [51].

This is related with a second central question, which is whether the two objectives of reproductive medicine involving germline genome editing, i.e., infertility treatment and disease prevention, are ethically the same [91]. We think that the reasons why the use of these techniques is generally wanted to be limited, for the time being, for disease prevention [23], are not ethical in nature, but have more to do with safety concerns. Limiting its application to these concrete cases implies that if things go wrong, the number of affected individuals will be much lower than if their use had also extended to cases of infertility, which are much more common. However, if the safety of these techniques is proven, there is no doubt that their use will be generalized to other fields, as this is already happening nowadays. Certainly, if their safety and efficacy were proven, it would be difficult to deny treatment to some and to allow it to others on ethical grounds, since in both cases it is not a matter of curing a patient but of producing a new individual.

3.3. Donor-Recipient Relationship

In terms of the donor–recipient relationship, as mitochondria have their own DNA, their donations have implications not seen in organ or tissue donation, since any children conceived will have a genetic link with three people: their parents and the donor. Today, only a few children have been born following ooplasmic transfer, a technique which consists of adding ooplasm (with its mitochondria) from a young healthy donor to the eggs of a woman with fertility problems. There is no evidence that these people have attempted to establish any type of relationship with their donors or vice versa. However, given the small number of cases, these data do not have great importance [14].

Thus, only assumptions can be made about the consequences that being genetically related to three people will have for the child. In 2014, BBC News introduced a girl, named Alana Saarinen, who was born from ooplasmic transfer, and stated that she would not consider the donor of her mitochondria a third parent [92]. In fact, taking into account the minimal proportion of DNA contributed by the mitochondria (0.1%), it does not seem reasonable to consider the donor as a third progenitor (or a second mother). Therefore, this term does not seem correct for referring to the donor. Moreover, calling the donor "mother" could be harmful for the child, since it could affect the development of their personal identity and their perception of the parental unit.

As regards the child's possible interest in contacting the donor or vice versa, this is something that could happen, as occurs in other cases of organ, tissue or gamete donation. Thus, mitochondrial donation techniques must be legally regulated so that matters of confidentiality and possible contact with the donors are guaranteed.

One potential way of avoiding this problem is to apply the new gene editing techniques (CRISPR and TALENs) to the mitochondria, in the unfertilised egg (which would be a preventive approach) or in the already created embryo (which would be a curative approach). This would involve correcting the mitochondrial mutation by eliminating the mutant gene and replacing it with the correct one. In this case, the contribution of a donor would not be necessary, so the child would have only its parent's DNA. Nevertheless, it still remains germline editing.

4. Conclusions

It is important to consider that not all the techniques used or proposed in the prevention of mitochondrial disease have the same ethical implications. Thus, within the mitochondrial replacement techniques, MST and PB1T act on the female gamete, while PNT and PB2T entail the destruction of one embryo for every healthy embryo produced, so from an ethical point of view, we consider that the first two are more advisable. This consideration is also effective in the case of the use of mitochondrial replacement to resolve fertility issues. Apart from these differences, the safety evidence up to now is far from reassuring in all cases.

Moreover, gene editing techniques do not require the intervention of a donor with healthy mitochondria, which avoids the problem of the genetic link of the individual with three persons, and with the legal and ethical problems that this entails.

From our perspective, studies must be conducted in animal models on mitochondria–nucleus communication, transmission of mtDNA to the offspring—especially as regards the variable transmission of heteroplasmy due to the bottleneck effect—and symptoms of mitochondrial diseases, before we can successfully undertake these techniques. Since animal studies have limitations in predicting outcomes in humans, subsequent research in humans would be necessary.

It therefore does not seem prudent to continue moving forward in the application of mitochondrial replacement techniques in humans, nor for infertility treatment nor for disease prevention, when there is still so much to discover about the biology of mtDNA, more so when the intention is not to treat sick people, but to produce new individuals in vitro. We therefore propose a moratorium on their use in humans until we have gained an in depth understanding of the biological mechanisms involved.

Acknowledgments: The authors would like to thank Samuel Martos Mínguez for his invaluable help with Figures 1–3 and with the graphic abstract. This research has been supported by a studentship from the Catholic University of Valencia San Vicente Mártir.

Author Contributions: All authors have substantially contributed to the design of the study and preparation of the manuscript. L.G-T. was responsible for reviewing the technical aspects of each method and preparing the manuscript. J.M.H-A. was responsible of the study conception and design. J.A. was responsible of the critical revision of the manuscript.

Conflicts of Interest: The authors declare no conflict of interest.

References

1. Van der Giezen, M.; Tovar, J. Degenerate mitochondria. *EMBO Rep.* **2005**, *6*, 525–530. [CrossRef] [PubMed]
2. Gorman, G.S.; Chinnery, P.F.; DiMauro, S.; Hirano, M.; Koga, Y.; McFarland, R.; Suomalainen, A.; Thorburn, D.R.; Zeviani, M.; Turnbull, D.M. Mitochondrial diseases. *Nat. Rev. Dis. Primers* **2016**. [CrossRef] [PubMed]
3. Nass, S.; Nass, M.M.K. Intramitochondrial fibers with DNA characteristics II. Enzymatic and Other Hydrolytic Treatments. *J. Cell Biol.* **1963**, *19*, 613–629. [CrossRef] [PubMed]
4. Andrews, R.M.; Kubacka, I.; Chinnery, P.F.; Lightowlers, R.N.; Turnbull, D.M.; Howell, N. Reanalysis and revision of the Cambridge reference sequence for human mitochondrial DNA. *Nat. Genet.* **1999**, *23*, 147. [PubMed]

5. Lapuente-Brun, E.; Moreno-Loshuertos, R.; Acín-Pérez, R.; Latorre-Pellicer, A.; Colás, C.; Balsa, E.; Perales-Clemente, E.; Quirós, P.M.; Calvo, E.; Rodríguez-Hernández, M.A.; et al. Supercomplex assembly determines electron flux in the mitochondrial electron transport chain. *Science* **2013**, *340*, 1567–1570. [CrossRef] [PubMed]

6. Chinnery, P.F.; Hudson, G. Mitochondrial genetics. *Br. Med. Bull.* **2013**, *106*, 135–159. [CrossRef] [PubMed]

7. Youle, R.J.; van der Bliek, A.M. Mitochondrial fission, fusion, and stress. *Science* **2012**, *337*, 1062–1065. [CrossRef] [PubMed]

8. Schapira, A.H. Mitochondrial diseases. *Lancet* **2012**, *379*, 1825–1834. [CrossRef]

9. McFarland, R.; Taylor, R.W.; Turnbull, D.M. A neurological perspective on mitochondrial disease. *Lancet Neurol.* **2010**, *9*, 829–840. [CrossRef]

10. DiMauro, S.; Schon, E.A.; Carelli, V.; Hirano, M. The clinical maze of mitochondrial neurology. *Nat. Rev. Neurol.* **2013**, *9*, 429–444. [CrossRef] [PubMed]

11. Elliott, H.R.; Samuels, D.C.; Eden, J.A.; Relton, C.L.; Chinnery, P.F. Pathogenic mitochondrial DNA mutations are common in the general population. *Am. J. Hum. Genet.* **2008**, *83*, 254–260. [CrossRef] [PubMed]

12. Chinnery, P.F.; Howell, N.; Lightowlers, R.N.; Turnbull, D.M. Molecular pathology of MELAS and MERRF. The relationship between mutation load and clinical phenotypes. *Brain* **1997**, *120*, 1713–1721. [PubMed]

13. White, S.L.; Collins, V.R.; Wolfe, R.; Cleary, M.A.; Shanske, S.; DiMauro, S.; Dahl, H.H.; Thorburn, D.R. Genetic counseling and prenatal diagnosis for the mitochondrial DNA mutations at nucleotide 8993. *Am. J. Hum. Genet.* **1999**, *65*, 474–482. [CrossRef] [PubMed]

14. Nuffield Council on Bioethics. Novel Techniques for the Prevention of Mitochondrial DNA Disorders: An Ethical Review. Available online: http://nuffieldbioethics.org/wp-content/uploads/2014/06/Novel_techniques_for_the_prevention_of_mitochondrial_DNA_disorders_compressed.pdf (accessed on 23 December 2016).

15. Wilson, I.J.; Carling, P.J.; Alston, C.L.; Floros, V.I.; Pyle, A.; Hudson, G.; Sallevelt, S.C.; Lamperti, C.; Carelli, V.; Bindoff, L.A.; et al. Mitochondrial DNA sequence characteristics modulate the size of the genetic bottleneck. *Hum. Mol. Genet.* **2016**, *25*, 1031–1041. [CrossRef] [PubMed]

16. Chinnery, P.F.; Howell, N.; Lightowlers, R.N.; Turnbull, D.M. Genetic counseling and prenatal diagnosis for mtDNA disease. *Am. J. Hum. Genet.* **1998**, *63*, 1908–1911. [CrossRef] [PubMed]

17. Thorburn, D.R.; Dahl, H.H. Mitochondrial disorders: Genetics, counseling, prenatal diagnosis and reproductive options. *Am. J. Med. Genet.* **2001**, *106*, 102–114. [CrossRef] [PubMed]

18. Brown, D.T.; Herbert, M.; Lamb, V.K.; Chinnery, P.F.; Taylor, R.W.; Lightowlers, R.N.; Craven, L.; Cree, L.; Gardner, J.L.; Turnbull, D.M. Transmission of mitochondrial DNA disorders: Possibilities for the future. *Lancet* **2006**, *368*, 87–89. [CrossRef]

19. Pfeffer, G.; Majamaa, K.; Turnbull, D.M.; Thorburn, D.; Chinnery, P.F. Treatment for mitochondrial disorders. *Cochrane Database Syst. Rev.* **2012**, *18*, CD004426.

20. HFEA Approves Licence Application to Use Gene Editing in Research. Available online: http://www.hfea.gov.uk/10187.html (accessed on 23 December 2016).

21. HFEA Reconvenes Independent Expert Panel and Launches Call for Evidence. Available online: http://www.hfea.gov.uk/10363.html (accessed on 23 December 2016).

22. Scientific Review of the Safety and Efficacy of Methods to Avoid Mitochondrial Disease through Assisted Conception: 2016 Update. Available online: http://www.hfea.gov.uk/docs/Fourth_scientific_review_mitochondria_2016.PDF (accessed on 23 December 2016).

23. HFEA Permits Cautious Use of Mitochondrial Donation in Treatment, Following Advice from Scientific Experts. Available online: http://www.hfea.gov.uk/10563.html (accessed on 23 December 2016).

24. Yamada, M.; Emmanuele, V.; Sanchez-Quintero, M.J.; Sun, B.; Lallos, G.; Paull, D.; Zimmer, M.; Pagett, S.; Prosser, R.W.; Sauer, M.V.; et al. Genetic Drift Can Compromise Mitochondrial Replacement by Nuclear Transfer in Human Oocytes. *Cell Stem Cell* **2016**, *18*, 749–754. [CrossRef] [PubMed]

25. Hyslop, L.A.; Blakeley, P.; Craven, L.; Richardson, J.; Fogarty, N.M.; Fragouli, E.; Lamb, M.; Wamaitha, S.E.; Prathalingam, N.; Zhang, Q.; et al. Towards clinical application of pronuclear transfer to prevent mitochondrial DNA disease. *Nature* **2016**, *534*, 383–386. [CrossRef] [PubMed]

26. Kang, E.; Wu, J.; Gutierrez, N.M.; Koski, A.; Tippner-Hedges, R.; Agaronyan, K.; Platero-Luengo, A.; MartinezRedondo, P.; Ma, H.; Lee, Y.; et al. Mitochondrial replacement in human oocytes carrying pathogenic mitochondrial DNA mutations. *Nature* **2016**, *540*, 270–275. [CrossRef] [PubMed]

27. Samuels, D.C.; Wonnapinij, P.; Chinnery, P.F. Preventing the transmission of pathogenic mitochondrial DNA mutations: Can we achieve long-term benefits from germ-line gene transfer? *Hum. Reprod.* **2013**, *28*, 554–559. [CrossRef] [PubMed]
28. Tachibana, M.; Sparman, M.; Sritanaudomchai, H.; Ma, H.; Clepper, L.; Woodward, J.; Li, Y.; Ramsey, C.; Kolotushkina, O.; Mitalipov, S. Mitochondrial gene replacement in primate offspring and embryonic stem cells. *Nature* **2009**, *461*, 367–372. [CrossRef] [PubMed]
29. Tachibana, M.; Amato, P.; Sparman, M.; Woodward, J.; Sanchis, D.M.; Ma, H.; Gutierrez, N.M.; Tippner-Hedges, R.; Kang, E.; Lee, H.S.; et al. Towards germline gene therapy of inherited mitochondrial diseases. *Nature* **2013**, *493*, 627–631. [CrossRef] [PubMed]
30. Craven, L.; Tuppen, H.A.; Greggains, G.D.; Harbottle, S.J.; Murphy, J.L.; Cree, L.M.; Murdoch, A.P.; Chinnery, P.F.; Taylor, R.W.; Lightowlers, R.N.; et al. Pronuclear transfer in human embryos to prevent transmission of mitochondrial DNA disease. *Nature* **2010**, *465*, 82–85. [CrossRef] [PubMed]
31. Wang, T.; Sha, H.; Ji, D.; Zhang, H.L.; Chen, D.; Cao, Y.; Zhu, J. Polar body genome transfer for preventing the transmission of inherited mitocondrial diseases. *Cell* **2014**, *157*, 1591–1604. [CrossRef] [PubMed]
32. National Academies of Sciences, Engineering, and Medicine. *Mitochondrial Replacement Techniques: Ethical, Social, and Policy Considerations*; National Academies Press: Washington, DC, USA, 2016.
33. Bacman, S.R.; Williams, S.L.; Pinto, M.; Peralta, S.; Moraes, C.T. Specific elimination of mutant mitochondrial genomes in patient-derived cells by mitoTALENs. *Nat. Med.* **2013**, *19*, 1111–1113. [CrossRef] [PubMed]
34. Gammage, P.A.; Rorbach, J.; Vincent, A.I.; Rebar, E.J.; Minczuk, M. Mitochondrially targeted ZFNs for selective degradation of pathogenic mitochondrial genomes bearing large-scale deletions or point mutations. *EMBO Mol. Med.* **2014**, *6*, 458–466. [CrossRef] [PubMed]
35. Hashimoto, M.; Bacman, S.R.; Peralta, S.; Falk, M.J.; Chomyn, A.; Chan, D.C.; Williams, S.L.; Moraes, C.T. MitoTALEN: A General Approach to Reduce Mutant mtDNA Loads and Restore Oxidative Phosphorylation Function in Mitochondrial Diseases. *Mol. Ther.* **2015**, *23*, 1592–1599. [CrossRef] [PubMed]
36. Minczuk, M.; Papworth, M.A.; Kolasinska, P.; Murphy, M.P.; Klug, A. Sequence-specific modification of mitochondrial DNA using a chimeric zinc finger methylase. *Proc. Natl. Acad. Sci. USA* **2006**, *103*, 19689–19694. [CrossRef] [PubMed]
37. Hellebrekers, D.M.; Wolfe, R.; Hendrickx, A.T.; de Coo, I.F.; de Die, C.E.; Geraedts, J.P.; Chinnery, P.F.; Smeets, H.J. PGD and heteroplasmic mitochondrial DNA point mutations: A systematic review estimating the chance of healthy offspring. *Hum. Reprod. Update* **2012**, *18*, 341–349. [CrossRef] [PubMed]
38. Zhang, J.; Zhuang, G.; Zeng, Y.; Grifo, J.; Acosta, C.; Shu, Y.; Liu, H. Pregnancy derived from human zygote pronuclear transfer in a patient who had arrested embryos after IVF. *Reprod. Biomed. Online* **2016**, *33*, 529–533. [CrossRef] [PubMed]
39. Zhang, J.; Liu, H.; Luo, S.; Chavez-Badiola, A.; Liu, Z.; Yang, M.; Munne, S.; Konstantinidis, M.; Wells, D.; Huang, T. First live birth using human oocytes reconstituted by spindle nuclear transfer for mitochondrial DNA mutation causing Leigh syndrome. *Fertil. Steril.* **2016**, *106*, e375–e376. [CrossRef]
40. Review of the Safety and Efficacy of Polar Body Transfer to Avoid Mitochondrial Disease. Available online: http://www.hfea.gov.uk/docs/2014--10--07_-_Polar_Body_Transfer_Review_-_Final.PDF (accessed on 23 December 2016).
41. Mojica, F.J.; Díez-Villaseñor, C.; García-Martínez, J.; Soria, E. Intervening sequences of regularly spaced prokaryotic repeats derive from foreign genetic elements. *J. Mol. Evol.* **2005**, *60*, 174–182. [CrossRef] [PubMed]
42. Jinek, M.; Chylinski, K.; Fonfara, I.; Hauer, M.; Doudna, J.A.; Charpentier, E. A programmable dual-RNA-guided DNA endonuclease in adaptive bacterial immunity. *Science* **2012**, *337*, 816–821. [CrossRef] [PubMed]
43. Ledford, H. CRISPR, the disruptor. *Nat. News* **2015**, *522*, 20–24. [CrossRef] [PubMed]
44. Baltimore, D.; Berg, P.; Botchan, M.; Carroll, D.; Charo, R.A.; Church, G.; Corn, J.E.; Daley, G.Q.; Doudna, J.A.; Fenner, M.; et al. A prudent path forward for genomic engineering and germline gene modification. *Science* **2015**, *348*, 36–38. [CrossRef] [PubMed]
45. Lanphier, E.; Urnov, F.; Haecker, S.E.; Werner, M.; Smolenski, J. Don't edit the human germ line. *Nature* **2015**, *519*, 410–411. [CrossRef] [PubMed]

46. Liang, P.; Xu, Y.; Zhang, X.; Ding, C.; Huang, R.; Zhang, Z.; Lv, J.; Xie, X.; Chen, Y.; Li, Y.; et al. CRISPR/Cas9-mediated gene editing in human tripronuclear zygotes. *Protein Cell* **2015**, *6*, 363–372. [CrossRef] [PubMed]

47. Kang, X.; He, W.; Huang, Y.; Yu, Q.; Chen, Y.; Gao, X.; Sun, X.; Fan, Y. Introducing precise genetic modifications into human 3PN embryos by CRISPR/Cas-mediated genome editing. *J. Assist. Reprod. Genet.* **2016**, *33*, 581–588. [CrossRef] [PubMed]

48. Jo, A.; Ham, S.; Lee, G.H.; Lee, Y.I.; Kim, S.; Lee, Y.S.; Shin, J.H.; Lee, Y. Efficient Mitochondrial Genome Editing by CRISPR/Cas9. *Biomed. Res. Int.* **2015**, *2015*, 305716:1–305716:10. [CrossRef] [PubMed]

49. Cermak, T.; Doyle, E.L.; Christian, M.; Wang, L.; Zhang, Y.; Schmidt, C.; Baller, J.A.; Somia, N.V.; Bogdanove, A.J.; Voytas, D.F. Efficient design and assembly of custom TALEN and other TAL effector-based constructs for DNA targeting. *Nucl. Acids Res.* **2011**, *39*, e82. [CrossRef] [PubMed]

50. Reddy, P.; Ocampo, A.; Suzuki, K.; Luo, J.; Bacman, S.R.; Williams, S.L.; Sugawara, A.; Okamura, D.; Tsunekawa, Y.; Wu, J.; et al. Selective elimination of mitochondrial mutations in the germline by genome editing. *Cell* **2015**, *23*, 161, 459–469. [CrossRef] [PubMed]

51. Baylis, F. The ethics of creating children with three genetic parents. *Reprod. Biomed. Online* **2013**, *26*, 531–534. [CrossRef] [PubMed]

52. Darnovsky, M. A slippery slope to human germline modification. *Nature* **2013**, *499*, 127. [CrossRef] [PubMed]

53. Newson, A.J.; Wrigley, A. Is Mitochondrial Donation Germ-Line Gene Therapy? Classifications and Ethical Implications. *Bioethics* **2017**, *31*, 55–67. [CrossRef] [PubMed]

54. Jones, D. The Other Woman: Evaluating the Language of "Three Parent" Embryos. *Clin. Eth.* **2015**, *10*, 97–106. [CrossRef]

55. Nisker, J. The Latest Thorn by Any Other Name: Germ-Line Nuclear Transfer in the Name of "Mitochondrial Replacement". *J. Obstet. Gynaecol. Can.* **2015**, *37*, 829–831. [CrossRef]

56. Baylis, F. Human Nuclear Genome Transfer (So-Called Mitochondrial Replacement): Clearing the Underbrush. *Bioethics* **2017**, *31*, 7–19. [CrossRef] [PubMed]

57. Kim, K.H.; Lee, K.A. Maternal effect genes: Findings and effects on mouse embryo development. *Clin. Exp. Reprod. Med.* **2014**, *41*, 47–61. [CrossRef] [PubMed]

58. Thorburn, D.; Group Leader of Mitchondrial Research, Murdoch Childrens Research Institute. Personal communication, 2017.

59. Bredenoord, A.L.; Pennings, G.; de Wert, G. Ooplasmic and nuclear transfer to prevent mitochondrial DNA disorders: Conceptual and normative issues. *Hum. Reprod. Update* **2008**, *14*, 669–678. [CrossRef] [PubMed]

60. Reinhardt, K.; Dowling, D.K.; Morrow, E.H. Mitochondrial replacement, evolution and the clinic. *Science* **2013**, *341*, 1345–1346. [CrossRef] [PubMed]

61. Roubertoux, P.L.; Sluyter, F.; Carlier, M.; Marcet, B.; Maarouf-Veray, F.; Chérif, C.; Marican, C.; Arrechi, P.; Godin, F.; Jamon, M.; et al. Mitochondrial DNA modifies cognition in interaction with the nuclear genome and age in mice. *Nat. Genet.* **2003**, *35*, 65–69. [CrossRef] [PubMed]

62. Lease, L.R.; Winnier, D.A.; Williams, J.T.; Dyer, T.D.; Almasy, L.; Mahaney, M.C. Mitochondrial genetic effects on latent class variables associated with susceptibility to alcoholism. *BMC Genet.* **2005**, *6*, S158. [CrossRef] [PubMed]

63. Latorre-Pellicer, A.; Moreno-Loshuertos, R.; Lechuga-Vieco, A.V.; Sánchez-Cabo, F.; Torroja, C.; Acín-Pérez, R.; Calvo, E.; Aix, E.; González-Guerra, A.; Logan, A.; et al. Mitochondrial and nuclear DNA matching shapes metabolism and healthy ageing. *Nature* **2016**, *535*, 561–565. [CrossRef] [PubMed]

64. Moreno-Loshuertos, R.; Acín-Pérez, R.; Fernández-Silva, P.; Movilla, N.; Pérez-Martos, A.; Rodriguez de Cordoba, S.; Gallardo, M.E.; Enríquez, J.A. Differences in reactive oxygen species production explain the phenotypes associated with common mouse mitochondrial DNA variants. *Nat. Genet.* **2006**, *38*, 1261–1268. [CrossRef] [PubMed]

65. Nagao, Y.; Totsuka, Y.; Atomi, Y.; Kaneda, H.; Lindahl, K.F.; Imai, H.; Yonekawa, H. Decreased physical performance of congenic mice with mismatch between the nuclear and the mitochondrial genome. *Genes Genet. Syst.* **1998**, *73*, 21–27. [CrossRef] [PubMed]

66. Innocenti, P.; Morrow, E.H.; Dowling, D.K. Experimental evidence supports a sexspecific selective sieve in mitochondrial genome evolution. *Science* **2011**, *332*, 845–848. [CrossRef] [PubMed]

67. Camus, M.F.; Clancy, D.J.; Dowling, D.K. Dowling, Mitochondria, maternal inheritance, and male aging. *Curr. Biol.* **2012**, *22*, 1717–1721. [CrossRef] [PubMed]

68. Clancy, D.J. Variation in mitochondrial genotype has substantial lifespan effects which may be modulated by nuclear background. *Aging Cell* **2008**, *7*, 795–804. [CrossRef] [PubMed]
69. Dowling, D.K.; Maklakov, A.A.; Friberg, U.; Hailer, F. Applying the genetic theories of ageing to the cytoplasm: Cytoplasmic genetic covariation for fitness and lifespan. *J. Evol. Biol.* **2009**, *22*, 818–827. [CrossRef] [PubMed]
70. Maklakov, A.A.; Friberg, U.; Dowling, D.K.; Arnqvist, G. Within-population variation in cytoplasmic genes affects female life span and aging in *Drosophila melanogaster*. *Evolution* **2006**, *60*, 2081–2086. [CrossRef] [PubMed]
71. Clancy, D.J.; Hime, G.R.; Shirras, A.D. Cytoplasmic male sterility in *Drosophila melanogaster* associated with a mitochondrial CYTB variant. *Heredity* **2011**, *107*, 374–376. [CrossRef] [PubMed]
72. Yee, W.K.; Sutton, K.L.; Dowling, D.K. In vivo male fertility is affected by naturally occurring mitochondrial haplotypes. *Curr. Biol.* **2013**, *23*, R55–R56. [CrossRef] [PubMed]
73. Montooth, K.L.; Meiklejohn, C.D.; Abt, D.N.; Rand, D.M. Mitochondrial-nuclear epistasis affects fitness within species but does not contribute to fixed incompatibilities between species of Drosophila. *Evolution* **2010**, *64*, 3364–3379. [CrossRef] [PubMed]
74. Sackton, T.B.; Haney, R.A.; Rand, D.M. Cytonuclear coadaptation in Drosophila: Disruption of cytochrome c oxidase activity in backcross genotypes. *Evolution* **2003**, *57*, 2315–2325. [CrossRef] [PubMed]
75. Dowling, D.K.; Abiega, K.C.; Arnqvist, G. Temperature-specific outcomes of cytoplasmic-nuclear interactions on egg-to-adult development time in seed beetles. *Evolution* **2007**, *61*, 194–201. [CrossRef] [PubMed]
76. Arnqvist, G.; Dowling, D.K.; Eady, P.; Gay, L.; Tregenza, T.; Tuda, M.; Hosken, D.J. Genetic architecture of metabolic rate: Environment specific epistasis between mitochondrial and nuclear genes in an insect. *Evolution* **2010**, *64*, 3354–3363. [CrossRef] [PubMed]
77. Dowling, D.K.; Nowostawski, A.L.; Arnqvist, G. Effects of cytoplasmic genes on sperm viability and sperm morphology in a seed beetle: Implications for sperm competition theory? *J. Evol. Biol.* **2007**, *20*, 358–368. [CrossRef] [PubMed]
78. Dowling, D.K.; Meerupati, T.; Arnqvist, G. Cytonuclear interactions and the economics of mating in seed beetles. *Am. Nat.* **2010**, *176*, 131–140. [CrossRef] [PubMed]
79. Burton, R.S.; Ellison, C.K.; Harrison, J.S. The sorry state of F2 hybrids: Consequences of rapid mitochondrial DNA evolution in allopatric populations. *Am. Nat.* **2006**, *168*, S14–S24. [CrossRef] [PubMed]
80. Hamzelou, J. Exclusive: World's First Baby Born with New "3 Parent" Technique. *New Sci.* **2016**. Available online: https://www.newscientist.com/article/2107219-exclusive-worlds-first-baby-born-with-new-3-parent-technique/ (accessed on 3 February 2017).
81. Schatten, H.; Sun, Q.Y.; Prather, R. The impact of mitochondrial function/dysfunction on IVF and new treatment possibilities for infertility. *Reprod. Biol. Endocrinol.* **2014**, *12*, 111. [CrossRef] [PubMed]
82. Cohen, J.; Scott, R.; Alikani, M.; Schimmel, T.; Munné, S.; Levron, J.; Wu, L.; Brenner, C.; Warner, C.; Willadsen, S. Ooplasmic transfer in mature human oocytes. *Mol. Hum. Reprod.* **1998**, *4*, 269–280. [CrossRef] [PubMed]
83. Yabuuchi, A.; Beyhan, Z.; Kagawa, N.; Mori, C.; Ezoe, K.; Kato, K.; Aono, F.; Takehara, Y.; Kato, O. Prevention of mitochondrial disease inheritance by assisted reproductive technologies: Prospects and challenges. *Biochim. Biophys. Acta* **2012**, *1820*, 637–642. [CrossRef] [PubMed]
84. Cohen, J.; Scott, R.; Schimmel, T.; Levron, J.; Willadsen, S. Birth of infant after transfer of anucleate donor oocyte cytoplasm into recipient eggs. *Lancet* **1997**, *350*, 186–187. [CrossRef]
85. Barritt, J.A.; Brenner, C.A.; Willadsen, S.; Cohen, J. Spontaneous and artificial changes in human ooplasmic mitochondria. *Hum. Reprod.* **2000**, *15*, 207–217. [CrossRef] [PubMed]
86. Barritt, J.A.; Brenner, C.A.; Malter, H.E.; Cohen, J. Rebuttal: Interooplasmic transfers in humans. *Reprod. Biomed. Online* **2001**, *3*, 47–88. [CrossRef]
87. Couzin-Frankel, J. Reproductive medicine. Eggs' power plants energize new IVF debate. *Science* **2015**, *348*, 14–15. [CrossRef] [PubMed]
88. The Incredible, Surprising, Controversial, New Way to Make a Baby. Available online: http://www.ovascience.com/files/TIME_Magazine_May_2015.pdf (accessed on 23 December 2016).
89. Coghlan, A. Exclusive: '3-Parent' Baby Method Already Used for Infertility. *New Sci.* **2016**. Available online: https://www.newscientist.com/article/2108549-exclusive-3-parent-baby-method-already-used-for-infertility/ (accessed on 23 December 2016).

90. Cook, M. Ethics Ignored in '3-Person Embryo' Technique. *BioEdge* **2016**. Available online: http://www.bioedge.org/bioethics/ethics-ignored-in-3-person-embryo-technique/12045 (accessed on 23 December 2016).
91. Ishii, T. Reproductive medicine involving genome editing: Clinical uncertainties and embryological needs. *Reprod. Biomed. Online* **2017**, *34*, 27–31. [CrossRef] [PubMed]
92. Pritchard, C. The girl with three biological parents. *BBC News Mag.* **2014**. Available online: http://www.bbc.com/news/magazine-28986843?OCID=fbasia&ocid=socialflow_facebook (accessed on 4 February 2017).

Journal of
Clinical Medicine

MDPI

Article

The Effect of Mitochondrial Supplements on Mitochondrial Activity in Children with Autism Spectrum Disorder

Leanna M. Delhey [1,2], Ekim Nur Kilinc [1], Li Yin [3], John C. Slattery [1,2], Marie L. Tippett [1,2], Shannon Rose [1,2], Sirish C. Bennuri [1,2], Stephen G. Kahler [1,2], Shirish Damle [4], Agustin Legido [4], Michael J. Goldenthal [4] and Richard E. Frye [1,2,*]

[1] Arkansas Children's Research Institute, Little Rock, AR 72202, USA; lmdelhey@uams.edu (L.M.D.); ekimkilinc@gmail.com (E.N.K.); jcslattery@uams.edu (J.C.S.); mltippett@uams.edu (M.L.T.); srose@uams.edu (S.R.); scbennutri@uams.edu (S.C.B.); kahlerstepheng@uams.edu (S.G.K.)
[2] Department of Pediatrics, University of Arkansas for Medical Sciences, Little Rock, AR 72202, USA
[3] Child and Adolescent Department, Mental Health Centre, West China Hospital of Sichuan University, Chengdu 610041, China; dr.yinli@hotmail.com
[4] Department of Pediatrics, Drexel University College of Medicine, Neurology Section, St. Christopher's Hospital for Children, Philadelphia, PA 19134, USA; shirish.damle@drexelmed.edu (S.D.); agustin.legido@drexelmed.edu (A.L.); michael.goldenthal@drexelmed.edu (M.J.G.)
* Correspondence: refrye@uams.edu; Tel.: +1-501-364-4662; Fax: +1-501-364-1648

Academic Editor: Ian P. Hargreaves
Received: 31 December 2016; Accepted: 6 February 2017; Published: 13 February 2017

Abstract: Treatment for mitochondrial dysfunction is typically guided by expert opinion with a paucity of empirical evidence of the effect of treatment on mitochondrial activity. We examined citrate synthase and Complex I and IV activities using a validated buccal swab method in 127 children with autism spectrum disorder with and without mitochondrial disease, a portion of which were on common mitochondrial supplements. Mixed-model linear regression determined whether specific supplements altered the absolute mitochondrial activity as well as the relationship between the activities of mitochondrial components. Complex I activity was increased by fatty acid and folate supplementation, but folate only effected those with mitochondrial disease. Citrate synthase activity was increased by antioxidant supplementation but only for the mitochondrial disease subgroup. The relationship between Complex I and IV was modulated by folate while the relationship between Complex I and Citrate Synthase was modulated by both folate and B12. This study provides empirical support for common mitochondrial treatments and demonstrates that the relationship between activities of mitochondrial components might be a marker to follow in addition to absolute activities. Measurements of mitochondrial activity that can be practically repeated over time may be very useful to monitor the biochemical effects of treatments.

Keywords: antioxidants; autism spectrum disorder; B12; Complex I; Complex IV; electron transport chain; fatty acids; folate; mitochondrial disease; mitochondrial dysfunction

1. Introduction

Primary mitochondrial disease, as well as secondary mitochondrial dysfunction, is becoming increasingly recognized [1]. Indeed, the contribution of the mitochondria to many diverse, common disorders such as diabetes, obesity, cancer and heart, neurologic and psychiatric disease is significant. What is less well known is the optimal treatment for mitochondrial disease and dysfunction. Several expert opinion papers provide insight into the recognized management of patients with mitochondrial disease. However, such expert opinion is based on a paucity of clinical evidence [2,3]. Although new novel therapies have undergone increasing investigation recently, most of the published information remains in the preclinical stage, isolated to evidence from model organisms [4]. Although clinical trials have been conducted, the rarity of certain mitochondrial diseases; small subject numbers; heterogeneity in symptoms; severity of specific mitochondrial diseases; short treatment and follow-up periods; variability in outcomes measures; and the use of measures that are not specifically designed to measure mitochondrial outcomes, are all factors which probably contribute to the lack of positive findings in clinical trials [5,6]. Thus, there has been a recent call to develop new biomarkers of mitochondrial function that can be used in future well-designed clinical trials [7].

Biochemical measurements of mitochondrial function can be variable or difficult to obtain. For example, laboratory measures are commonly very sensitive to collection techniques and laboratory processing, resulting in significant variability. Magnetic resonance spectroscopy is a promising technique to non-invasively measure energy metabolism in muscle and brain tissues, but is limited to centers with specialized equipment. In addition, to date, none of these markers have been found to be systematically altered in high-quality clinical trials [7]. Direct measurement of mitochondrial function by enzymology typically requires biopsies that are somewhat invasive, limiting their ability to be repeated to follow the disease status. In 2012, Goldenthal et al. developed and validated the non-invasive buccal swab technique, demonstrating the correspondence between enzymology measurements in buccal tissue and muscle biopsy in individuals with mitochondrial disease [8]. The buccal swab technique has been used to measure mitochondrial function in individuals with mitochondrial disease [8–10], specific genetic syndromes [10,11] and Autism Spectrum Disorder (ASD) [12,13].

ASD is a behaviorally defined disorder which now affects ~2% of children [14]. Recent studies suggest that ASD is linked to mitochondrial dysfunction [13,15,16], although the exact nature of mitochondrial abnormalities in ASD appears to be complicated. For example, classic mitochondrial disease is found in 5% of children with ASD [16], yet up to 50% of children with ASD may have biomarkers of mitochondrial dysfunction [16,17] and a higher rate of abnormal electron transport chain (ETC) activity is found in immune cells [18,19] and post-mortem brain tissue [20]. Perhaps more unique is the fact that ETC activity in muscle [21,22], skin [23], buccal cells [11–13] and the brain [20] has been documented to be significantly increased, rather than decreased, in individuals with ASD, consistent with in vitro data showing elevated mitochondrial respiration in cell lines derived from children with ASD [24,25]. More recently, mitochondrial respiration in cell lines has been shown to be related to the stereotyped behaviors and restricted interests subscale on the Autism Diagnostic Observation Scale (ADOS) with elevated respiratory rates corresponding to worse behavior [26].

Individuals with ASD are a particularly important group of patients that would benefit from a biomarker of mitochondrial dysfunction as well as a marker of the effect of treatments on mitochondrial function. First, the great majority of children with ASD do not have genetic mutations to explain their mitochondrial dysfunction [16], making diagnosis complicated. Second, many children with ASD are treated with supplements that potentially target the mitochondrial but it is unclear whether such treatments influence mitochondrial function [27]. Understanding which treatments would be most helpful and effective for children with ASD, especially on an individual basis, would be tremendously helpful for guiding treatment in a personalized medicine fashion.

In this study, we aimed to ask whether the functional effect of common treatments that target the mitochondria can be measured with a non-invasive buccal swab technique and what are the measures that might be sensitive to the effect of treatment. To this end, we measured the activity of ETC Complex I and IV as well as Citrate Synthase. We not only examined the absolute level of activity of mitochondrial components, but also the relationship between the components, to better understand whether treatments not only modulated the activity level but how the mitochondrial components work together. To this end, we utilized the data from our study of the natural history of mitochondrial function in children with ASD to examine the mitochondrial function on individuals taking and abstaining from common treatments that affect the mitochondrial. Since specific supplements were not systematically manipulated, it is not possible to equate the findings from this study to a clinical trial of specific supplements. Rather, this study is designed to answer the question of whether the technique and measurements used in the study show promise for future research.

2. Material and Methods

The study was approved by the Institutional Review Board at the University of Arkansas for Medical Sciences (Little Rock, AR, USA) under two protocols (#137162 originally approved on August 7th 2012 and #136272 originally approved on May 25th 2012). Parents of participants provided written informed consent.

2.1. Participants

2.1.1. Autism Spectrum Disorder

Individuals with ASD who met the inclusion and exclusion criteria had mitochondrial function measured up to four times using the buccal swab technique described below. Inclusion criteria included: (i) age 3 to 14 years of age and (ii) ASD diagnosis. Exclusion criteria included prematurity.

The ASD diagnosis was defined by one of the following: (i) a gold-standard diagnostic instrument such as the ADOS and/or Autism Diagnostic Interview-Revised; (ii) the state of Arkansas diagnostic standard, defined as the agreement of a physician, psychologist and speech therapist; and/or (iii) Diagnostic Statistical Manual (DSM) diagnosis by a physician along with standardized validated questionnaires and diagnosis confirmation by the Principal Investigator.

2.1.2. Mitochondrial Disease

Individuals included in this study were screened for mitochondrial disease through a standard clinical protocol [23,28]. Mitochondrial disease was diagnosed in a portion of the individuals using a combination of biochemical, enzymology and genetic testing. In general, the modified Walkers criterion was used to diagnose mitochondrial disease, although in some cases with clear repeated biochemical abnormalities with clinical symptomatology that lacked an identifiable genetic component, the Morava criterion was used [15].

2.1.3. Historical Healthy Controls

Controls of similar age and gender included 68 healthy individuals without neurological disease as described in previous studies [12]. Controls ranged in age from 3 to 21 years of age [mean (Standard Deviation (SD)) 10.1 years (4.6 years)] with 33 (49%) being female. In a previous report, it was found that there was no correlation between enzyme activities and age and no difference in protein activities across ethnicity or race in both controls and mitochondrial disease patients [8].

2.2. Measures of Mitochondrial Function

The buccal cells were collected using Catch-All Buccal Collection Swabs (Epicentre Biotechnologies, Madison, WI, USA). Four swabs were collected by firmly pressing a swab against the inner cheek while twirling for 30 s. Swabs were clipped and placed in 1.5 mL microcentrifuge tubes that were labeled and placed on dry ice for overnight transportation to the Goldenthal laboratory.

Buccal extracts were prepared using an ice-cold buffered solution (Buffer A, ABCAM, Cambridge, MA, USA) containing protease inhibitor cocktail and membrane solubilizing non-ionic detergent and cleared of insoluble cellular material by high speed centrifugation at 4 °C. Duplicate aliquots of the protein extract were analyzed for protein concentration using the bicinchoninic acid method (Pierce Biotechnology, Rockford, IL, USA). Samples were typically stored at −80 °C for up to 1 week prior to enzymatic analysis.

Dipstick immunocapture assays measured ETC Complex I activity using 50 μg extracted protein [8–10]. Signals were quantified using a Hamamatsu immunochromato MS 1000 Dipstick reader (ABCAM, Cambridge, MA, USA). Raw mABS (milliAbsorbance) results were corrected for protein concentration and data were expressed as percentages of the values obtained with control extracts run on the same assay. ETC Complex IV and Citrate Synthase (CS) activity was assessed using standard spectrophotometric procedures in 0.5 mL reaction volume. Specific activities of respiratory complexes and citrate synthase were initially expressed as nanomoles/min/mg protein. This activity was then normalized to control values so that the final value represented a z-score. This allowed for the direct comparison of activities across complexes and citrate synthase.

2.3. Statistical Analysis

Analyses were performed using SAS 9.4 (SAS Institute Inc., Cary, NC, USA). Graphs were produced using Excel version 14.0 (Microsoft Corp, Redmond, WA, USA). Normal control values for mitochondrial function were based upon the established controls from the Goldenthal laboratory [12]. A mixed-model linear regression was used to account for both within-subject variation from repeated measurements on the same individual as well as between-subject variation such as mitochondrial disease subgroup. The module "glimmix" in SAS was used with an $p \leq 0.05$.

A series of analyses first examined the effect of specific supplements on overall normalized mitochondrial activities including an interaction the with mitochondrial disease subgroup (mitochondrial disease vs no mitochondrial disease). Main effects and interactions in the model are F-distributed so they were evaluated using a F-test. If the interaction was significant, post-hoc orthogonal contrasts were used to determine whether the effect of the supplement was specific to one subgroup. Post-hoc orthogonal contrasts are t-distributed and thus were evaluated using a t-distribution. The supplements that were found to have a significant effect were then entered into a stepwise backward mixed-model regression (with mitochondrial disease subgroup interaction if significant in individuals regressions) with a criteria of $p \leq 0.05$ to keep in the model. Essentially, at each step, the variable with the highest p-value was eliminated and the model was recalculated until all of the variables in the model were significant at the $p \leq 0.05$ level. Of course, variables that were dependents of an interaction were kept in the model irrespective of their significance.

Similarly, a series of analyses examined the effect of specific supplements on the relationship between the normalized mitochondrial component activities, including an interaction with the mitochondrial disease subgroup. If the interaction was significant, post-hoc orthogonal contrasts were used to determine whether the effect of the supplement on the relationship between the mitochondrial components (i.e., the slope of the regression) was specific to one subgroup. The supplements that were found to be significant were then entered into a stepwise backward mixed-model regression (with mitochondrial disease subgroup interaction if significant in the single supplement models). As before, at each step, the variable with the highest p-value was eliminated and the model was recalculated until all of the variables in the model were significant at the $p \leq 0.05$ level. Variables that were dependents of an interaction were kept in the model irrespective of their significance.

3. Results

3.1. Participants

A total of 127 individuals with ASD who met the inclusion and exclusion criteria had mitochondrial function measured. Of the 127 participants, 38 had mitochondrial function measured twice, seven had mitochondrial function measured three times and one participant had mitochondrial function measured four times. The mean age at the first mitochondrial function measurement was 8.3 years (SD = 4.0 years) with 77% being male. Age and gender were entered into the regressions initially but were found not to be significant so they were not included in the subsequent analyses. A total of 15% of the sample was clinically diagnosed with mitochondrial disease. Mitochondrial disease was found to be a significant factor in the regression analyses so it was included in most analyses. The only exceptions were for the analysis of multivitamin (MVI) and herbal supplements where there were too few individuals with mitochondrial disease taking these supplements for a valid analysis of this effect by subgroup.

3.2. Supplements

Participants were not on any supplements for 118 measurements. Of those measurements, 26 participants were naïve to supplements, whereas for 92 measurements, supplements were held for an average of 21.3 days (SD 15.4 days; Range 1–61 days) before the measurements. For 55 measurements, supplements were given as scheduled. The specific supplements taken by the participants are outlined in Table 1. The factors of (a) time since taking the supplement when supplements were held, and (b) naivety to supplementation, were entered into the regressions initially but were found not to be significant so they were not included in the subsequent analyses.

Table 1. Supplements taken by participants.

Supplement	% Taking Regularly	% Taking During Mitochondrial Testing	% Holding During Mitochondrial Testing
Amino Acids	23%	4%	19%
B12	38%	13%	25%
B Vitamins	36%	10%	26%
Carnitine	42%	13%	29%
Coenzyme Q10	36%	9%	27%
Fatty Acids	45%	16%	29%
Folate	54%	16%	38%
Herbal	17%	6%	11%
Multivitamin	26%	12%	14%
Antioxidants	46%	14%	32%
Other Vitamins	49%	14%	35%

3.3. Supplement Effect on Mitochondrial Complexes and Citrate Synthase Activity

3.3.1. Normalized Complex I Activity

Carnitine, antioxidant and other vitamin supplementation were associated with significantly higher Complex I activity [$F_{(1.44)} = 7.58$, $p < 0.01$, $F_{(1.44)} = 6.54$, $p = 0.01$, $F_{(1.44)} = 6.70$, $p = 0.01$, respectively] (See Table 2).

Table 2. Means (Standard Error) of Normalized Complex I activity on and off supplements.

Supplement	Off Supplement	On Supplement
Carnitine	−0.1 (0.28)	1.4 (0.46)
Antioxidants	0.1 (0.26)	1.5 (0.50)
Other Vitamins	0.1 (0.27)	1.4 (0.45)

Fatty acids and folate supplementation influenced Complex I activity with this influence different for the mitochondrial disease subgroup (See Table 3). Fatty acid supplementation significantly increased Complex I activity overall [$F_{(1.44)} = 16.86$, $p < 0.0005$] but also interacted with mitochondrial disease subgroup [$F_{(1.44)} = 4.53$, $p < 0.05$]. This interaction resulted from this increase being more marked for the mitochondrial disease subgroup when the subgroups were analyzed separately despite the fact that both the no mitochondrial disease [$t_{(44)} = 2.05$, $p < 0.05$] and the mitochondrial disease [$t_{(44)} = 3.56$, $p < 0.001$] subgroups demonstrated a significant effect of supplementation. Folate supplementation significantly increased Complex I activity overall [$F_{(1.44)} = 11.15$, $p < 0.005$] but there was an interaction with mitochondrial disease subgroup [$F_{(1.44)} = 5.61$, $p < 0.05$]. This interaction resulted from folate only significantly influencing Complex I activity in the mitochondrial disease group [$t_{(44)} = 3.28$, $p < 0.005$] when the subgroups were analyzed separately.

Table 3. Means (Standard Error) of Normalized Complex I activity on and off supplements by Mitochondrial Disease group. Supplements that are confirmed to be significant in the stepwise regression are bolded and italicized.

Supplement	No Mitochondrial Disease		Mitochondrial Disease	
	Off Supplement	On Supplement	Off Supplement	On Supplement
Fatty Acids	*0.1 (0.20)*	*1.2 (0.48)*	*−0.3 (0.47)*	*3.1 (0.84)*
Folate	0.2 (0.21)	0.7 (0.5)	−0.3 (0.49)	2.7 (0.79)

Complex I was not significantly influenced by amino acid, B12, MVI, B vitamins, Coenzyme Q10 (CoQ10), or herbal supplementation.

To determine which supplements were driving the effect on Complex I activity, the supplements that demonstrated a significant effect on Complex I activity were entered into a stepwise backwards regression. The regression demonstrated that fatty acids supplementation significantly increased Complex I activity [$F(1.43) = 6.39$, $p < 0.05$] without an interaction between subgroups and that the effect of folate on Complex I activity was influenced by the subgroup [$F(1.43) = 5.94$, $p < 0.05$] since the effect of folate was isolated to the mitochondrial disease group [$t(43) = 2.42$, $p < 0.05$].

3.3.2. Normalized Citrate Synthase Activity

Fatty acids, folate and antioxidant supplementation influenced Citrate Synthase activity with this influence being different for the mitochondrial disease subgroup (See Table 4). Fatty acid supplementation significantly increased Citrate Synthase activity [$F(1.44) = 9.58$, $p < 0.005$] but also interacted with the mitochondrial disease subgroup [$F(1.44) = 4.69$, $p < 0.05$]. This interaction resulted from fatty acids significantly influencing Citrate Synthase activity only in the mitochondrial disease group [$t(44) = 3,02$, $p < 0.005$] when the subgroups were analyzed separately. Folate supplementation significantly increased Citrate Synthase activity [$F(1.44) = 7.00$, $p = 0.01$] but also interacted with the mitochondrial disease subgroup [$F(1.44) = 7.56$, $p < 0.01$]. This interaction resulted from folate significantly influencing Citrate Synthase activity only in the mitochondrial disease group [$t(44) = 3.13$, $p < 0.005$] when the subgroups were analyzed separately. Antioxidant supplementation significantly increased Citrate Synthase activity [$F(1.44) = 7.37$, $p < 0.01$] but also interacted with the mitochondrial disease subgroup [$F(1.44) = 8.30$, $p < 0.01$]. This interaction resulted from antioxidants significantly influencing Citrate Synthase activity only in the mitochondrial disease group [$t(44) = 3.22$, $p = 0.01$] when the subgroups were analyzed separately.

Table 4. Means (Standard Error) of Normalized Citrate Synthase activity on and off supplements by Mitochondrial Disease group. Supplements that are confirmed to be significant in the stepwise regression are bolded and italicized.

Supplement	No Mitochondrial Disease		Mitochondrial Disease	
	Off Supplement	On Supplement	Off Supplement	On Supplement
Fatty Acids	0.8 (0.17)	1.2 (0.40)	0.3 (0.40)	2.6 (0.70)
Folate	0.9 (0.18)	0.8 (0.40)	0.2 (0.42)	2.4 (0.66)
Antioxidants	0.9 (0.17)	0.8 (0.42)	*0.2 (0.41)*	*2.7 (0.71)*

To determine which supplements were driving the effect on Complex I activity, the supplements that demonstrated a significant effect on Citrate Synthase activity were entered into a stepwise backwards elimination regression. The regression only selected antioxidant supplementation for improvement in Citrate Synthase activity (results same as above).

Citrate Synthase was not significantly influenced by amino acid, B12, B vitamins, multivitamin, CoQ10, carnitine, other vitamins or herbal supplementation.

3.3.3. Normalized Complex IV Activity

Normalized Complex IV activity was not significantly influenced by amino acid, B12, B vitamins, CoQ10, carnitine, other vitamins, herbal, fatty acids, folate, multivitamin or antioxidant supplementation.

3.4. Supplement Effect on Relationship between Mitochondrial Complexes and Citrate Synthase Activity

3.4.1. The Relationship between Normalized Complex I and Complex IV Activity

The relationship between Normalized Complex I and Complex IV activity was not influenced by amino acids, multivitamin, B12, herbal, other vitamins or CoQ10. B vitamins, fatty acids, folate, antioxidants and carnitine all influenced the relationship between Normalized Complex I and Complex IV activity with this relationship influenced by whether or not the participant was in the mitochondrial disease subgroup.

Most of the supplements predominantly influenced the mitochondrial disease subgroup. B vitamins significantly influenced the relationship between complexes [$F(1.40) = 5.88$, $p < 0.05$] with this effect interacting with the mitochondrial disease subgroup [$F(1.40) = 7.45$, $p < 0.01$] since the effect was significant only in the mitochondrial disease subgroup [$t(40) = 3.09$, $p < 0.005$]. Fatty acids significantly influenced the relationship between complexes [$F(1.40) = 12.64$, $p = 0.001$] with this effect interacting with the mitochondrial disease subgroup [$F(1.40) = 7.45$, $p < 0.01$] since the effect was significant only in the mitochondrial disease subgroup [$t(40) = 3.17$, $p < 0.005$]. Antioxidants significantly influenced the relationship between complexes [$F(1.40) = 14.34$, $p = 0.0005$] with this effect interacting with the mitochondrial disease subgroup [$F(1.40) = 10.01$, $p < 0.005$] because the effect was only significant in the mitochondrial disease subgroup [$t(40) = 3.83$, $p < 0.0005$]. Carnitine significantly influenced the relationship between complexes [$F(1.40) = 7.64$, $p < 0.01$] with this effect interacting with the mitochondrial disease subgroup [$F(1.40) = 4.49$, $p < 0.05$] because the effect was only significant in the mitochondrial disease subgroup [$t(40) = 2.71$, $p < 0.01$].

Folate influenced the relationship between Normalized Complex I and Complex IV activity [$F(1.40) = 18.13$, $p = 0.0001$] with the effect of folate being influenced by the mitochondrial disease subgroup [$F(1.40) = 7.04$, $p = 0.01$]. The effect of folate was more marked in the mitochondrial disease subgroup [$t(40) = 3.75$, $p < 0.001$] than the non-mitochondrial disease subgroup [$t(40) = 2.07$, $p < 0.05$] but folate did influence mitochondrial function for both those with and without mitochondrial disease.

To determine which supplements were driving the effect, the supplements that demonstrated significant effects were entered into a stepwise backwards regression. The regression demonstrated that folate supplementation significantly improved the relationship between Normalized Complex I and Complex IV activity. Figure 1 depicts the effect of folate supplementation on the relationship between Normalized Complex I and Complex IV activity. Those on folate supplementation had a stronger relationship between Complex I and Complex IV activity as compared to individuals not on folate supplementation. The regression coefficients suggest that each increase in Complex IV activity results in a 2.4 times increase in Complex I activity if an individual was on folate supplementation while this increase was only 0.9 times if an individual was not on folate supplementation.

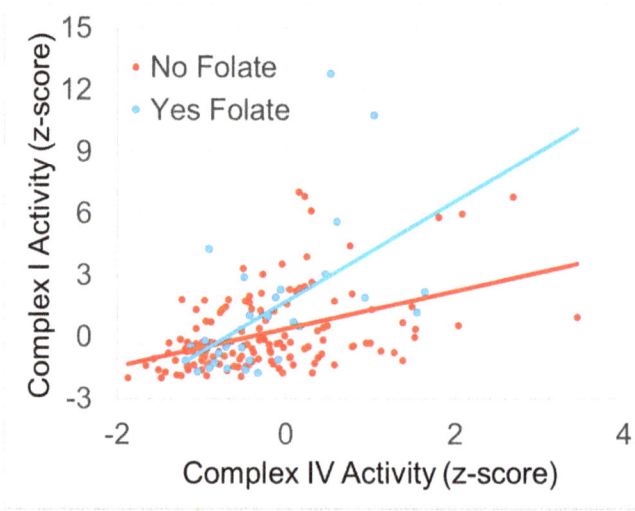

Figure 1. The relationship between Normalized Complex I and IV activity. Folate supplementation is associated with a significantly greater slope in the relationship between complex activities.

3.4.2. The Relationship between Normalized Complex I and Citrate Synthase Activity

The relationship between Normalized Complex I and Citrate Synthase activity was not influenced by amino acids, MVI, fatty acids, herbal, other vitamins or CoQ10. B vitamins, B12, folate, antioxidants and carnitine all influenced the relationship between Normalized Complex I and Complex IV activity.

Some supplements predominately influenced the mitochondrial disease subgroup. The effect of B vitamins on the relationship between mitochondrial components was significantly influenced by the mitochondrial disease subgroup [$F_{(1.40)} = 5.44$, $p < 0.05$] because the effect was isolated to the mitochondrial disease subgroup [$t_{(40)} = 2.43$, $p < 0.05$]. B12 significantly influenced the relationship between mitochondrial components [$F_{(1.40)} = 4.27$, $p < 0.05$] with this effect being influenced by the mitochondrial disease subgroup [$F_{(1.40)} = 6.24$, $p = 0.01$] because the effect was only significant for the mitochondrial disease subgroup [$t_{(40)} = 2.53$, $p < 0.05$].

Folate, antioxidants and carnitine appear to influence the relationship between Normalized Complex I and Citrate Synthase activity without a difference in this effect across subgroups [$F_{(1.40)} = 21.36$, $p < 0.0001$, $F_{(1.40)} = 7.09$, $p = 0.01$ and $F_{(1.40)} = 4.74$, $p < 0.05$, respectively].

To determine which supplementation was driving the effect, the supplements that demonstrated significant effects were entered into a stepwise backwards elimination regression. The regression demonstrated that folate and B12 supplementation significantly altered the relationship between Normalized Complex I and Citrate Synthase activity [$F_{(1.41)} = 28.23$, $p < 0.0001$ and $F_{(1.41)} = 8.35$, $p < 0.005$, respectively] without an interaction between subgroups. Figure 2 depicts these relationships. Folate increased the slope of the relationship between Normalized Complex I and Citrate Synthase activity such that any increase in Citrate Synthase resulted in a 1.5 times increase in Normalized Complex I activity if an individual was on folate, whereas it resulted in only a 0.5 times increase in Normalized Complex I activity if the individual was not on folate. For B12 supplementation, an increase in Citrate Synthase resulted in a 0.8 times increase in Complex I activity if an individual was supplementing with B12, whereas this was only 0.6 times if an individual was not supplementing with B12.

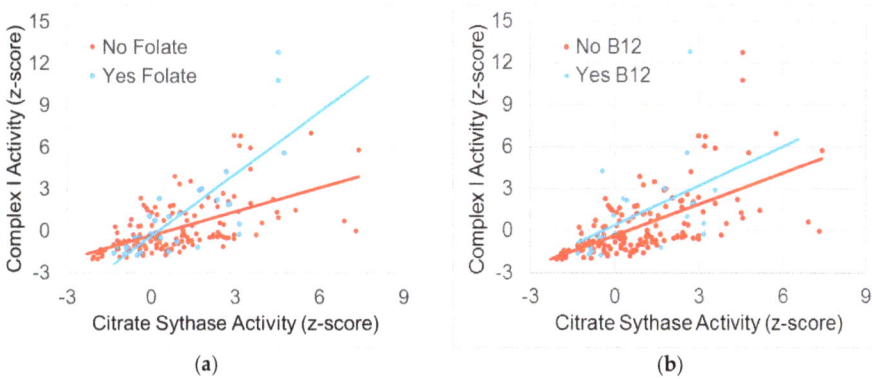

Figure 2. The relationship between normalized Complex I and Citrate Synthase activity. (**a**) Folate and (**b**) B12 supplementation are associated with a significantly greater slope in the relationship between Complex I and Citrate Synthase.

3.4.3. The Relationship between Normalized Complex IV and Citrate Synthase Activity

None of the supplements were found to influence the relationship between Normalized Complex IV and Citrate Synthase activity.

4. Discussion

This study examined the effect of common mitochondrial treatments on specific mitochondrial components in a group of children diagnosed with ASD, some of which also were diagnosed with co-morbid mitochondrial disease. Measurement of mitochondrial function is important in ASD since many children with ASD appear to have mitochondrial dysfunction even if they are not diagnosed with classic mitochondrial disease. Furthermore, the influence of mitochondrial treatment in ASD is important as randomized controlled clinical trials have demonstrated that common treatments for mitochondrial disease, such as L-carnitine, improve ASD symptoms, suggesting that such treatments may have a role in the treatment of ASD [27]. However, what remains unclear is whether these treatments are targeting mitochondrial function per se.

In addition, in this study, we examined not only whether common mitochondrial supplements affect the absolute levels of activity of three mitochondrial components, but whether the treatments alter the relationship between the components. This may be important, as optimal coupling of the various mitochondrial components is essential for the mitochondria to function optimally.

Results from this study suggested that several common mitochondrial supplements such as fatty acids and antioxidants appeared to influence Complex I and Citrate Synthase activity, respectively, with this influence being more marked for the mitochondrial disease subgroup. This is not unexpected as such treatments are sometimes recommended for individuals with mitochondrial disease, particularly antioxidants. Fatty acids are not always recommended for individuals with mitochondrial disease. Several studies have suggested that omega 3 fatty acids, which are the most fatty acids prescribed to children with ASD, have positive effects on behavior [29]. Interestingly, recent research has highlighted the role of fatty acids in preserving mitochondrial function in such diseases as stroke [30] and cancer [31] as well as improving muscle health through modulation of mitochondrial function [32]. The findings that these effects were more marked in the mitochondrial disease subgroup suggest that these treatments are indeed targeting and improving mitochondrial function and further suggest that certain treatments may be best targeted to subpopulations of individuals with ASD.

Folate was found to be potentially important in modulating the relationship between both Complex I and IV and Complex I and Citrate Synthase, while B12 appeared to be potentially important in the relationship between Complex I and Citrate Synthase. In a clinical trial on individuals with ASD, the combination of B12 and folate has been shown to improve cognitive development [28] and glutathione [33], the major intrinsic antioxidant that is essential for protecting the mitochondrial. In another clinical trial, B12 alone has been shown to improve methylation in individuals with ASD [34]. Folate is also essential for mitochondrial function as one-carbon metabolism is highly compartmentalized [35]. Most notable in the context of mitochondrial disease, is that mitochondria often replicate to compensate for poorly functioning mitochondria. Since mitochondria contain their own DNA, folate is needed for the synthesis of purines and pyrimidine nucleotides [35]. Thus, given the important role of folate in many critical cellular processes, it should not be surprising that it was found to be important in mitochondrial function.

This study has many limitations, including the lack of systematically manipulating the treatment studied and simultaneous treatments with multiple supplements in many cases. In addition, the subgroups of individuals with mitochondrial disease were not diagnosed with one specific mitochondrial disease. Nevertheless, this study provides a novel framework to build upon in order to consider the development of alternative methods for monitoring mitochondrial function.

5. Conclusions

This study provides empirical support for common mitochondrial treatments and demonstrates that the relationship between activities of mitochondrial components might be a marker to follow in addition to absolute activities. In addition, measurements of mitochondrial activity that can be practically repeated over time, especially those that are non-invasive such as the buccal swab technique, may be very useful to monitor the biochemical effects of mitochondrial targeted treatments.

Acknowledgments: This study has not been published previously although data from the same patient have been used in other publications. This trial is registered on clinicaltrials.gov as NCT02000284. This research was supported, in part, by the Autism Research Institute (San Diego, CA, USA), the Arkansas Biosciences Institute (Little Rock, AR, USA), the Jane Botsford Johnson Foundation (New York, NY, USA) and the Arkansas Children's Research Institute (Little Rock, AR, USA). None of the sponsors were involved with the design or conduct of the study, collection, management, analysis, or interpretation of the data; or preparation, review, approval of the manuscript or decision to submit the manuscript for publication. The first two and last authors, Richard E Frye, John Slattery and Leanna Delhey, had full access to all of the data in the study and take responsibility for the integrity of the data and the accuracy of the data analysis. Funds for covering the costs to publish in open access were provided by the Gupta family.

Author Contributions: Leanna M. Delhey, John C. Slattery, Stephen G. Kahler and Richard E. Frye designed the study and wrote the protocol, collected the data, analyzed the data, drafted the manuscript. Ekim Nur Kilinc, Li Yin and Marie L. Tippett analyzed and collected the data. Shannon Rose, Sirish C. Bennuri, Shirish Damle and Agustin Legido performed laboratory analysis. Michael J. Goldenthal designed the study and supervised the laboratory analysis of mitochondrial function. All authors contributed to and have approved the final manuscript.

Conflicts of Interest: The authors declare no conflict of interest. The sponsors and funders had no role in the design of the study; in the collection, analyses, or interpretation of data; in the writing of the manuscript, and in the decision to publish the results.

References

1. Niyazov, D.M.; Kahler, S.G.; Frye, R.E. Primary mitochondrial disease and secondary mitochondrial dysfunction: Importance of distinction for diagnosis and treatment. *Mol. Syndromol.* **2016**, *7*, 122–137. [CrossRef] [PubMed]
2. Camp, K.M.; Krotoski, D.; Parisi, M.A.; Gwinn, K.A.; Cohen, B.H.; Cox, C.S.; Enns, G.M.; Falk, M.J.; Goldstein, A.C.; Gopal-Srivastava, R.; et al. Nutritional interventions in primary mitochondrial disorders: Developing an evidence base. *Mol. Genet. Metab.* **2016**, *119*, 187–206. [CrossRef] [PubMed]
3. Parikh, S.; Goldstein, A.; Koenig, M.K.; Scaglia, F.; Enns, G.M.; Saneto, R.; Anselm, I.; Cohen, B.H.; Falk, M.J.; Greene, C.; et al. Diagnosis and management of mitochondrial disease: A consensus statement from the mitochondrial medicine society. *Genet. Med. Off. J. Am. Coll. Med. Genet.* **2015**, *17*, 689–701. [CrossRef] [PubMed]

4. Viscomi, C. Toward a therapy for mitochondrial disease. *Biochem. Soc. Trans.* **2016**, *44*, 1483–1490. [CrossRef] [PubMed]

5. Chinnery, P.; Majamaa, K.; Turnbull, D.; Thorburn, D. Treatment for mitochondrial disorders. *Cochrane Database Syst. Rev.* **2006**. [CrossRef]

6. Pfeffer, G.; Majamaa, K.; Turnbull, D.M.; Thorburn, D.; Chinnery, P.F. Treatment for mitochondrial disorders. *Cochrane Database Syst. Rev.* **2012**. [CrossRef]

7. Pfeffer, G.; Horvath, R.; Klopstock, T.; Mootha, V.K.; Suomalainen, A.; Koene, S.; Hirano, M.; Zeviani, M.; Bindoff, L.A.; Yu-Wai-Man, P.; et al. New treatments for mitochondrial disease-no time to drop our standards. *Nat. Rev. Neurol.* **2013**, *9*, 474–481. [CrossRef] [PubMed]

8. Goldenthal, M.J.; Kuruvilla, T.; Damle, S.; Salganicoff, L.; Sheth, S.; Shah, N.; Marks, H.; Khurana, D.; Valencia, I.; Legido, A. Non-invasive evaluation of buccal respiratory chain enzyme dysfunction in mitochondrial disease: Comparison with studies in muscle biopsy. *Mol. Genet. Metab.* **2012**, *105*, 457–462. [CrossRef] [PubMed]

9. Yorns, W.R., Jr.; Valencia, I.; Jayaraman, A.; Sheth, S.; Legido, A.; Goldenthal, M.J. Buccal swab analysis of mitochondrial enzyme deficiency and DNA defects in a child with suspected myoclonic epilepsy and ragged red fibers (merrf). *J. Child. Neurol.* **2012**, *27*, 398–401. [CrossRef] [PubMed]

10. Ezugha, H.; Goldenthal, M.; Valencia, I.; Anderson, C.E.; Legido, A.; Marks, H. 5q14.3 deletion manifesting as mitochondrial disease and autism: Case report. *J. Child. Neurol.* **2010**, *25*, 1232–1235. [CrossRef] [PubMed]

11. Frye, R.E.; Cox, D.; Slattery, J.; Tippett, M.; Kahler, S.; Granpeesheh, D.; Damle, S.; Legido, A.; Goldenthal, M.J. Mitochondrial dysfunction may explain symptom variation in phelan-mcdermid syndrome. *Sci. Rep.* **2016**, *6*, 19544. [CrossRef] [PubMed]

12. Goldenthal, M.J.; Damle, S.; Sheth, S.; Shah, N.; Melvin, J.; Jethva, R.; Hardison, H.; Marks, H.; Legido, A. Mitochondrial enzyme dysfunction in autism spectrum disorders; a novel biomarker revealed from buccal swab analysis. *Biomark. Med.* **2015**, *9*, 957–965. [CrossRef] [PubMed]

13. Legido, A.; Jethva, R.; Goldenthal, M.J. Mitochondrial dysfunction in autism. *Semin. Pediatr. Neurol.* **2013**, *20*, 163–175. [CrossRef] [PubMed]

14. Zablotsky, B.; Black, L.I.; Maenner, M.J.; Schieve, L.A.; Blumberg, S.J. Estimated prevalence of autism and other developmental disabilities following questionnaire changes in the 2014 national health interview survey. *Nat. Health Stat. Rep.* **2015**, *87*, 1–20.

15. Frye, R.E.; Rossignol, D.A. Mitochondrial dysfunction can connect the diverse medical symptoms associated with autism spectrum disorders. *Pediatr. Res.* **2011**, *69*, 41R–47R. [CrossRef] [PubMed]

16. Rossignol, D.A.; Frye, R.E. Mitochondrial dysfunction in autism spectrum disorders: A systematic review and meta-analysis. *Mol. Psychiatry* **2012**, *17*, 290–314. [CrossRef] [PubMed]

17. Frye, R.E. Biomarkers of abnormal energy metabolism in children with autism spectrum disorder. *N. Am. J. Med. Sci.* **2012**, *5*, 141–147. [CrossRef]

18. Giulivi, C.; Zhang, Y.F.; Omanska-Klusek, A.; Ross-Inta, C.; Wong, S.; Hertz-Picciotto, I.; Tassone, F.; Pessah, I.N. Mitochondrial dysfunction in autism. *JAMA* **2010**, *304*, 2389–2396. [CrossRef] [PubMed]

19. Napoli, E.; Wong, S.; Hertz-Picciotto, I.; Giulivi, C. Deficits in bioenergetics and impaired immune response in granulocytes from children with autism. *Pediatrics* **2014**, *133*, e1405–e1410. [CrossRef] [PubMed]

20. Palmieri, L.; Papaleo, V.; Porcelli, V.; Scarcia, P.; Gaita, L.; Sacco, R.; Hager, J.; Rousseau, F.; Curatolo, P.; Manzi, B.; et al. Altered calcium homeostasis in autism-spectrum disorders: Evidence from biochemical and genetic studies of the mitochondrial aspartate/glutamate carrier agc1. *Mol. Psychiatry* **2010**, *15*, 38–52. [CrossRef] [PubMed]

21. Frye, R.E. Novel cytochrome b gene mutations causing mitochondrial disease in autism. *J. Pediatr. Neurol.* **2012**, *10*, 35–40.

22. Frye, R.E.; Naviaux, R.K. Autistic disorder with complex iv overactivity: A new mitochondrial syndrome. *J. Pediatr. Neurol.* **2011**, *9*, 427–434.

23. Frye, R.E.; Melnyk, S.; Macfabe, D.F. Unique acyl-carnitine profiles are potential biomarkers for acquired mitochondrial disease in autism spectrum disorder. *Transl. Psychiatry* **2013**, *3*, e220. [CrossRef] [PubMed]

24. Rose, S.; Frye, R.E.; Slattery, J.; Wynne, R.; Tippett, M.; Pavliv, O.; Melnyk, S.; James, S.J. Oxidative stress induces mitochondrial dysfunction in a subset of autism lymphoblastoid cell lines in a well-matched case control cohort. *PLoS ONE* **2014**, *9*, e85436. [CrossRef] [PubMed]

25. Rose, S.; Frye, R.E.; Slattery, J.; Wynne, R.; Tippett, M.; Melnyk, S.; James, S.J. Oxidative stress induces mitochondrial dysfunction in a subset of autistic lymphoblastoid cell lines. *Transl. Psychiatry* **2014**, *4*, e377. [CrossRef] [PubMed]

26. Rose, S.; Bennuri, S.C.; Wynne, R.; Melnyk, S.; James, S.J.; Frye, R.E. Mitochondrial and redox abnormalities in autism lymphoblastoid cells: A sibling control study. *FASEB J.* **2016**. [CrossRef] [PubMed]

27. Frye, R.E.; Rossignol, D.A. Treatments for biomedical abnormalities associated with autism spectrum disorder. *Front. Pediatr.* **2014**, *2*, 66. [CrossRef] [PubMed]

28. Frye, R.E.; Delatorre, R.; Taylor, H.; Slattery, J.; Melnyk, S.; Chowdhury, N.; James, S.J. Redox metabolism abnormalities in autistic children associated with mitochondrial disease. *Transl. Psychiatry* **2013**, *3*, e273. [CrossRef] [PubMed]

29. Frye, R.E.; Rossignol, D.; Casanova, M.F.; Brown, G.L.; Martin, V.; Edelson, S.; Coben, R.; Lewine, J.; Slattery, J.C.; Lau, C.; et al. A review of traditional and novel treatments for seizures in autism spectrum disorder: Findings from a systematic review and expert panel. *Front. Public Health* **2013**, *1*, 31. [CrossRef] [PubMed]

30. Berressem, D.; Koch, K.; Franke, N.; Klein, J.; Eckert, G.P. Intravenous treatment with a long-chain omega-3 lipid emulsion provides neuroprotection in a murine model of ischemic stroke—A pilot study. *PLoS ONE* **2016**, *11*, e0167329. [CrossRef] [PubMed]

31. Agnihotri, N.; Sharma, G.; Rani, I.; Renuka; Bhatnagar, A. Fish oil prevents colon cancer by modulation of structure and function of mitochondria. *Biomed. Pharmacother.* **2016**, *82*, 90–97. [CrossRef] [PubMed]

32. Yoshino, J.; Smith, G.I.; Kelly, S.C.; Julliand, S.; Reeds, D.N.; Mittendorfer, B. Effect of dietary n-3 pufa supplementation on the muscle transcriptome in older adults. *Physiol. Rep.* **2016**. [CrossRef] [PubMed]

33. James, S.J.; Melnyk, S.; Fuchs, G.; Reid, T.; Jernigan, S.; Pavliv, O.; Hubanks, A.; Gaylor, D.W. Efficacy of methylcobalamin and folinic acid treatment on glutathione redox status in children with autism. *Am. J. Clin. Nutr.* **2009**, *89*, 425–430. [CrossRef] [PubMed]

34. Hendren, R.L.; James, S.J.; Widjaja, F.; Lawton, B.; Rosenblatt, A.; Bent, S. Randomized, placebo-controlled trial of methyl b12 for children with autism. *J. Child Adolesc. Psychopharmacol.* **2016**, *26*, 774–783. [CrossRef] [PubMed]

35. Desai, A.; Sequeira, J.M.; Quadros, E.V. The metabolic basis for developmental disorders due to defective folate transport. *Biochimie* **2016**, *126*, 31–42. [CrossRef] [PubMed]

Journal of
Clinical Medicine

MDPI

Review

Biomarkers for Detecting Mitochondrial Disorders

Josef Finsterer [1],*,[†] and Sinda Zarrouk-Mahjoub [2],[†]

[1]   Krankenanstalt Rudolfstiftung, Postfach 20, 1180 Vienna, Austria
[2]   El Manar and Genomics Platform, Pasteur Institute of Tunis, University of Tunis, Tunis 1068, Tunisia;
       sinda.z.m@gmail.com
*   Correspondence: fifigs1@yahoo.de; Tel.: +43-171-1659-2085
†   These authors contributed equally to this work.

Received: 9 December 2017; Accepted: 19 January 2018; Published: 30 January 2018

Abstract: (1) Objectives: Mitochondrial disorders (MIDs) are a genetically and phenotypically heterogeneous group of slowly or rapidly progressive disorders with onset from birth to senescence. Because of their variegated clinical presentation, MIDs are difficult to diagnose and are frequently missed in their early and late stages. This is why there is a need to provide biomarkers, which can be easily obtained in the case of suspecting a MID to initiate the further diagnostic work-up. (2) Methods: Literature review. (3) Results: Biomarkers for diagnostic purposes are used to confirm a suspected diagnosis and to facilitate and speed up the diagnostic work-up. For diagnosing MIDs, a number of dry and wet biomarkers have been proposed. Dry biomarkers for MIDs include the history and clinical neurological exam and structural and functional imaging studies of the brain, muscle, or myocardium by ultrasound, computed tomography (CT), magnetic resonance imaging (MRI), MR-spectroscopy (MRS), positron emission tomography (PET), or functional MRI. Wet biomarkers from blood, urine, saliva, or cerebrospinal fluid (CSF) for diagnosing MIDs include lactate, creatine-kinase, pyruvate, organic acids, amino acids, carnitines, oxidative stress markers, and circulating cytokines. The role of microRNAs, cutaneous respirometry, biopsy, exercise tests, and small molecule reporters as possible biomarkers is unsolved. (4) Conclusions: The disadvantages of most putative biomarkers for MIDs are that they hardly meet the criteria for being acceptable as a biomarker (missing longitudinal studies, not validated, not easily feasible, not cheap, not ubiquitously available) and that not all MIDs manifest in the brain, muscle, or myocardium. There is currently a lack of validated biomarkers for diagnosing MIDs.

Keywords: biomarker; diagnosis; mitochondrial disorder; mtDNA; oxidative phosphorylation; ATP

1. Introduction

Mitochondrial disorders (MIDs) are metabolic disorders due to impaired metabolic pathways within mitochondria [1]. Limiting MIDs to the respiratory chain is a narrow horizon, everything that goes wrong in the mitochondrion can become a MID (e.g., a beta-oixidation defect is a MID) [1]. MIDs carry substantial morbidity and are associated with excess premature death [2]. Due to their phenotypic and genetic heterogeneity and their intra-familial and inter-familial variability, MIDs are frequently missed or wrongly diagnosed. To ascertain the suspicion of an MID, the application of simple and widely available diagnostic tests is warranted. However, there is currently a lack of validated biomarkers for diagnosing MIDs [3]. This review aims at summarising current knowledge about biomarkers in the diagnostic work-up of MIDs.

2. Methods

Data for this review were identified by searches of MEDLINE for references of relevant articles. Search terms used were all acronyms known for specific MIDs ($n = 50$) and the terms "mitochondrial disorder", "mtDNA", "encephalomyopathy", and "mitochondrion" in individual combination with the terms "biomarker", "diagnostic test", "work-up", and "diagnosis". The results of the searches were screened for potentially relevant studies by the application of inclusion and exclusion criteria for the full texts of relevant studies. Only original articles about humans, published between 1966 and 2017, were included. Only randomised controlled trials (RCTs), observational studies with controls, case series, and case reports were included. Reviews, editorials, and letters were excluded. Additionally, reference lists of retrieved studies were checked for reports of studies not detected on the electronic search. Websites checked for additional information with regard to possible biomarkers for diagnosing MIDs were MITOMAP Neuromuscular Disease Center Database, and MitoTools.

3. Results

3.1. Biomarkers

3.1.1. Definition

Biomarkers (short for biological markers) are biological measures of a biological state. By definition, a biomarker is "a characteristic that is objectively measured and evaluated as an indicator of normal biological processes, pathogenic processes or pharmacological responses to a therapeutic intervention" [4]. Thus, two main groups of biomarkers are generally differentiated: disease-related biomarkers and drug-related biomarkers [3]. Disease-related biomarkers reflect the presence or absence of disease, aid in disease stratification, guide prognosis, and can inform about disease natural history [3]. The current review focuses only on disease-related biomarkers for diagnosing or suspecting MIDs.

3.1.2. Requirements

General requirements which a biomarker must meet include a continuous change of the process that is measured (changes as a function of the process being monitored), and thus a linear correlation between the measurement and the represented process, registration of quick changes (translational marker), stability without diurnal or seasonal variations, presence in detectable amounts in easily accessible biological fluids/tissues, cheap and easily feasible tests with widely available equipment, reliability when applied by different examiners and repeatedly, independence of age, sex, environmental, or climate conditions, pre-existing training condition, food, hydration, and proven usefulness, effectivity, and validation [5]. Biomarkers need to shorten the time necessary for diagnosing a condition and need to be cost-effective.

3.1.3. Classification

Biomarkers can be classified according to various different criteria. In addition to separation into disease-related and drug-related biomarkers, biomarkers may be classified according to the organ or tissue they refer to or being investigated, or according to the type of tissue investigated as wet, dry, or volatile biomarkers [6]. Furthermore, biomarkers can be categorised as invasive or non-invasive or as validated or non-validated [7].

3.2. Dry Biomarkers

3.2.1. History and Clinical Examination

Recently, an attempt has been undertaken to determine if clinical parameters from the individual or family history and findings on the clinical exam, in association with findings on easily available instrumental investigations, could raise the suspicion of an MID and could facilitate their diagnostic work-up [8]. According to this study, the so called "mitochondrial multiorgan disorder syndrome (MIMODS)" score suggests the presence of an MID if exceeding a limit of 10 points. Among 36 patients with a genetically or biochemically confirmed MID, the organs most frequently affected were the muscle (97%), the central nervous system (CNS) (72%), endocrine glands (69%), the heart (58%), intestines (55%), and the peripheral nerves (50%) [8]. MIDs manifested most frequently in the CNS as leukoencephalopathy, prolonged visually-evoked potentials or atrophy, in the endocrine organs as thyroid dysfunction, short stature, or diabetes, and in the heart as arrhythmias, heart failure, or hypertrophic cardiomyopathy [8]. Key clinical features suggesting an MID are short stature, facial dysmorphism, hypoacusis, epilepsy, migraine, cognitive impairment, diabetes, thyroid dysfunction, hypogonadism, hypertrophic cardiomyopathy, arterial hypertension, atrial fibrillation, hepatopathy, diverticulosis, nephrolithiasis, renal insufficiency, anaemia, neuropathy, and myopathy of extra-ocular, facial, bulbar, axial, respiratory, or the limb muscles. The score has not yet been validated in diseased or healthy controls. Though some MIDs may follow a pattern of organ involvement, the phenotypic heterogeneity is of such a degree that hardly a single biomarker may encompass all abnormalities developing during the course.

3.2.2. Imaging

Structural Imaging

(1) Muscle

Structural alterations of the skeletal muscles in MIDs can be easily determined by ultrasound, computed tomography (CT), or magnetic resonance imaging (MRI) by measuring muscle volume, amount of connective tissues, or amount of fat. In a study of nine patients with chronic progressive external ophthalmoplegia (CPEO), the range of eye movements (ROEM) correlated with the degree of atrophy of extra-ocular eye muscles on 3 Tesla magnetic resonance imaging (3T-MRI) [9]. There was a negative correlation between ROEM and the amount of T2-hyperintensities in the extra-ocular muscles [9]. This is why the authors proposed that ROEM could serve as a marker for assessing disease severity in these patients [9]. In a study of 10 patients with CPEO due to single-scale or multiple mtDNA deletions, atrophy of extra-ocular muscles was found in all of them [10]. Though imaging with ultrasound, CT, or MRI of skeletal muscles is increasingly applied, no longitudinal studies in a large number of patients have been carried out so far.

(2) Brain

Structural abnormalities on imaging of the brain in patients with MIDs are manifold, and may be different between early-onset and late-onset MIDs. Structural abnormalities found in paediatric MID patients include diffuse, patchy, periventricular, subcortical, or semioval white or grey matter lesions (WMLs, GMLs), stroke-like lesions (SLLs, the morphological equivalent of stroke-like episodes, SLEs), cerebral atrophy, calcifications, or optic atrophy [11]. Some of these lesions remain stable for years, whereas others are dynamic (e.g., SLLs and GMLs) [11]. Cerebral lesions reported in adult MID patients include SLLs, laminar cortical necrosis, basal ganglia necrosis, focal or diffuse WMLs, focal or diffuse atrophy, intra-cerebral calcifications, cysts, lacunas, haemorrhages, cerebral hypo- or hyperperfusion, intra-cerebral artery stenoses, or moyamoya syndrome [12]. Since cerebral lesions in paediatric and adult MIDs may go along with or without clinical manifestations, it is important to prospectively screen patients with an MID for cerebral involvement. Most of the CNS lesions in MIDs are non-specific and are thus unsuitable for serving as biomarkers of MID with CNS involvement.

(3) Heart

Since the heart is frequently involved in MIDs, screening for myocardial abnormalities could be an option to prematurely detect cardiac involvement. In a study of 64 MID patients undergoing cardiac MRI, 53% had at least one cardiac abnormality [13]. Late gadolinium enhancement was found in 33%, reduced systolic function in 28%, and left ventricular hypertrophy in 22% [13]. Thickness of the left ventricular wall was generally increased in MID patients as compared to controls [13]. Among the specific MIDs, myocardial thickening was most frequent among MELAS-like patients

(91% of patients), followed by patients with CPEO/ Kearns-Sayre syndrome (KSS) (late gadolinium enhancement, 80% of patients). Interestingly, more cardiac abnormalities were detected on cardiac MRI than on electrocardiogram (ECG) in this study. However, longitudinal studies to assess cardiac abnormalities over a longer period of time are warranted.

Functional Imaging

(1) Muscle

MR-spectroscopy (MRS) of skeletal muscles in MIDs allows measurement of muscle metabolites by means of ^{31}P or ^{1}H spectra [3]. ^{1}H-spectra reflect concentrations of lactate, choline, or N-acetyl-aspartate (NAA), whereas ^{31}P spectra reflect concentrations of phosphorus metabolites and thus the oxidative capacity [3]. In the majority of cases, the phospho-creatine recovery time—which correlates with the amount of ATP production—is measured by ^{31}P-MRS after phosphor-creatine depletion by exercise [3]. In a recent study on the amount of intramyocyte lipid accumulation by 7T-MRS, it turned out that the heteroplasmy rate correlated with the intramuscular lipid content in 10 patients with MELAS [14]. The authors concluded that intramyocyte lipid accumulation could serve as a novel biomarker for MELAS [14]. Another novel MRI technique allows measurement of the intramyocyte creatine content by means of creatine chemical exchange saturation transfer (CrCEST) MRI [15]. Intramyocyte creatine levels in this study correlated significantly with the capacity of oxidative phosphorylation (OXPHOS), as determined by means of ^{31}P-MRS [15]. It was concluded that CrCEST allows determination of the muscle creatine content, which can be disturbed in MID patients with muscle involvement [15]. In a study of 11 patients with MELAS or CPEO, ^{31}P-MRS showed an increased inorganic phosphate (iP)-to-phosphor-creatine (PCr) ratio and a decreased ATP/PCr ratio during exercise in MELAS patients [16]. Additionally, recovery to normal values of Pi/PCr and ATP/PCr was delayed in MELAS patients [16]. Energy failure as detected by ^{31}P-MRS correlated with the number of COX-positive ragged-red fibres in the MELAS patients of this study [16].

(2) Brain

Though increased lactate production in the cerebrospinal fluid (CSF) is well appreciated among MID patients with cerebral involvement, there is little published data available investigating the capability of cerebral MRS to monitor disease progression [3]. In a study of 45 MELAS patients, the lactate peak was increased and the NAA peak was decreased compared to healthy controls [17]. Patients who developed MELAS during follow-up (converters) had elevated NAA peaks, elevated total choline peaks, elevated lactate peaks, and elevated total creatine peaks compared to healthy controls [17]. The authors concluded that CSF lactate and choline could serve as biomarkers for predicting the risk of individual mutation carriers to develop a MELAS phenotype [17]. In a study of 14 patients with nonspecific MIMODS with cerebral lesions on MRI, 86% had a lactate peak on single-voxel ^{1}H-MRS of the brain [18]. Only eight patients had elevated serum lactate levels, and CSF lactate did not correlate with serum lactate [18].

Using PET studies, it has been shown that the cerebral oxygen metabolic rate is reduced, that the cerebral blood flow is increased, and that the glucose metabolic rate is increased in MELAS patients [19]. In a study of five patients with Leigh syndrome, the glucose uptake was decreased in the cerebellum and basal ganglia on 18-fluor-deoxi-glucose positron emission tomography (18FDG-PET) [20]. There are also studies which showed decreased oxygen extraction from blood during passage through the capillary bed in MID patients [21,22]. In these studies, the cerebral metabolic rate of oxygen was significantly decreased in the grey as well as the white matter in patients carrying the m.3243A>G variant, and thus it was concluded that the m.3243A>G variant results in a global decrease of oxygen consumption [22].

A promising future technique to be applied as a dry biomarker could be dynamic nuclear polarisation (DNP) MRI, which uses 13C-MRS to provide real-time functional imaging and allows determination of substrates and metabolites in low concentrations [3]. Novel PET-ligands such as

18F-BCPP-EF appear to be promising for quantification of the activity of complex-I of the respiratory chain [3].The disadvantage of all imaging biomarkers, however, is that every MID does not manifest in the brain, myocardium, or the muscle. Thus, MIDs without cerebral, myocardial, or muscle involvement may be missed by cerebral, muscle, or cardiac imaging.

3.2.3. Cutaneous Respirometry

Cutaneous respirometry is conceptualised to measure respiratory chain functions in vivo [23]. The method relies on the optical properties of proto-porphyrin-IX, a heme precursor synthesised in mitochondria, and is capable of measuring mito-pO_2 (pO_2: O_2 partial pressure) and mito-VO_2 (VO_2: O_2 volume) [23]. Though appealing, the method has not been applied to MIDs so far; it can be speculated that MID patients with clinical or subclinical cutaneous involvement may have increased mito-pO_2 or increased mito-VO_2. However, it needs to be confirmed that MID patients with involvement of the skin indeed show abnormal mito-pO_2 or mito-VO_2 on cutaneous respirometry. In a study of 30 healthy controls, reference limits of mito-pO_2 are given as 44 ± 17 mm Hg, and those of mito-VO_2 as 5.8 ± 2.3 mm Hg at 34 degrees Celsius [24]. The study showed that cutaneous respirometry allows measurement of mitochondrial oxygenation and oxygen consumption in humans [24].

3.3. Wet Biomarkers

3.3.1. Lactate, Pyruvate, Creatine-Kinase, Amino A2cids, Organic Acids, Carnitines, Oxidative Stress Parameters

The determination of lactate, pyruvate, creatine-kinase (CK), amino acids, organic acids, carnitines, and oxidative stress parameters is frequently carried out in fluids such as blood, urine, saliva, or CSF, but diagnostic accuracy is limited. This may be due to the fact that the muscle, myocardium, and cerebrum are not affected in every MID patient, may be unaffected at the time of the investigation, and that some of these parameters depend on whether these fluids are collected at rest or during exercise. However, these parameters are attractive since the collection of appropriate fluids can be carried out non-invasively (urine, saliva) or minimally invasively (blood). Recently, it has been shown that determination of the parameters retinol-binding protein (RBP) and albumin in the urine allows the delineation of patients with specific or non-specific MIDs from healthy controls [25]. RBP (respectively, albumin) was increased in 29/75 (respectively, 23/75) patients carrying the variant m.3243A>G [25]. In a study of fibroblasts from 16 patients with Leber's hereditary optic neuropathy (LHON) and from eight healthy volunteers, amino acids, spermidine, putrescine, isovaleryl-carnitine, propionyl-carnitine, and five sphingomyelin species were decreased, whereas ten phosphatidyl-choline species were increased [26]. Increase of sphingomyelins and decrease of phosphatidyl-choline together with decreased amino acids suggests involvement of the endoplasmic reticulum in MIDs [26].

3.3.2. Circulating Cytokines (FGF21, GDF15)

The cytokines FGF-21 and GDF-15 have been recently identified as potential biomarkers of MIDs [27,28]. Since FGF-21 is mainly produced in the skeletal muscle, a main disadvantage of FGF-21 as a biomarker of MIDs is that it may not be useful in MID patients without myopathy. In MIDs which do not manifest with mitochondrial myopathy [27,28], FGF-21 may be normal. Though FGF-21 and GDF-15 have been found elevated in a significant number of MID patients, their specificity is low. This is because FGF-21 and GDF-15 have also been found elevated in other conditions, such as diabetes, hepatopathy, renal insufficiency, malignancy, or obesity [3]. Additionally, FGF-21 levels may increase with stress, steatosis hepatis, or in metabolic syndrome [29]. Whether FGF-21 levels are associated with disease severity or disease progression is under debate, since conflicting results have been reported on this issue [30,31]. Though FGF-21 concentrations correlated with disease severity in a study of 99 carriers of the m.3243A>G variant, no significant correlation was found between disease severity and the heteroplasmy rate in urinary epithelial cells or leukocytes [30]. A weak correlation was found between FGF-21 concentrations and the severity of myopathy and between FGF-21 concentrations and the severity of the encephalopathy [30]. It has recently been reported that FGF-21 and GDF-15 levels have the highest specificity in MID patients due to mtDNA translation or maintenance defects [32]. The specificity of FGF-21 (respectively, GDF-15) to detect patients with mitochondrial myopathy was 89.3% (respectively, 86.4%), and the sensitivity was 67.3% (76.0%, respectively) [32].

3.3.3. microRNAs

microRNAs represent highly-conserved non-coding RNAs of 21–23 nucleotides in length (although some may reach >100 nucleotides in length), which control gene expression by silencing the transcription. Micro-RNAs regulate gene expression with high specificity on the post-transcriptional level. Micro-RNAs bind to the 3'-untranslated region (3'-UTR) of the mRNA. Due to this binding, mRNA are hindered in the translation or the mRNA is cut. Distinctive patterns of micro-RNAs are associated with various disorders and at least in cybrid cells carrying the m.3243A>G variant it has been shown that the micro-RNA 9/9* pattern is associated with the MELAS or myoclonic epilepsy with ragged-red fibers (MERRF) phenotype [33]. The micro-RNA pattern 9/9* acts as a post-transcriptional down-regulator of the mt-tRNA-modification enzymes GTPBP3, MTO1, and TRMU [33]. Down-regulation of these enzymes by microRNA-9/9* affects the U34 modification status of non-mutant tRNAs, and thus contributes to the MELAS phenotype [33].

3.3.4. Biopsy of Solid Tissues

Though biopsy of affected tissues is invasive, logistically demanding, time-consuming, and cost-intensive, it is still one of the best instruments to diagnose MIDs. This is particularly the case for biopsies of the skeletal muscle, the myocardium, liver, or the skin. Biopsies can not only be analysed with regard to histological or immunohistological features, but also with regard to ultrastructural and biochemical abnormalities of mitochondria and the respiratory chain in particular. The tissue most easily accessible for biopsy is the skin. It has been shown to be of diagnostic help in Leigh syndrome with cutaneous manifestations (cutis laxa) due to a mitochondrial β-oxidation defect [34]. Skin biopsy was also of diagnostic support in a patient with MELAS and skin manifestations such as scaly, pruritic, diffuse erythema, reticular pigmentation, moderate hypertrichosis, seborrheic eczema, atopy, and vitiligo [35]. Skin biopsy may be also helpful in MIDs without skin manifestations. In MELAS patients without clinical skin manifestations, investigations of skin fibroblasts revealed decreased membrane potential of fibroblast mitochondria [36].

3.3.5. Exercise Tests

Exercise tests for the diagnostic work-up of MIDs are applied to assess the aerobic capacity of mitochondria [37]. It has been shown in exercise tests of MIDs that oxygen consumption (peakVO$_2$) is decreased, that peak power (Wmax) is decreased, and that also peak arterio-venous oxygen difference is decreased [37]. In a similar study, peakVO$_2$ correlated with the heteroplasmy rate of the m.3243A>G variant as determined in the skeletal muscle but not in lymphocytes [38]. A second type of exercise test for diagnostic purposes is the lactate stress test [39]. It relies on the determination of serum lactate before, during, and after constant exercise below the anaerobic threshold [39]. In patients with an MID, lactate increases significantly during exercise, while in healthy subjects such an increase cannot be observed [40]. The sensitivity of the lactate stress test was calculated as 66%, and the specificity as 84% [39]. However, it is under debate as to whether the workload used should be adapted to the maximal individual workload, and if constant or incremental exercise should be carried out during the test.

3.3.6. Small Molecule Reporters

Small molecule reporters are tailor-made probes administered intravenously to react with a substrate of interest and consecutively accumulate in mitochondria of an organ of interest [3].

After reaction with the substrate, probes are modified such that they produce an exogenous marker, which can be extracted to undergo quantitative analysis. Exogenous markers also allow inferences about the reacting substrate [3]. In a cell study, it has been shown that concentrations of reactive oxidative species (ROS) such as H_2O_2 can be quantified by the so-called SNAP-tag technique, which relies on the small molecule reporter SNAP-peroxy-green [41]. Though small-molecule reporters have not been applied to MID patients thus far, they may enable the measurement of mitochondrial function, mitochondrion-specific metabolites, and the generation of ROS in vivo [3].

4. Conclusions

This review shows that there is currently no single biomarker available with which all different subtypes of MIDs could be detected. Particularly MIDs without cerebral, cardiac, or skeletal muscle involvement may be missed by application of the biomarkers presented above. The best "biomarker" for suspecting and diagnosing MIDs is still the individual and family history and the clinical exam. Since MIDs are frequently multisystem diseases, it is essential that all organs potentially affected in an MID are prospectively investigated, irrespective of whether there are clinical manifestations. However, if patients predominantly present with cerebral, muscle, or cardiac manifestations, imaging techniques can be helpful to confirm the suspicion of an MID. Concerning the wet biomarkers from blood, urine, saliva, or CSF, lactate, pyruvate, CK, amino acids, carnitines, and organic acids are frequently elevated in various MIDs, but do not meet all criteria to serve as a biomarker. This is also the case for circulating cytokines (FGF21, GDF15) and markers of oxidative stress. Biopsies from various tissues can be extremely helpful, but are usually invasive and cost-intensive, thus not fulfilling two main criteria of a biomarker. The role of exercise tests, microRNAs, and small-molecule reporters need to be further evaluated before a decision about their role in the work-up of MIDs can be finally made. Generally, there are few biomarkers reported which have been systematically investigated for their suitability to serve as a biomarker for diagnosing an MID. Currently-available biomarkers are not appropriate to distinguish between primary and secondary MIDs. Since most of the parameters so far applied to screen for MIDs failed to meet the criteria for a biomarker, effort needs to be increased to find more global MID parameters encompassing all subtypes of a MID.

Author Contributions: J.F.: design, literature search, discussion, first draft, S.Z.-M.: literature search, discussion, critical comments.

Conflicts of Interest: The authors declare no conflict of interest.

References

1. Gorman, G.S.; Chinnery, P.F.; DiMauro, S.; Hirano, M.; Koga, Y.; McFarland, R.; Suomalainen, A.; Thorburn, D.R.; Zeviani, M.; Turnbull, D.M. Mitochondrial diseases. *Nat. Rev. Dis. Prim.* **2016**, *2*, 16080. [CrossRef] [PubMed]
2. Kaufmann, P.; Engelstad, K.; Wei, Y.; Kulikova, R.; Oskoui, M.; Sproule, D.M.; Battista, V.; Koenigsberger, D.Y.; Pascual, J.M.; Shanske, S.; et al. Natural history of MELAS associated with mitochondrial DNA m.3243A>G genotype. *Neurology* **2011**, *77*, 1965–1971. [CrossRef] [PubMed]
3. Steele, H.E.; Horvath, R.; Lyon, J.J.; Chinnery, P.F. Monitoring clinical progression with mitochondrial disease biomarkers. *Brain* **2017**, *140*, 2530–2540. [CrossRef] [PubMed]
4. Strimbu, K.; Tavel, J.A. What are biomarkers? *Curr. Opin. HIV AIDS* **2010**, *5*, 463–466. [CrossRef] [PubMed]
5. Finsterer, J. Biomarkers of peripheral muscle fatigue during exercise. *BMC Musculoskelet. Disord.* **2012**, *13*, 218. [CrossRef] [PubMed]
6. Finsterer, J.; Drory, V.E. Wet, volatile, and dry biomarkers of exercise-induced muscle fatigue. *BMC Musculoskelet. Disord.* **2016**, *17*, 40. [CrossRef] [PubMed]
7. Pennuto, M.; Greensmith, L.; Pradat, P.F.; Sorarù, G.; European SBMA Consortium. 210th ENMC International Workshop: Research and Clinical Management of Patients with Spinal and Bulbar Muscular Atrophy, 27–29 March, 2015, Naarden, The Netherlands. *Neuromuscul. Disord.* **2015**, *25*, 802–812. [CrossRef] [PubMed]

8. Finsterer, J.; Zarrouk-Mahjoub, S. Mitochondrial multiorgan disorder syndrome score generated from definite mitochondrial disorders. *Neuropsychiatr. Dis. Treat.* **2017**, *13*, 2569–2579. [CrossRef] [PubMed]

9. Pitceathly, R.D.; Morrow, J.M.; Sinclair, C.D.; Woodward, C.; Sweeney, M.G.; Rahman, S.; Plant, G.T.; Ali, N.; Bremner, F.; Davagnanam, I.; et al. Extra-ocular muscle MRI in genetically-defined mitochondrial disease. *Eur. Radiol.* **2016**, *26*, 130–137. [CrossRef] [PubMed]

10. Yu-Wai-Man, C.; Smith, F.E.; Firbank, M.J.; Guthrie, G.; Guthrie, S.; Gorman, G.S.; Taylor, R.W.; Turnbull, D.M.; Griffiths, P.G.; Blamire, A.M.; et al. Extraocular muscle atrophy and central nervous system involvement in chronic progressive external ophthalmoplegia. *PLoS ONE* **2013**, *8*, e75048. [CrossRef] [PubMed]

11. Finsterer, J.; Zarrouk-Mahjoub, S. Cerebral imaging in pediatric mitochondrial disorders. *J. Neurol. Sci.* **2017**, submitted.

12. Finsterer, J. Central nervous system imaging in mitochondrial disorders. *Can. J. Neurol. Sci.* **2009**, *36*, 143–153. [CrossRef] [PubMed]

13. Florian, A.; Ludwig, A.; Stubbe-Dräger, B.; Boentert, M.; Young, P.; Waltenberger, J.; Rösch, S.; Sechtem, U.; Yilmaz, A. Characteristic cardiac phenotypes are detected by cardiovascular magnetic resonance in patients with different clinical phenotypes and genotypes of mitochondrial myopathy. *J. Cardiovasc. Magn. Reson.* **2015**, *17*, 40. [CrossRef] [PubMed]

14. Golla, S.; Ren, J.; Malloy, C.R.; Pascual, J.M. Intramyocellular lipid excess in the mitochondrial disorder MELAS: MRS determination at 7T. *Neurol. Genet.* **2017**, *3*, e160. [CrossRef] [PubMed]

15. DeBrosse, C.; Nanga, R.P.; Wilson, N.; D'Aquilla, K.; Elliott, M.; Hariharan, H.; Yan, F.; Wade, K.; Nguyen, S.; Worsley, D.; et al. Muscle oxidative phosphorylation quantitation using creatine chemical exchange saturation transfer (CrCEST) MRI in mitochondrial disorders. *JCI Insight* **2016**, *1*, e88207. [CrossRef] [PubMed]

16. Liu, A.H.; Niu, F.N.; Chang, L.L.; Zhang, B.; Liu, Z.; Chen, J.Y.; Zhou, Q.; Wu, H.Y.; Xu, Y. High cytochrome c oxidase expression links to severe skeletal energy failure by ^{31}P-MRS spectroscopy in mitochondrial encephalomyopathy, lactic acidosis, and stroke-like episodes. *CNS Neurosci. Ther.* **2014**, *20*, 509–514. [CrossRef] [PubMed]

17. Weiduschat, N.; Kaufmann, P.; Mao, X.; Engelstad, K.M.; Hinton, V.; DiMauro, S.; De Vivo, D.; Shungu, D. Cerebral metabolic abnormalities in A3243G mitochondrial DNA mutation carriers. *Neurology* **2014**, *82*, 798–805. [CrossRef] [PubMed]

18. Chi, C.S.; Lee, H.F.; Tsai, C.R.; Chen, W.S.; Tung, J.N.; Hung, H.C. Lactate peak on brain MRS in children with syndromic mitochondrial diseases. *J. Chin. Med. Assoc.* **2011**, *74*, 305–309. [CrossRef] [PubMed]

19. Nariai, T.; Ohno, K.; Ohta, Y.; Hirakawa, K.; Ishii, K.; Senda, M. Discordance between cerebral oxygen and glucose metabolism, and hemodynamics in a mitochondrial encephalomyopathy, lactic acidosis, and strokelike episode patient. *J. Neuroimaging* **2001**, *11*, 325–329. [CrossRef] [PubMed]

20. Haginoya, K.; Kaneta, T.; Togashi, N.; Hino-Fukuyo, N.; Kobayashi, T.; Uematsu, M.; Kitamura, T.; Inui, T.; Okubo, Y.; Takezawa, Y.; et al. FDG-PET study of patients with Leigh syndrome. *J. Neurol. Sci.* **2016**, *362*, 309–313. [CrossRef] [PubMed]

21. Frackowiak, R.S.; Herold, S.; Petty, R.K.; Morgan-Hughes, J.A. The cerebral metabolism of glucose and oxygen measured with positron tomography in patients with mitochondrial diseases. *Brain* **1988**, *111*, 1009–1024. [CrossRef] [PubMed]

22. Lindroos, M.M.; Borra, R.J.; Parkkola, R.; Virtanen, S.M.; Lepomäki, V.; Bucci, M.; Virta, J.R.; Rinne, J.O.; Nuutila, P.; Majamaa, K. Cerebral oxygen and glucose metabolism in patients with mitochondrial m.3243A>G mutation. *Brain* **2009**, *132*, 3274–3284. [CrossRef] [PubMed]

23. Harms, F.A.; Bodmer, S.I.; Raat, N.J.; Mik, E.G. Cutaneous mitochondrial respirometry: Non-invasive monitoring of mitochondrial function. *J. Clin. Monit. Comput.* **2015**, *29*, 509–519. [CrossRef] [PubMed]

24. Harms, F.A.; Stolker, R.J.; Mik, E.G. Cutaneous respirometry as novel technique to monitor mitochondrial function: A feasibility study in healthy volunteers. *PLoS ONE* **2016**, *11*, e0159544.

25. Hall, A.M.; Vilasi, A.; Garcia-Perez, I.; Lapsley, M.; Alston, C.L.; Pitceathly, R.D.; McFarland, R.; Schaefer, A.M.; Turnbull, D.M.; Beaumont, N.J.; et al. The urinary proteome and metabonome differ from normal in adults with mitochondrial disease. *Kidney Int.* **2015**, *87*, 610–622. [CrossRef] [PubMed]

26. Chao de la Barca, J.M.; Simard, G.; Amati-Bonneau, P.; Safiedeen, Z.; Prunier-Mirebeau, D.; Chupin, S.; Gadras, C.; Tessier, L.; Gueguen, N.; Chevrollier, A.; et al. The metabolomic signature of Leber's hereditary optic neuropathy reveals endoplasmic reticulum stress. *Brain* **2016**, *139*, 2864–2876. [CrossRef] [PubMed]

27. Tranchant, C.; Anheim, M. Movement disorders in mitochondrial diseases. *Rev. Neurol. (Paris)* **2016**, *172*, 524–529. [CrossRef] [PubMed]

28. Rasool, N.; Lessell, S.; Cestari, D.M. Leber Hereditary Optic Neuropathy: Bringing the Lab to the Clinic. *Semin. Ophthalmol.* **2016**, *31*, 107–116. [CrossRef] [PubMed]

29. Morovat, A.; Weerasinghe, G.; Nesbitt, V.; Hofer, M.; Agnew, T.; Quaghebeur, G.; Sergeant, K.; Fratter, C.; Guha, N.; Mirzazadeh, M.; et al. Use of FGF-21 as a biomarker of mitochondrial disease in clinical practice. *J. Clin. Med.* **2017**, *6*, 80. [CrossRef] [PubMed]

30. Koene, S.; de Laat, P.; van Tienoven, D.H.; Vriens, D.; Brandt, A.M.; Sweep, F.C.; Rodenburg, R.J.; Donders, A.R.; Janssen, M.C.; Smeitink, J.A. Serum FGF21 levels in adult m.3243A>G carriers: Clinical implications. *Neurology* **2014**, *83*, 125–133. [CrossRef] [PubMed]

31. Suomalainen, A.; Elo, J.M.; Pietiläinen, K.H.; Hakonen, A.H.; Sevastianova, K.; Korpela, M.; Isohanni, P.; Marjavaara, S.K.; Tyni, T.; Kiuru-Enari, S.; et al. FGF-21 as a biomarker for muscle-manifesting mitochondrial respiratory chain deficiencies: A diagnostic study. *Lancet Neurol.* **2011**, *10*, 806–818. [CrossRef]

32. Lehtonen, J.M.; Forsström, S.; Bottani, E.; Viscomi, C.; Baris, O.R.; Isoniemi, H.; Höckerstedt, K.; Österlund, P.; Hurme, M.; Jylhävä, J.; et al. FGF21 is a biomarker for mitochondrial translation and mtDNA maintenance disorders. *Neurology* **2016**, *87*, 2290–2299. [CrossRef] [PubMed]

33. Meseguer, S.; Martínez-Zamora, A.; García-Arumí, E.; Andreu, A.L.; Armengod, M.E. The ROS-sensitive microRNA-9/9* controls the expression of mitochondrial tRNA-modifying enzymes and is involved in the molecular mechanism of MELAS syndrome. *Hum. Mol. Genet.* **2015**, *24*, 167–184. [CrossRef] [PubMed]

34. Balasubramaniam, S.; Riley, L.G.; Bratkovic, D.; Ketteridge, D.; Manton, N.; Cowley, M.J.; Gayevskiy, V.; Roscioli, T.; Mohamed, M.; Gardeitchik, T.; et al. Unique presentation of cutis laxa with Leigh-like syndrome due to *ECHS₁* deficiency. *J. Inherit. Metab. Dis.* **2017**, *40*, 745–747. [CrossRef] [PubMed]

35. Carmi, E.; Defossez, C.; Morin, G.; Fraitag, S.; Lok, C.; Westeel, P.F.; Canaple, S.; Denoeux, J.P. MELAS syndrome (mitochondrial encephalopathy with lactic acidosis and stroke-like episodes). *Ann. Dermatol. Venereol.* **2001**, *128*, 1031–1035. [PubMed]

36. James, A.M.; Wei, Y.H.; Pang, C.Y.; Murphy, M.P. Altered mitochondrial function in fibroblasts containing MELAS or MERRF mitochondrial DNA mutations. *Biochem. J.* **1996**, *318*, 401–407. [CrossRef] [PubMed]

37. Jeppesen, T.D.; Schwartz, M.; Olsen, D.B.; Wibrand, F.; Krag, T.; Dunø, M.; Hauerslev, S.; Vissing, J. Aerobic training is safe and improves exercise capacity in patients with mitochondrial myopathy. *Brain* **2006**, *129*, 3402–3412. [CrossRef] [PubMed]

38. Jeppesen, T.D.; Schwartz, M.; Frederiksen, A.L.; Wibrand, F.; Olsen, D.B.; Vissing, J. Muscle phenotype and mutation load in 51 persons with the 3243A>G mitochondrial DNA mutation. *Arch. Neurol.* **2006**, *63*, 1701–1706. [CrossRef] [PubMed]

39. Finsterer, J.; Milvay, E. Stress lactate in mitochondrial myopathy under constant, unadjusted workload. *Eur. J. Neurol.* **2004**, *11*, 811–816. [CrossRef] [PubMed]

40. Finsterer, J.; Milvay, E. Lactate stress testing in 155 patients with mitochondriopathy. *Can. J. Neurol. Sci.* **2002**, *29*, 49–53. [CrossRef] [PubMed]

41. Srikun, D.; Albers, A.E.; Nam, C.I.; Iavarone, A.T.; Chang, C.J. Organelle-targetable fluorescent probes for imaging hydrogen peroxide in living cells via SNAP-Tag protein labeling. *J. Am. Chem. Soc.* **2010**, *132*, 4455–4465. [CrossRef] [PubMed]

Journal of
Clinical Medicine

MDPI

Review

Oxidative Stress: Mechanistic Insights into Inherited Mitochondrial Disorders and Parkinson's Disease

Mesfer Al Shahrani [1,2,3], Simon Heales [1,2,4], Iain Hargreaves [1,5] and Michael Orford [2,*]

1   Neurometabolic Unit. National Hospital for Neurology and Neurosurgery, Queen Square,
    London WC1N 3BG, UK; mesfer.shahrani.14@ucl.ac.uk (M.A.S.); s.heales@ucl.ac.uk (S.H.);
    i.p.hargreaves@ljmu.ac.uk (I.H.)
2   Department of Genetics and Genomic Medicine, UCL Great Ormond Street Institute of Child Health,
    London WC1N 1EH, UK
3   College of Applied Medical Sciences, King Khalid University, Abha 61481, Saudi Arabia
4   Chemical Pathology, Great Ormond Street for Children Hospital NHS Foundation Trust,
    London WC1N 3JH, UK
5   School of Pharmacy and Biomolecular Sciences, Liverpool John Moores University, Liverpool L2 2AZ, UK
*   Correspondence: m.orford@ucl.ac.uk

Received: 12 October 2017; Accepted: 23 October 2017; Published: 27 October 2017

Abstract: Oxidative stress arises when cellular antioxidant defences become overwhelmed by a surplus generation of reactive oxygen species (ROS). Once this occurs, many cellular biomolecules such as DNA, lipids, and proteins become susceptible to free radical-induced oxidative damage, and this may consequently lead to cellular and ultimately tissue and organ dysfunction. Mitochondria, as well as being a source of ROS, are vulnerable to oxidative stress-induced damage with a number of key biomolecules being the target of oxidative damage by free radicals, including membrane phospholipids, respiratory chain complexes, proteins, and mitochondrial DNA (mt DNA). As a result, a deficit in cellular energy status may occur along with increased electron leakage and partial reduction of oxygen. This in turn may lead to a further increase in ROS production. Oxidative damage to certain mitochondrial biomolecules has been associated with, and implicated in the pathophysiology of a number of diseases. It is the purpose of this review to discuss the impact of such oxidative stress and subsequent damage by reviewing our current knowledge of the pathophysiology of several inherited mitochondrial disorders together with our understanding of perturbations observed in the more commonly acquired neurodegenerative disorders such as Parkinson's disease (PD). Furthermore, the potential use and feasibility of antioxidant therapies as an adjunct to lower the accumulation of damaging oxidative species and hence slow disease progression will also be discussed.

Keywords: mitochondria; oxidative stress; reactive oxygen species; antioxidant

---

1. Introduction

Up to 90% of cellular metabolic energy is generated by mitochondria via the oxidative phosphorylation pathway [1]. In concert with glycolysis, the tricarboxylic acid (TCA) cycle additionally generates a small amount of energy via substrate level phosphorylation, although the vast proportion of metabolic energy is harnessed via the generation of reducing power and subsequent donation of high energy electron pairs through the electron carriers NADH and FADH2, which ultimately feed directly into mitochondrial respiratory chain (MRC). The MRC is composed of four multi-subunit proteins; complex I (NADH: ubiquinone reductase; EC 1.6.5.3), complex II (succinate: ubiquinone reductase; EC 1.3.5.1), complex III (ubiquinol: cytochrome c reductase; EC 1.10.2.2), and complex IV (cytochrome c oxidase; EC 1.9.3.1) [2], each of which contain a variety of cofactors such as hemes, flavins, and iron–sulphur clusters. In addition to these redox cofactors, two mobile electron carriers, namely coenzyme Q_{10} (ubiquinone) and cytochrome c are involved in transferring electrons between the complexes. As a result of the passage of electrons between chains, protons are pumped out of the mitochondrial matrix and into the intermembrane space, creating a proton-motive force. It is the subsequent dissipation of these protons through the mitochondrial ATPase enzyme which results in the direct phosphorylation of ADP to ATP [3].

It has been well established that the formation of reactive oxygen species (ROS) is a significant component produced during the generation of ATP. Under normal conditions, approximately 1% of total oxygen utilized by the MRC is converted to ROS, although under pathological conditions this may increase dramatically. Mitochondrial ROS, particularly in the form of the superoxide radical ($O^{\bullet-}_2$) is mostly generated either in the matrix from complex I or both in the intermembrane space and matrix from complex III [4]. Mitochondria are additionally a site of nitric oxide (NO) synthesis which in turn may form the peroxynitrite ion ($ONOO^-$) when NO reacts with $O^{\bullet-}_2$, leading to the generation of equally undesirable reactive nitrogen species (RNS) [5]. $O^{\bullet-}_2$ is rapidly removed by conversion to hydrogen peroxide (H_2O_2) either by a manganese-dependent superoxide dismutase (Mn-SOD) or a copper, zinc-dependent superoxide dismutase (Cu, Zn-SOD), and then ultimately reduced to water by glutathione peroxidase (GPx) utilizing the active and reduced form of glutathione (GSH) as a cofactor (Figure 1) [6,7]. A large number of biomolecules have over the years been recognised as potent antioxidants. GSH itself, besides being a cofactor for the enzymatic antioxidant GPx, also serves as non-enzymatic antioxidant by directly removing free radicals as well as other oxidative agents [7]. Similarly, other powerful non-enzymatic antioxidants known to act as potent free radical scavengers include ascorbate (vitamin C) [8], α-tocopherol (vitamin E) [9], ubiquinol-10, the reduced form coenzyme Q_{10} [10], α-lipoic acid (ALA) [11], and carotenoids (β-carotene) [12].

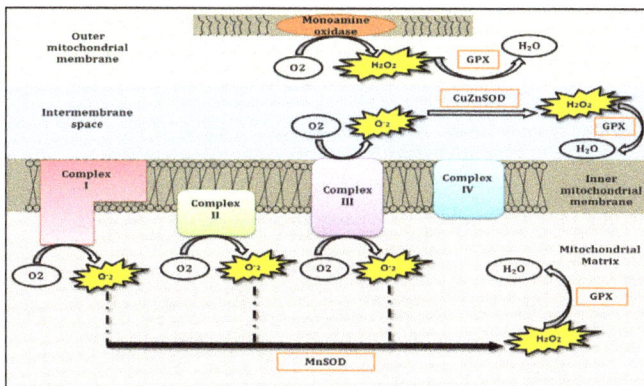

Figure 1. A schematic underlies the pathway of mitochondrial free radical generation and their enzymatic antioxidant defences. The mitochondrial $O^{\bullet-}_2$ is subsequently converted to H_2O_2 either by manganese-dependent superoxide dismutase (Mn-SOD) or copper, zinc-dependent superoxide dismutase (Cu, Zn-SOD), and then ultimately reduced to water by glutathione peroxidase (GPx).

It is the uncontrolled or overproduction of ROS (oxidative stress) or RNS (nitrosative stress) which can indiscriminately cause damage to cellular molecules, including DNA, proteins and lipids [13]. Furthermore, it is believed that the accumulation of these free radical species, resulting in oxidative/nitrosative stress, could lead to impaired MRC function and this in turn may be a major contributory factor to the pathophysiology of various inherited and acquired disorders [14,15].

In this review, the potential impact of oxidative stress and subsequent molecular and cellular damage will be discussed, including lessons learnt from our knowledge of the pathophysiology of a number of inherited mitochondrial disorders together with our growing understanding of perturbations observed in the more commonly acquired neurodegenerative disorders. Furthermore, we will consider the potential use of antioxidant therapies as an adjunct to standard pharmacological care as a means to limit free radical accumulation and thereby attempt to slow disease progression.

2. Inherited Mitochondrial Disorders

2.1. Inherited Mitochondrial DNA (mtDNA) Disorders

Inherited mitochondrial disorders are generally believed to be one of the most common inborn errors of metabolism, with an overall birth prevalence of about 1:5000 [3], with those resulting from mitochondrial DNA (mtDNA) mutations estimated at about 1:8000 [16]. Furthermore, mtDNA mutations are rare in children, accounting for less than 10% of all mitochondrial disorders affecting infants [17]. At least 200 pathogenic point mutations affecting the mtDNA-encoded MRC complexes I, III, and IV as well as tRNAs have recently been reported [18]. In comparison with nuclear-DNA (nDNA), mtDNA is particularly vulnerable to oxidative damage since it lacks protective histones and has limited repair mechanisms, as well as being located in close proximity to the MRC which is known to be the major source of ROS generation in the cell [19,20]. Therefore, mtDNA has a potentially higher mutation rate than nDNA. A unifying hypothesis, known as "mitochondrial catastrophe", postulates that the accumulation of mtDNA lesions results in a decline in MRC function, which in turn, leads to the generation of further ROS, and eventually cell death [21]. This phenomenon therefore provides an insightful working hypothesis that oxidative stress could be considered as a major cause of rather than as a consequence of mtDNA disorders.

Leber's hereditary optic neuropathy (LHON) (OMIM 540000) is one of the most well-known inherited mtDNA disorders. It is caused in most cases by three mtDNA point mutations within MRC complex I subunits [22] and results predominantly in visual loss as the main clinical feature [23]. The aetiology of oxidative stress in the mechanism of LHON disorder has been described [24,25]. It is worth emphasizing that MRC complex I is one of the major sources of ROS generation, predominately in the form of $O^{\bullet}_2$, and it is this reactive species that is implicated to have significant effects in some, if not all of LHON disorders [26]. The inhibition of MRC complex I causes a significant increase in oxidative stress, which in turn promotes apoptosis and cell death. [27]. A recent study of patients with LHON demonstrated an increase in plasma free radical formation as well as a reduction in antioxidant levels compared to controls [28]. In addition to a reduction of MRC complex I activity and consequential increased $O^{\bullet}_2$ levels, increases in protein carbonyl, and lipid peroxidation have also been reported in mutant mitochondrially encoded NADH dehydrogenase 6 (MT-ND6) subunit of MRC complex I. However, the mitochondrial antioxidant enzymes Mn-SOD and GPX were not altered in this study, suggesting that the mutation threshold might not be significant [29]. Interestingly, increased lipid peroxidation and raised levels of the potent oxidant hydroxyl radical (OH$^{\bullet}$) together with an elevation in the activity of both Mn-SOD and Cu, Zn-SOD have also been observed [30]. Consistent with this, increased levels of chemical ROS markers have been demonstrated in LHON neurons [31] as well as the marker of oxidative DNA damage, 8-hydroxy-2'-deoxyguanosine (8-OHDG), being elevated in white blood cells of LHON patients [32]. It should be further noted that in addition to endogenous ROS production, exogenous ROS sources such as those contained in tobacco smoke have also been linked to the onset of LHON disorders [33], thereby strengthening the evidence of ROS-mediated events.

2.2. An Inherited Mitochondrial Lipid Disorder

Barth syndrome (BTHS) (OMIM 302060), is a rare X-linked genetic disorder, characterized by cardiomyopathy, neutropenia, skeletal weakness, and growth disorders [34]. In fact, it has been previously described as a mitochondrial disorder as BTHS patients show symptoms consistent with known mitochondrial disorders [35]. It is mainly caused by a mutation in the tafazzin (*TAZ*) gene, which encodes a putative enzyme acyltransferase, an enzyme largely responsible for enzymatic remodelling of cardiolipin (CL) [36]. CL is a phospholipid component found exclusively within the inner mitochondrial membrane (IMM) and constitutes approximately 25% of the total lipid contents in mitochondria [37]. It plays a crucial role in aspects of maintaining the functional properties of mitochondrial components [38]. For example, it is required for the enhancement of the enzymatic function of the MRC complexes following CL binding [39] and its molecular interaction with all individual respiratory complexes is necessary for their assembly into super-complexes [40]. CL also plays an essential role in the retention of cytochrome *c* which protects against apoptosis [41].

The biochemical findings following MRC enzyme studies in 1983 by Barth together with other groups have indicated evidence of multiple MRC defects [42]. However, results are somewhat conflicting, suggesting the possibility that primary MRC deficiency may result in a secondary loss of other MRC activities. This latter possibility was previously been investigated in human astrocytoma cells by Hargreaves et al. in 2007 where a pharmacologically-induced MRC complex IV deficiency was found to result in a secondary loss of MRC complex II–III activity due to the progressive nature of MRC defects [43]. Interestingly, loss of CL content has been associated with mtDNA instability [44] suggesting another possible mechanism that the dysfunctional MRC encoded by mtDNA may be the consequence of oxidative damage as mtDNA structurally lacks protective histones [45]. Increased ROS levels have evidently been implicated in the *TAZ* mutation seen in cardiomyopathy which is a hallmark clinical feature of BTHS syndrome [46]. It is worth highlighting that CL is a susceptible target for oxidative damage for the following reasons: (1) CL has a naturally high unsaturated content, which is easily attacked by free radical species; (2) It is involved in the structural assembly of the MRC, a major intracellular site for ROS production; and (3) In addition to CL peroxidation, calcium-mediated detachment of cytochrome *c* from CL is induced by generating further ROS levels and this results in apoptotic cell death.

2.3. An Inherited Mitochondrial Protein Disorder

Friedreich ataxia (FRDA) (OMIM 229300), is a progressive neurodegenerative disorder with an autosomal recessive mode of inheritance, affecting roughly 1:50,000 live births [47]. In addition to neuronal injury in the dorsal root ganglia (DRG) and sensory peripheral nerves, FRDA patients also manifest with non-neurological symptoms including diabetes, cardiomegaly, and muscle weakness [48]. FRDA is caused by a GAA expansion in the frataxin gene, the product of which is predominantly located in mitochondria [49]. The exact role of the frataxin protein is not yet fully understood. However, it has been proposed to play crucial roles primarily in regulating iron machinery, and functioning as a mitochondrial Fe–S cluster chaperone [50,51]. In this regard, increased iron capacity and the loss of activity of mitochondrial Fe–S cluster-containing enzymes has been observed in FRDA patients, highlighting the important function of frataxin in iron metabolism [47,49,52]. In addition to its well-established role in iron metabolism, frataxin can protect against iron-mediated oxidative stress [53]. In a previous study, exposure of fibroblast obtained from patients with FRDA to ferrous ions and H_2O_2 reduced the viability of the cells compared to control patients [54]. The most direct evidence of the critical function of frataxin in protecting against oxidative stress however comes from the observation of a combined reduction in activity of nuclear factor E2-related factor 2 (Nrf2) and GSH levels in the YG8R mouse model of FRDA [55]. In contrast, an increased resistance to oxidative stress induced by the overexpression of mitochondrial frataxin has been reported in *Drosophila* [56]. Since the discovery of the gene in 1996, dysfunction of mitochondrial Fe–S cluster-containing enzymes including MRC complexes I and III as well as aconitase resulting in oxidative stress has been found to make a major contribution to the pathophysiology of FRDA [57].

Aconitase (EC 4.2.1.3) is a multi-domain enzyme, containing a closely associated iron–sulphur cluster, and exists in two slightly different structural forms: an active $[4Fe-4S]^{2+}$ and an inactive $[3Fe-4S]^{1+}$ cluster [58]. The active form of aconitase is highly sensitive to oxidation by the superoxide anion, which in turn, converts it to the inactive form. This oxidation reaction is accompanied by the release of a ferrous ion, which subsequently contributes to the generation of OH$^•$ via the Fenton reaction [59]. As a consequence, oxidative damage to mtDNA, lipids, and proteins may occur [60]. Since the aconitase enzyme is susceptible to direct attack by free radicals, it has been recognized as an oxidative stress marker in mitochondria, suggesting it may function as a mitochondrial redox sensor [61]. Aconitase exists in two isoenzyme forms in mammalian cells: the mitochondrial aconitase (m-aconitase), and cytosolic aconitase (c-aconitase), which both enzymatically catalyse the isomerization of citrate to isocitrate. In addition to its role in the TCA cycle, c-aconitase, also known as iron-responsive protein-1 (IRP1) additionally performs a dual role in the regulation of iron homeostasis through binding to iron-responsive elements (IREs) and controlling cellular iron levels [62]. Despite the m-aconitase being identical in function (with 25% sequence homology identity) to that of c-aconitase, it is clearly not recognized to have role as an IRP [63]. However, the brain is highly dependent on m-aconitase activity [64], and is regulated by a 5′IRE in its mRNA [65]. As a consequence of inactivation of m-aconitase; neurons could be highly vulnerable to free radical attack and subsequent iron overload, resulting in a dramatic increase in oxidative stress [66]. Due to its important role in TCA cycle energy metabolism, dysfunction of aconitase may consequently lead to TCA cycle impairment, a deficit in MRC activity, and a decline in ATP production, which in turn, could lead to subsequent accumulation of ROS generation, and resultant oxidative damage (Figure 2) [67].

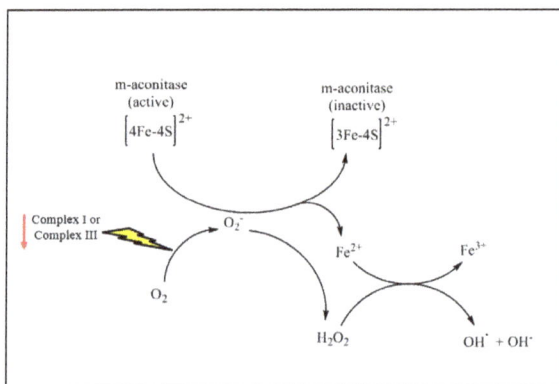

Figure 2. A potential mechanism for the oxidative inactivation of m-aconitase by mitochondrial O$^•_2$. This oxidation reaction is accompanied by the release of a ferrous ion, which subsequently contributes to the generation of OH$^•$ via Fenton reaction. This scenario could consequently lead to an impairment of tricarboxylic acid (TCA) cycle capacity, a deficit in mitochondrial respiratory chain (MRC) activity, and a decline in ATP production, which in turn, leads to further oxidative damage.

3. Parkinson's Disease (PD)

PD is a chronic and progressive neurological disorder. It is currently ranked as the second most common neuromuscular disorder after Alzheimer's disease, affecting roughly 1% of people almost exclusively in the over 60 age group [68]. Furthermore, the male sex is particularly susceptible to this disorder, with larger proportion of men being affected than women [69]. Clinically, PD patients commonly experience motor symptoms such as bradykinesia, tremor (particularly in the hands and/or arms), muscle stiffness (rigidity), and postural instability [70]. During later stages of the disorder, non-motor symptoms may manifest such as depression, sleep disturbances, anxiety, constipation, and in some cases dementia. Despite receiving intensive research interest over several decades and in particular with a large focus on the mechanistic aspects of PD, the exact aetiological mechanism is still poorly understood. Under normal conditions, the neurotransmitter dopamine (DA) is produced in the substantia nigra pars compacta (SNpc). However, a typical and major characteristic feature of PD is a significant depletion in its levels. The formation of intracytoplasmic eosinophilic inclusions, known as Lewy bodies (LBs), is another pathological sign of PD observed in a majority, but not all PD cases [71]. For several decades, it was postulated that PD is likely caused by environmental factors, until 1997 when the autosomal dominant mutation in the alpha-synculein (*SNKA*) gene was discovered [72]. Since this discovery, at least five other mutant genes that are linked to familial PD, including *parkin*, PTEN-induced putative kinase 1 (*PINK1*), *DJ-1*, High temperature requirement protein A2 (*HTRA2*), and leucine-rich-repeat kinase 2 (*LRRK2*) have also been identified [73]. These gene products appear to be in part localized to mitochondria and therefore may contribute towards mitochondrial dysfunction and oxidative stress [73].

Evidence of MRC dysfunction in PD emerged in the early 1980s following the intravenous injection of 1-methyl-4-phenyl-1,2,3,4-tetrahydropyridine (MPTP) by drug abusers, producing the neurotoxin 1-methyl-4-phenylpyridinium (MPP$^+$) via the monoamine oxidase-B (MAO-B) enzyme, which consequently induced Parkinson-like symptoms [72]. Similarly, in animal models, rats and primates were shown to share Parkinson-like symptoms following the administration of MPTP [74]. In parallel with MPTP, a chronic low-dose infusion of rotenone to rats additionally induced similar features of Parkinson disorders [75]. Taken together, these MRC complex I inhibitors have become widely used to create PD models to investigate the pathogenesis and therapeutic approaches for this disorder. Further studies conducted to support the role of mitochondrial dysfunctions have shown strong links to the aetiopathogenesis of PD. Studies of post-mortem PD patient brain tissue demonstrated MRC complex I deficiency in the substantia nigra and frontal cortex [76]. Consistent with these findings, MRC complex I deficiency was also shown in platelets [77] and skeletal muscle [78] from individuals with PD. In this context, it seems that a reduction of MRC complex I activity is systemic, thereby simultaneously affecting many tissues. In addition to a decrease in the activity of MRC complex I, a reduction in MRC complex III activity was further demonstrated in lymphocytes and platelets in patients with PD [79]. Remarkably, loss of MRC complex III activity may contribute to the impairment in function of MRC complex I since the stability of MRC complex I is evidently dependent on a correctly assembled MRC complex III [80]. Taken together, inhibition of MRC complex I and III can have devastating consequences, leading to excessive free radical species generation, oxidative stress and subsequent depletion of ATP levels, elevated intracellular calcium levels, excitotoxicity, and ultimately enhanced cell death (Figure 3) [81].

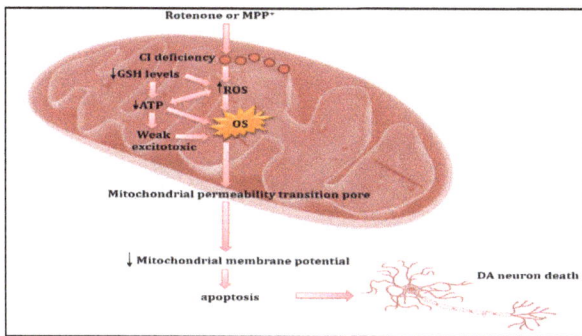

Figure 3. A schematic showing the role of mitochondrial dysfunction in the pathogenesis of Parkinson's disease (PD). Neurotoxins, such as rotenone or 1-methyl-4-phenylpyridinium (MPP$^+$) elicit MRC complex I deficiency, and subsequently generate reactive oxygen species (ROS), reducing levels of the antioxidant glutathione (GSH), with resulting oxidative stress. Oxidative stress induces mitochondrial permeability by transiently opening a pore, which subsequently causes depolarization of the mitochondrial membrane potential. These events ultimately lead to neural cell death via the release of pro-apoptotic mitochondrial proteins, including cytochrome *c* and apoptosis-initiating factor. DA: dopamine.

Any discussion would be incomplete without a reference to the interplay of iron and its contribution to mitochondrial dysfunction. Amongst all transition metals, iron is considered to be the most abundant metal in the brain, predominately in the basal ganglia [82]. It significantly contributes to the proper functioning of neurotransmitters, myelination, and mitochondria [83,84]. Brain iron metabolism is primarily regulated by transferrin and ferritin [85]. It is commonly conjugated into iron–sulphur clusters in many proteins, which have the potential ability to accept or donate electrons, particularly in the MRC pathway [86]. The evidence supporting the alteration of the iron metabolism in the neuropathology of PD is also overwhelming [87–90]. In fact, the potential mitochondrial ROS toxicity due to a defect in MRC complex I activity has been widely demonstrated in PD models [91]. Nevertheless, the exact mechanism of whether an enhanced production of ROS-induced neuronal injury is yet to be fully elucidated. Neurotoxins such the product of MPTP metabolism, commonly used to create PD models [72,75,91], have been utilized to demonstrate the potential harmful effects of the inactivation m-aconitase and high amounts of iron content on dopaminergic neurons [92]. In mice, this neurotoxin has been linked to the inactivation of m-aconitase, an increase in iron content, and a depletion of DA level [93].

It appears that excess ROS production is a common denominator of these cascades. Iron and iron derivatives contribute to the generation of the most active $OH^\bullet$ via the Fenton reaction, which in conjunction with DA autoxidation may further enhance oxidative stress, leading to degeneration of dopaminergic neurons (Figure 4) [94–96]. Post-mortem brain tissue from PD patients exhibited accumulation of iron content, which together with a reduction in the glutathione redox ratio of reduced glutathione/oxidized glutathione (GSH/GSSG) is a potential indicator of oxidative stress [96,97]. In the glutathione-depleted Δgsh1 cell model, on the other hand, the mitochondrial (Fe–S) cluster was unaffected, suggesting that m-aconitase is resistant to oxidative stress [98]. Furthermore, accumulation of iron was found to potentiate LB formation in the substantia nigra of PD patients, supporting the link between iron-mediated oxidative stress and the degeneration of dopaminergic neurons in PD [99].

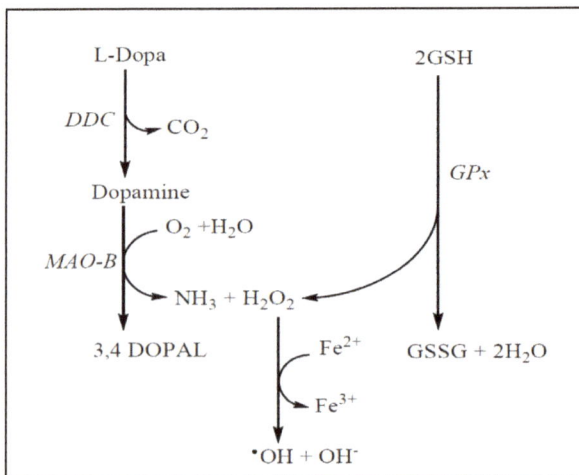

Figure 4. A potential mechanism of dopamine metabolism and $OH^\bullet$ radical formation in the striatum of PD patients as a consequence of iron accumulation and decline in GSH levels. DDC: dopa decarboxylase; 3,4 DOPAL: 3,4-dihydroxyphenylacetaldehyde.

4. The Role of Antioxidants in the Prevention of Oxidative Damage

Despite extensive research to elucidate the underlying mechanism of mitochondrial dysfunction in various conditions, there is no currently satisfactory treatment available. According to our recent understanding and knowledge regarding the mechanisms of mitochondrial dysfunction, it is however without doubt that mitochondrial free radical-induced oxidative damage is a plausible pathogenic facilitator in both inherited and acquired mitochondrial disorders. Alleviation of ROS/RNS free radical-mediated oxidative stress and increased availability of ATP by antioxidants could be effective therapeutic approaches to restore mitochondrial function, or at least to limit the progression of symptoms in a tremendous number of patients with mitochondrial dysfunction.

To limit free radical-induced oxidative stress, the human body is endowed with a variety of enzymatic and non-enzymatic antioxidant defence mechanisms. The two major antioxidants that protect the cell from ROS and RNS are GSH and coenzyme Q_{10} [10,100,101]. By cooperative actions, the primary function of the antioxidants is to scavenge and eliminate harmful ROS/RNS free radicals, thereby minimizing or delaying mitochondrial damage and enhancing mitochondrial bioenergetics.

4.1. Glutathione (GSH)

The tripeptide GSH, is a major intracellular thiol-dependent antioxidant, which protects the cellular components from free radical-induced oxidative damage [102]. Consequently, a compromised cellular GSH status results in increased production of ROS and RNS [100,101]. Despite being predominantly localised in the cytosol, GSH is also present in other intracellular organelles including, mitochondria, the nucleus, and the endoplasmic reticulum [102]. It exists in two forms: a reduced (GSH) and oxidized disulphide form (GSSG), with the ratio of reduced to oxidized forms being a key indicator of OS. In addition to its vital antioxidant role, GSH also serves as a substrate for other antioxidant defences including GPx, glutaredoxin (GRX), and thioredoxin (Trx) as well as maintaining vitamins C and E to be functionally active [7]. Accumulating evidence suggests that the depletion of GSH is associated with MRC defects [103,104]. The reduction of MRC complex I activity, followed by a depletion in GSH, has been reported previously [105]. Interestingly, results from our group have recently shown that GSH levels were significantly decreased in skeletal muscle from patients with MRC defects, compared to the control group [106]. Furthermore, patients with multiple MRC defects exhibited marked reductions in GSH levels, suggesting that oxidative stress may contribute to the pathophysiology of MRC disorders. In neurological disorders, particularly PD, it is thought that GSH depletion could be an early common event in PD pathogenesis before any significant impairment of MRC complex I and iron metabolism occur [107]. With regards to the latter however, it is uncertain whether this depletion occurs as a consequence of decreased ATP availability (required for GSH biosynthesis) or is due to increased ROS levels. Thus, the replenishment of cellular GSH could hold a promising therapeutic avenue for patients with inherited or acquired mitochondrial disorders. In rat brain, the GSH ethyl ester (GEE) derivative has been subcutaneously administered to enhance GSH levels. However, elevated brain levels were only evident post-administration directly to the left cerebral ventricle [108]. Furthermore, following co-administration with neurotoxin MPP$^+$, GGE has been demonstrated to partially protect dopaminergic neuron against neurotoxicity. However, complete protection was only achieved only after pre-treating with GEE [108]. As cysteine is a major component in GSH, it hinders GSH passage across the blood–brain barrier (BBB). For this reason, the modified *N*-acetyl cysteine (NAC) form, has been effectively utilized due to it is increased ability to penetrate the BBB [109]. As such, it has also been shown to restore GSH level and consequently ameliorate free radical-induced oxidative stress [110]. Encouragingly, lesions in dopaminergic tissue have been reduced by approximately 30%, following administration with NAC, suggesting it is able to provide neuroprotection [111].

4.2. Coenzyme Q_{10}

For many years, the clinical use of coenzyme Q_{10} or ubiquinone and its quinone analogues has been proven to be effective for treatment of mitochondrial disorders due to their capacity to augment electron transfer in the MRC, increase ATP, and enhance mitochondrial antioxidant activity, which in turn, can ameliorate the harmful effects of ROS [112]. In addition to its function as an electron carrier in the MRC pathway, Coenzyme Q_{10} also serves as a powerful free radical-scavenging antioxidant. The reduced ubiquinol form of coenzyme Q_{10} serves this function [113].

The therapeutic potential of coenzyme Q_{10} in the treatment of mitochondrial disorders took the spotlight in 1985 after Ogashara and colleagues reported sustained improvements in the clinical phenotype of patients with Kearns–Sayre syndrome (KSS) following administration with coenzyme Q_{10} [114]. More recently, Maldergem also reported that coenzyme Q_{10} therapy was beneficial to two sisters diagnosed with Leigh's encephalopathy [115]. Remarkably, the beneficial effects of CoQ_{10} in two patients with KSS and hypoparathyroidism were also shown to help maintain calcium levels in the serum of both patients, suggesting that treatment with coenzyme Q_{10} restored the capacity of calcitriol, a hormone located in the mitochondria of proximal renal tubules [116]. Some degree of sustained improvement has been noted with some patients whose clinical features can be associated with mitochondrial disorders, such as ataxia, muscle stiffness, and exercise intolerance following implementation of coenzyme Q_{10} therapy [114]. Despite the oral coenzyme Q_{10} supplementation being significantly effective in patients with all forms of coenzyme Q_{10} deficiency, it has been shown to be only partially effective in patients who present with neurological symptoms, suggesting that these sequelae may be somewhat refractory to coenzyme Q_{10} supplementation [117]. The efficacy of the synthetic ubiquinone analogues such as idebenone has been reported in patients with mitochondrial disorders including, LOHN, FRDA, and MELAS (mitochondrial encephalomyopathy, lactic acidosis and stroke like episodes) [118,119]. It has also been recommended that patients with deficient levels of coenzyme Q_{10} should be given coenzyme Q_{10} supplementation rather than idebenone since the synthetic analogue is not a potential replacement for coenzyme Q_{10} in the MRC [120]. However, in addition to its beneficial effects, idebenone may reduce MRC complex I activity, thereby affecting the mitochondrial bioenergetics function [121]. Hence, further clinical studies regarding the overall benefits of idebenone need to be conducted to address this issue.

Both the impairment of mitochondrial function and oxidative stress have been potentially linked to neurodegenerative pathogenesis, particularly in PD. Strategies to enhance mitochondrial function and suppress oxidative stress may therefore contribute to the development of novel therapies for PD. As coenzyme Q_{10} performs two roles, one in mitochondrial energy metabolism and the other as a free-radical scavenger, low levels of coenzyme Q_{10} may therefore result in the impairment of the MRC activities as well as in the accumulation of ROS levels, and thereby contribute towards the pathogenesis of PD. Coenzyme Q_{10} deficiency associated with PD has been previously described [122,123]. A reduction in coenzyme Q_{10} level was demonstrated in the plasma [124] and platelets [125] in patients with PD, thereby suggesting that systemic effects may be important. For the first time, a UK study demonstrated that coenzyme Q_{10} levels were lower in the brain cortex of patients with PD [123]. The neuroprotective role of coenzyme Q_{10} has also been investigated in both animal and human cell models [126–128]. Using in vitro models of PD, coenzyme Q_{10} has been reported to protect dopaminergic neurons against neurotoxin-induced PD symptoms using either rotenone, paraquat or MPP^+ [129]. Another study has shown that coenzyme Q_{10} treatment improved both MRC complex I and complex IV activities in skin fibroblast from PD patients [128]. To investigate the neuroprotective potential of coenzyme Q_{10} treatment in PD, 80 patients with early stage PD were randomly allocated to participate in a 16-month multicentre clinical trial [130]. Results showed that participants who received high doses of coenzyme Q_{10} had a large improvement in their motor functions, whilst lower doses only provided mild benefits. It was therefore concluded that the beneficial effect of coenzyme Q_{10} treatment may contribute to a reduction in the progression of PD.

4.3. Other Antioxidants

There is a considerable body of scientific literature which focuses on the beneficial effects of other antioxidants, including vitamins C and E, creatine, α-lipoic acid, urate, melatonin, and their derivatives as potential mediators in treating mitochondrial disorders [131–134]. Many patients with mitochondrial dysfunction, however, have not shown any significant clinical improvements when treated with theses antioxidants alone. However, it could be hypothesized that in combination, a cocktail therapy may improve mitochondrial conditions to an extent not seen previously with a single antioxidant agent. Several recent studies have also demonstrated some initial promise in the use of mitochondrial-targeted antioxidants with different modes of action to provide additional beneficial effects, such as mitoquinone (MitoQ) and mitotocopherol (MitoVitE) [135]. However, these compounds need to be further investigated to evaluate their full efficacy and safety as potential therapeutic treatments for mitochondrial disorders.

4.4. Ketogneic Diet (KD)

The KD, in its various forms, has successfully been used to treat patients with pharmacoresistant epilepsy [136–138]. Whilst the exact mechanism with regards to how the diet exerts its efficacy is not known, there is growing evidence that, in part, this may occur as result of stimulation of mitochondrial biogenesis [139]. This raises the possibility of use in patients with acquired and inherited mitochondrial disorders. Recently, we have shown that a component of the medium chain triglyceride KD, decanoic acid (C10), stimulates mitochondrial biogenesis and increases MRC complex I activity and antioxidant status in neuronal cells [140]. Furthermore, cells from patients with MRC complex I deficiency have, in some cases, been shown to respond positively to C10 exposure [141].

5. Conclusion Remarks

As highlighted in this review, free radical-induced oxidative damage to the biomolecules of the mitochondria are intrinsically linked to the pathophysiology of a number of disorders (see summary in Figure 5). Despite the number of markers available to determine evidence of oxidative stress together with its pathological consequences, few clinical centres as yet include the determination of this parameter as part of the diagnostic algorithm of patient evaluation. Furthermore, in view of the vulnerability of mitochondrial biomolecules to oxidative damage by ROS or RNS, therapeutic strategies should be targeted the free radical threshold [13] and the molecular structure targeted by these radicals. Recent advances using mitochondrial-targeted antioxidants and dietary modification may hold potential promise to provide therapeutic benefit for patients with oxidative stress-associated disorders.

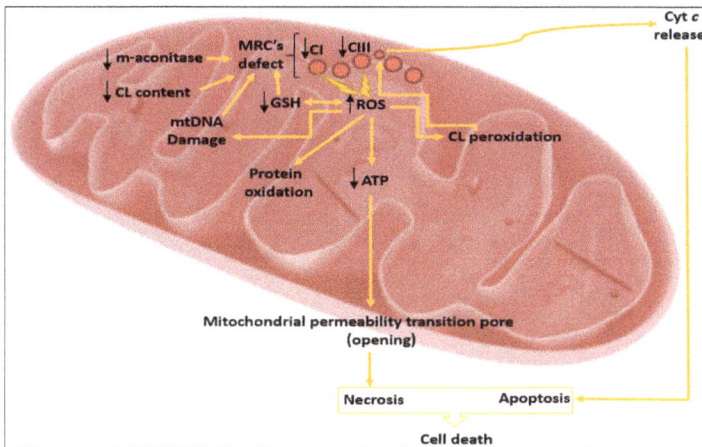

Figure 5. A summary of oxidative stress-induced mitochondrial damage is a common mechanistic link in the pathogenesis of inherited mitochondrial disorders and PD. CL: cardiolipin; mtDNA: mitochondrial DNA.

Acknowledgments: M.A.S. acknowledges support from the King Khalid University (Abha, Saudi Arabia). M.O. is in receipt of funding from the National Institute for Health Research UK (NIHR). M.O. and S.J.R.H. both acknowledge support from Great Ormond Street NIHR Research Centre.

Conflicts of Interest: The authors declare no conflict of interest.

Abbreviations

ROS	Reactive oxygen species
PD	Parkinson's disease
MRC	Mitochondrial respiratory chain
GPx	Glutathione peroxidase
GSH	Reduced glutathione
CL	Cardiolipin
mtDNA	Mitochondrial DNA
LHON	Leber's hereditary optic neuropathy
BTHS	Barth syndrome
FRDA	Friedreich ataxia
MPTP	1-methyl-4-phenyl-1,2,3,4-tetrahydropyridine

References

1. Brealey, D.; Brand, M.; Hargreaves, I.; Heales, S.; Land, J.; Smolenski, R.; Dnavies, N.; Cooper, C.; Singer, M. Association between Mitochondrial Dysfunction and Severity and Outcome of Septic Shock. *Lancet* **2002**, *360*, 219–223. [CrossRef]
2. Land, J.; Hughes, J.; Hargreaves, I.; Heales, S. Mitochondrial disease: A historical, biochemical, and London perspective. *Neurochem. Res.* **2004**, *29*, 483–491. [CrossRef] [PubMed]
3. Hargreaves, I.; Al Shahrani, M.; Wainwright, L.; Heales, S. Drug-Induced Mitochondrial Toxicity. *Drug Saf.* **2016**, *39*, 661–674. [CrossRef] [PubMed]
4. Chen, O.; Vazquez, E.; Moghaddas, S.; Hoppel, C.; Lesnefsky, E. Production of Reactive Oxygen Species by Mitochondria: Central Role of Complex III. *J. Biol. Chem.* **2003**, *278*, 36027–36031. [CrossRef] [PubMed]
5. Heales, S.; Bolanos, J.; Stewart, V.; Brookes, P.; Land, J.; Clark, J. Nitric Oxide, Mitochondria and Neurological Disease. *Biochim. Biophys. Acta Bioenerg.* **1999**, *1410*, 215–228. [CrossRef]
6. Turrens, J. Mitochondrial Formation of Reactive Oxygen Species. *J. Physiol.* **2003**, *552*, 335–344. [CrossRef] [PubMed]
7. Birben, E.; Sahiner, U.; Sackesen, C.; Erzurum, S.; Kalayci, O. Oxidative Stress and Antioxidant Defense. *World Allergy Organ. J.* **2012**, *5*, 9–19. [CrossRef] [PubMed]
8. Bunker, V. Free radicals, antioxidants and ageing. *Med. Lab. Sci.* **1992**, *49*, 299–312. [PubMed]
9. White, E.; Shannon, J.; Patterson, R. Relationship between Vitamin and Calcium Supplement Use and Colon Cancer. *Cancer Epidemiol. Biomark. Prev.* **1997**, *6*, 769–774.
10. Duberley, K.; Heales, S.; Abramov, A.; Chalasani, A.; Land, J.; Rahman, S.; Hargreaves, I. Effect of Coenzyme Q_{10} Supplementation on Mitochondrial Electron Transport Chain Activity and Mitochondrial Oxidative Stress in Coenzyme Q_{10} Deficient Human Neuronal Cells. *Int. J. Biochem. Cell Biol.* **2014**, *50*, 60–63. [CrossRef] [PubMed]
11. Smith, A.; Shenvi, S.; Widlansky, M.; Suh, J.; Hagen, T. Lipoic Acid as a Potential Therapy for Chronic Diseases Associated with Oxidative Stress. *Curr. Med. Chem.* **2004**, *11*, 1135–1146. [CrossRef] [PubMed]
12. Fiedor, J.; Burda, K. Potential Role of Carotenoids as Antioxidants in Human Health and Disease. *Nutrients* **2014**, *6*, 466–488. [CrossRef] [PubMed]
13. Jacobson, J.; Duchen, M.; Hothersall, J.; Clark, J.; Heales, S. Induction of Mitochondrial Oxidative Stress in Astrocytes by Nitric Oxide Precedes Disruption of Energy Metabolism. *J. Neurochem.* **2005**, *95*, 388–395. [CrossRef] [PubMed]
14. Hayashi, G.; Cortopassi, G. Oxidative Stress in Inherited Mitochondrial Disease. *Free Radic. Biol. Med.* **2015**, *88*, 10–17. [CrossRef] [PubMed]
15. Stewart, V.; Heales, S. Nitric Oxide-Induced Mitochondrial Dysfunction: Implication for Neurodegeneration. *Free Radic. Biol. Med.* **2003**, *34*, 287–303. [CrossRef]

16. Chinnery, P.; Johnson, M.; Wardell, T.; Singh-Kler, R.; Hayes, C.; Taylor, R.; Bindoff, L.; Turnbull, D. The Epidemiology of Pathogenic Mitochondrial DNA Mutations. *Ann. Neurol.* **2000**, *48*, 188–193. [CrossRef]

17. Lamont, P.; Surtees, R.; Woodward, C.; Leonard, J.; Wood, N.; Harding, A. Clinical and Laboratory Findings in Referrals for Mitochondrial DNA Analysis. *Arch. Dis. Child.* **1998**, *79*, 22–27. [CrossRef] [PubMed]

18. Gillis, L.; Sokol, R. Gastrointestinal Manifestation of Mitochondrial Disease. *Gastroenterol. Clin. N. Am.* **2003**, *32*, 789–817. [CrossRef]

19. Sykora, P.; Wilson, D.; Bohr, V. Repair of Persistent Strand Breaks in the Mitochondrial Genome. *Mech. Aging Dev.* **2012**, *133*, 169–175. [CrossRef] [PubMed]

20. Cline, S. Mitochondrial DNA Damage and Its Consequences for Mitochondrial Gene Expression. *Biochim. Biophys. Acta Bioenerg.* **2012**, *1819*, 979–991. [CrossRef] [PubMed]

21. Leeuwenburgh, C.; Hiona, A. The Role Mitochondrial DNA Mutations in Aging and Sarcopenia. *Exp. Gerontol.* **2008**, *43*, 24–33.

22. Huoponen, K.; Vilkki, J.; Aula, P.; Nikoskelainen, E.; Savontaus, M. A New mtDNA Mutation Associated with Leber Hereditary Optic Neuroretinopathy. *Am. J. Hum. Genet.* **1991**, *48*, 1147–1153. [PubMed]

23. Sadun, A.; Morgia, C.; Carelli, V. Leber Hereditary Optic Neuropathy. *Curr. Treat. Options Neurol.* **2011**, *13*, 109–117. [CrossRef] [PubMed]

24. Battisti, C.; Formichi, P.; Cardaioli, E.; Bianchi, S.; Mangiavacchi, P.; Tripodi, S.; Tosi, P.; Federico, A. Cell Response to Oxidative Stress Induced Apoptosis in Patients with Leber Hereditary Optic Neuropathy. *J. Neurol. Neurosurg. Psychiatry* **2004**, *75*, 1731–1736. [CrossRef] [PubMed]

25. Lin, C.; Sharpley, M.; Fan, W.; Waymire, K.; Sadun, A.; Carelli, F.; Ross-Cisneros, F.; Baciu, P.; Sung, E.; McManus, M.; et al. Mouse mtDNA Mutant Model of Leber Hereditary Optic Neuropathy. *Proc. Natl. Acad. Sci. USA* **2012**, *109*, 20065–20070. [CrossRef] [PubMed]

26. Howell, N. Leber Hereditary Optic Neuropathy: Respiratory Chain Dysfunction and Degeneration of the Optic Nerve. *Vis. Res.* **1998**, *38*, 1495–1504. [CrossRef]

27. Kirches, E. LHON: Mitochondrial Mutation and More. *Curr. Genom.* **2011**, *12*, 44–54. [CrossRef] [PubMed]

28. Wang, Y.; Gu, Y.; Wang, J.; Tong, Y. Oxidative Stress in Chinese Patients with Leber Hereditary Optic Neuropathy. *J. Int. Med. Res.* **2008**, *36*, 544–550. [CrossRef] [PubMed]

29. Gonzalo, R.; Arumi, E.; Llige, D.; Marti, R.; Solano, A.; Montoya, J.; Arenas, J.; Andreu, A. Free Radicals-mediated Damage in Transmitochondrial Cells Harboring the T14487C Mutation in the ND6 Gene of mtDNA. *FEBS Lett.* **2005**, *579*, 6909–6913. [CrossRef] [PubMed]

30. Luo, X.; Pitkanen, S.; Kassovska, S.; Robinson, B.; Lehotay. Excessive Formation of Hydroxyl Radicals and Aldehydic Lipid Peroxidation Products in Cultured Skin Fibroblasts From Patients with Complex I Deficiency. *J. Clin. Investig.* **1997**, *99*, 2877–2882. [CrossRef] [PubMed]

31. Wong, A.; Cavelier, L.; Collins-Schramm, H.; Seldin, M.; McGrogan, M.; Savontaus, M.; Cortopassi, G. Differentiation-specific Effects of LHON Mutation Introduced into Neuronal NT2 Cells. *Hum. Mol. Genet.* **2002**, *11*, 431–438. [CrossRef] [PubMed]

32. Yen, M.; Kao, S.; Wang, A.; Wei, Y. Increased 8 hydroxy 2′ deoxyguanosine in Leukocyte DNA in Leber Hereditary Optic Neuropathy. *Investig. Ophthalmol. Vis. Sci.* **2004**, *45*, 1688–1691. [CrossRef]

33. Sadun, A.; Carelli, V.; Salomao, S.; Berezovsky, A.; Quiros, P.; Sadun, F.; DeNegri, A.; Andrade, R.; Moraes, M.; Passos, A.; et al. Extensive Investigation of a Large Brazilian Pedigree of 11778/haplogroup J Leber Hereditary Optic Neuropathy. *Am. J. Ophthalmol.* **2004**, *136*, 231–238. [CrossRef]

34. Barth, P.; Scholte, H.; Berden, J.; Moorsel, J.; Luyt-Houwen, I.; Veer-Korthof, E.; Harten, J.; Sobotka-Plojhar, M. An X-linked Mitochondrial Disease Affecting Cardiac Muscle, Skeletal Muscle and Neutrophil Leucocytes. *J. Neurol. Sci.* **1983**, *62*, 327–355. [CrossRef]

35. Jefferies, J. Barth Syndrome. *Am. J. Med. Genet. C Semin. Med. Genet.* **2013**, *163*, 198–205. [CrossRef] [PubMed]

36. Barth, P.; Valianpour, F.; Bowen, V.; Lam, J.; Duran, M.; Vaz, F.; Wanders, R. X-linked Cardioskeletal Myopathy and Neutropenia (Barth Syndrome): An Update. *Am. J. Med. Genet.* **2004**, *126*, 349–354. [CrossRef] [PubMed]

37. Pope, S.; Land, J.; Heales, J. Oxidative Stress and Mitochondrial Dysfunction in Neurodegeneration; Cardiolipin a Critical Target? *Biochim. Biophys. Acta Bioenerg.* **2008**, *1777*, 794–799. [CrossRef] [PubMed]

38. Schlame, M.; Greenberg, R. The role of cardiolipin in The Structural Organization of Mitochondrial Membranes. *Biochim. Biophys. Acta Bioenerg.* **2009**, *1788*, 2080–2083. [CrossRef] [PubMed]

39. Paradies, G.; Paradies, V.; Benedictis, V.; Ruggiero, F.; Petrosillo, G. Functional Role of Cardiolipin in Mitochondrial Bioenergetics. *Biochim. Biophys. Acta Bioenerg.* **2014**, *1837*, 408–417. [CrossRef] [PubMed]

40. Pfeiffer, K.; Gohil, V.; Stuart, R.; Hunte, C.; Brandt, U.; Greenberg, M.; Schagger, H. Cardiolipin Stabilizes Respiratory Chain Supercomplexes. *J. Biol. Chem.* **2003**, *278*, 52873–52880. [CrossRef] [PubMed]

41. Vladimir, G.; Sten, O.; Boris, Z. Multiple Pathways of Cytochrome *c* Release from Mitochondria in Apoptosis. *Biochim. Biophys. Acta Bioenerg.* **2006**, *1757*, 639–647.

42. Barth, P.; Wanders, R.; Vreken, P.; Janssen, E.; Lam, J.; Baas, F. X-linked Cardioskeletal Myopathy and Neutropenia (Barth Syndrome) (MIM 302060). *J. Inherit. Metab. Dis.* **1999**, *22*, 555–567. [CrossRef] [PubMed]

43. Hargreaves, I.; Duncan, A.; Wu, L.; Agrawal, A.; Land, J.; Heales, S. Inhibition of Mitochondrial Complex IV Leads to Secondary Loss Complex II–III Activity: Implication for the Pathogenesis and Treatment of Mitochondrial Encephalomyopathies. *Mitochondrion* **2007**, *7*, 284–287. [CrossRef] [PubMed]

44. Martinez, L.; Forni, M.; Santos, V.; Pinto, N.; Kowaltowski, A. Cardiolipin is a Key Determinant for mtDNA Stability and Segregation during Mitochondrial Stress. *Biochim. Biophys. Acta Bioenerg.* **2015**, *1847*, 587–598. [CrossRef] [PubMed]

45. Alexeyev, M.; Shokolenko, I.; Wilson, G.; LeDoux, S. The Maintenance of Mitochondrial DNA Integrity—Critical Analysis and Update. *Cold Spring Harb. Perspect. Biol.* **2013**, *5*, a012641. [CrossRef] [PubMed]

46. Saric, A.; Andreau, K.; Armand, A.; Moller, I.; Petit, P. Barth Syndrome: From Mitochondrial Dysfunctions Associated with Aberrant Production of Reactive Oxygen Species to Pluripotent Stem Cell Studies. *Front. Genet.* **2016**, *6*, 359. [CrossRef] [PubMed]

47. Rotig, A.; Lonlay, P.; Chretien, D.; Foury, F.; Koenig, M.; Sidi, D.; Munnich, A.; Rustin, P. Aconitase and Mitochondrial Iron-sulphur Protein Defeciencey in Friedreich Ataxia. *Nat. Genet.* **1997**, *17*, 215–217. [CrossRef] [PubMed]

48. Koeppen, A. Friedreich Ataxia: Pathology, Pathogenesis, and Molecular Genetics. *J. Neurol. Sci.* **2011**, *303*, 1–12. [CrossRef] [PubMed]

49. Abeti, R.; Parkinson, M.; Hargreaves, I.; Angelova, P.; Sandi, C.; Pook, M.; Giunti, P.; Abramov, A. Mitochondrial Energy Imbalance and Lipid Peroxidation Cause Cell Death in Friedreich Ataxia. *Cell Death Dis.* **2016**, *7*, e2237. [CrossRef] [PubMed]

50. Babcock, M.; Silva, D.; Oaks, R.; Davis-Kaplan, S.; Jiralerspong, S.; Montermini, L.; Pandolfo, M.; Kaplan, J. Regulation of Mitochomdrial Iron Accumulation by Yfh1p, a putative Homolog of Frataxin. *Science* **1997**, *276*, 1709–1712. [CrossRef] [PubMed]

51. Gerber, J.; Muhlenhoff, U.; Lill, R. An Interaction between Frataxin and Isu1/Nfs1 That is Crucial for Fe/S Cluster Synthesis on Isu1. *EMBO Rep.* **2003**, *4*, 906–911. [CrossRef] [PubMed]

52. Kaplan, J. Friedreich Ataxia is a Mitochondrial Disorder. *Proc. Natl. Acad. Sci. USA* **1999**, *96*, 10948–10949. [CrossRef] [PubMed]

53. Bresgen, N.; Eckl, P. Oxidative Stress and the Homoeodynamics of Iron Metabolism. *Biomolecules* **2015**, *5*, 808–847. [CrossRef] [PubMed]

54. Wong, A.; Yang, J.; Cavadini, P.; Gellera, C.; Lonnerdal, B.; Taroni, F.; Cortopassi, G. The Friedreich Ataxia Mutation Confers Cellular Sensitivity to Oxidant Stress Which is Rescued by Chelators of Iron and Calcium and Inhibitors of Apoptosis. *Hum. Mol. Genet.* **1999**, *8*, 6. [CrossRef]

55. Yuxi, S.; Schoenfeld, R.; Hayashi, G.; Napoli, E.; Akiyama, T.; Carstens, M.; Carstens, E.; Pook, M.; Cortopa, G. Frataxin Deficiency Leads to Defect in Expression of Antioxidants and Nrf2 Expression in Dorsal Root Ganglia of the Friedreich Ataxia YG8R Mouse Model. *Antioxid. Redox Signal.* **2013**, *19*, 1481–1493.

56. Runko, A.; Griswold, A.; Kyung-Tai, M. Overexpression of Frataxin in the Mitochondria Increases Resistance to Oxidative Stress and Extends Lifespan in *Drosophila. FEBS Lett.* **2008**, *582*, 715–719. [CrossRef] [PubMed]

57. Koenig, M.; Mandel, J. Deciphering the Cause of Friedreich Ataxia. *Curr. Opin. Neurobiol.* **1997**, *7*, 689–694. [CrossRef]

58. Beinert, H.; Kennedy, M. Aconitase, a Two-faced Protein: Enzyme and Iron Regulatory Factor. *FASEB J.* **1993**, *15*, 1442–1449.

59. Vasquez-Vivar, J.; Kalyanaraman, B.; Kennedyl, M. Mitochondrial Aconitase is a Source of Hydroxyl Radical: An Electron Spin Resonance Investigation. *J. Biol. Chem.* **2000**, *275*, 14064–14069. [CrossRef] [PubMed]

60. Houten, B.; Woshner, V.; Santos, J. Role of Mitochondrial DNA in Toxic Responses to Oxidative Stress. *DNA Repair (Amst.)* **2006**, *5*, 145–152. [CrossRef] [PubMed]

61. Gardner, P.; Fridovich, I. Superoxide Sensitivity of the Escherichia Coli Aconitase. *J. Biol. Chem.* **1991**, *266*, 19328–19333. [PubMed]

62. Haile, D.; Rouault, T.; Harford, J.; Kennedy, M.; Blondin, G.; Klausner, R. Cellualr Regulation of the Iron-responsive Element Binding Protein: Disassembly of the Cubane Iron-sulfur Cluster Results in High Affinity RNA Binding. *Proc. Natl. Acad. Sci. USA* **1992**, *89*, 11735–11739. [CrossRef] [PubMed]
63. Liang, L.; Ho, Y.; Patel, M. Mitochondrial Superoxide Production in Kainate-induced Hippocampal Damage. *Neuroscience* **2000**, *101*, 563–570. [CrossRef]
64. Kim, H.; LaVaute, T.; Iwai, K.; Klausner, R.; Rouault, T. Identification of a Conserved and Functional Iron-responsive Element in the 5′-Untranslated Region of Mammalian Mitochondrial Aconitase. *J. Biol. Chem.* **1996**, *271*, 24226–24230. [CrossRef] [PubMed]
65. Stankiewics, J.; Brass, S. Role of Iron in Neurotoxicity: A cause for Concern in the Elderly? *Curr. Opin. Clin. Nutr. Metab. Care* **2009**, *12*, 22–29. [CrossRef] [PubMed]
66. Fariss, M.; Chan, C.; Patel, M.; Houten, B.; Orrenius, S. Role of Mitochondria in Toxic Oxidative Stress. *Mol. Interv.* **2005**, *5*, 94–111. [CrossRef] [PubMed]
67. Reeve, A.; Simcox, E.; Turnbull, D. Aging and Parkinson's disease: Why is Advancing Age the Biggest Risk Factor? *Aging Res. Rev.* **2014**, *14*, 19–30. [CrossRef] [PubMed]
68. Eden, S.; Tanner, M.; Bernstein, A.; Fross, R.; Leimpeter, A.; Bloch, D.; Nelson, L. Incidence of Parkinson's Disease: Variation by Age, Gender, and Race/Ethnicity. *Am. J. Epidemiol.* **2003**, *157*, 1015–1022. [CrossRef]
69. Jankovic, J. Parkinson's Disease: Clinical Features and Diagnosis. *J. Neurol. Neurosurg. Psychiatry* **2008**, *79*, 368–376. [CrossRef] [PubMed]
70. Stefanis, L. α-Synuclein in Parkinson's Disease. *Cold Spring Harb. Perspect. Med.* **2012**, *2*, a009399. [CrossRef] [PubMed]
71. Langston, J.; Ballard, P.; Tetrud, J.; Irwin, I. Chronic Parkinsonism in Humans due to a Product of Meperidine-analog Synthesis. *Science* **1983**, *219*, 979–980. [CrossRef] [PubMed]
72. Abou-Sleiman, P.; Muqit, M.; Wood, N. Expanding Insights of Mitochondrial Dysfunction in Parkinson's Disease. *Nat. Rev. Neurosci.* **2006**, *7*, 207–219. [CrossRef] [PubMed]
73. Dauer, W.; Przedborski, S. Parkinson Disease: Mechanisms and Models. *Neuron* **2003**, *39*, 889–909. [CrossRef]
74. Blesa, J.; Phani, S.; Jackson-Lewis, V.; Przedborski, S. Classic and New Animal Models of Parkinson's Disease. *J. Biomed. Biotechnol.* **2012**, *2012*, 1–10. [CrossRef] [PubMed]
75. Parker, W.; Parks, J.; Swerdlow, R. Complex I Deficiency in Parkinson's Disease Frontal Cortex. *Brain Res.* **2008**, *1189*, 215–218. [CrossRef] [PubMed]
76. Blandini, F.; Nappi, G.; Greenamyre, J. Quantitative Study of Mitochondrial Complex I in Platelets of Parkinsonian Patients. *Mov. Disord.* **1998**, *13*, 11–15. [CrossRef] [PubMed]
77. Blin, O.; Desnuelle, C.; Rascol, O.; Borg, M.; Peyro, H.; Azulay, J.; Bille, F.; Figarella, D.; Coulom, F.; Pellissier, J. Mitochondrial Respiratory Failure in Skeletal Muscle from Patients with Parkinson's Disease and Multiple System Atrophy. *J. Neurol. Sci.* **1994**, *125*, 95–101. [CrossRef]
78. Haas, R.; Nasirian, F.; Nakano, K.; Ward, D.; Pay, M.; Hill, R.; Shults, C. Low platelet Mitochondrial complex I and Complex II/III Activity in Early Untreated Parkinson's Disease. *Ann. Neurol.* **1995**, *37*, 714–722. [CrossRef] [PubMed]
79. Acín-Pérez, R.; Bayona-Bafaluy, M.; Fernández-Silva, P.; Moreno-Loshuertos, R.; Pérez-Martos, A.; Bruno, C.; Moraes, C.; Enríquez, J. Respiratory Complex III Is Required to Maintain Complex I in Mammalian Mitochondria. *Mol. Cell* **2004**, *13*, 805–815. [CrossRef]
80. Keane, P.; Kurzawa, M.; Blain, P.; Morris, C. Mitochondrial Dysfunction in Parkinson's Disease. *Parkinsons Dis.* **2011**, *2011*, 1–18. [CrossRef] [PubMed]
81. Dominic, H.; Scott, A.; Ashley, B.; Peng, L. A Delicate Balance: Iron Metabolism and Disease of the Brain. *Front. Aging Neurosci.* **2013**, *5*, 34.
82. Moos, T.; Nielsen, T.; Skjorringe, T.; Morgan, E. Iron Trafficking Inside the Brain. *J. Neurochem.* **2007**, *103*, 1730–1740. [CrossRef] [PubMed]
83. Winter, W.; Bazydlo, L.; Harris, N. The Molecular Biology of Human Iron Metabolism. *Lab. Med.* **2014**, *45*, 92–102. [CrossRef] [PubMed]
84. Conner, J.; Benkovic, S. Iron Regulation in the Brain: Histochemical, Biochemical, and Molecular Considerations. *Ann. Neurol.* **1992**, *32*, 51–61. [CrossRef]
85. Brzoska, K.; Meczynska, S.; Kruszewski, M. Iron-sulfur Cluster Proteins: Electron Transfer and Beyond. *Acta Biochim. Pol.* **2006**, *53*, 685–691. [PubMed]
86. Hirsch, E.; Faucheux, B. Iron Metabolism and Parkison's Disease. *Mov. Disord.* **1998**, *13*, 39–45. [PubMed]

87. Dexter, D.; Wells, F.; Lees, A.; Javoy-Agid, F.; Agid, Y.; Jenner, P.; Marsden, C. Increased Nigral Iron Content and Alterations in Other Metal Ions Occurring in Brain in Parkinson's Disease. *J. Neurochem.* **1989**, *52*, 1830–1836. [CrossRef] [PubMed]

88. Dexter, D.; Wells, F.; Javoy-Agid, F.; Agid, Y.; Lees, A.; Jenner, P.; Marsden, C. Incresed Nigral Iron Content in Postmortem Parkinsonian Brain. *Lancet* **1987**, *330*, 1219–1220. [CrossRef]

89. Hirsch, E.; Brandel, J.; Galle, P.; Javoy-Agid, F.; Agid, Y. Iron and Aluminum Increase in the Substantia Nigra of Patients with Parkinson's Disease: An X-Ray Microanalysis. *J. Neurochem.* **1991**, *56*, 446–451. [CrossRef] [PubMed]

90. Mazzio, E.; Soliman, K. Effects of Enhancing Mitochondrial Oxidative Phoshporylation with Reducing Equivalents and Ubiquinone on 1-methyl-4-phenylpyridinium Toxicity and Complex I-IV Damage in Neuroblastoma Cells. *Biochem. Pharmacol.* **2004**, *67*, 1167–1184. [CrossRef] [PubMed]

91. Shang, T.; Kotamraiu, S.; Kalivendi, S.; Hillard, C.; Kalyanaraman, B. 1-Methyl-4-phenylpyridinium-induced Apoptosis in Cerebellar Granule Neurons is Mediated by Transferrin Receptor Iron-dependent Depletion of Tetrahydrobiopterin and Neuronal Nitric Oxide Synthase-derived Superoxide. *J. Biol. Chem.* **2004**, *279*, 19099–19112. [CrossRef] [PubMed]

92. Li-Ping, L.; Patel, M. Iron-sulfur Enzyme Mediated Mitochondrial Superoxide Toxicity in Experimental Parkinson's Disease. *J. Neurochem.* **2004**, *90*, 1076–1084.

93. Ebadi, M.; Srinivasan, K.; Baxi, M. Oxidative Stress and Antioxidant Therapy in Parkinson's Disease. *Prog. Neurobiol.* **1996**, *48*, 1–19. [CrossRef]

94. Obata, T. Dopamine Efflux by MPTP and Hydroxyl Radical Generation. *J. Neural Trans.* **2002**, *109*, 1159–1180. [CrossRef] [PubMed]

95. Jomova, K.; Valko, M. Advances in Metal-induced Oxidative Stress and Human Disease. *Toxicology* **2011**, *283*, 65–87. [CrossRef] [PubMed]

96. Nunez, T.; Urrutia, P.; Mena, N.; Aguirre, P.; Tapia, V.; Salazar, J. Iron Toxicity in Neurodegeneration. *Biometals* **2012**, *25*, 761–776.

97. Owen, J.; Butterfield, D. Measurment of Oxidized/Reduced Glutathione Ratio. *Methods Mol. Biol.* **2010**, *648*, 269–277. [PubMed]

98. Sipos, K.; Lange, H.; Fekete, Z.; Ullmann, P.; Lill, R.; Kispal, G. Maturation of Cytosolic Iron-sulfur Proteins Requires Glutathione. *J. Biol. Chem.* **2002**, *277*, 26944–26949. [CrossRef] [PubMed]

99. Dias, V.; Junn, E.; Mouradian, M. The Role Oxidative Stress in Parkinson's Disease. *J. Parkinsons Dis.* **2013**, *3*, 461–491. [PubMed]

100. Heales, S.; Bolanos, J. Impirment of Brain Mitochondrial Function by Reactive Nitrogen Species: The Role of Glutathione in Dictating Susceptibility. *Neurochem. Int.* **2002**, *40*, 469–474. [CrossRef]

101. Dimonte, D.; Chan, P.; Sandy, M. Glutathione in Parkinson's Disease: A link Between Oxidative Stress and Mitochondrial Damage? *Ann. Neurol.* **1992**, *32*, 111–115. [CrossRef]

102. Mari, M.; Morales, A.; Colell, A.; Garcia-Ruiz, C.; Fernandez-Checa, J. Mitochondrial Glutathione, a Key Survival Antioxidant. *Antioxid. Redox Signal.* **2009**, *11*, 2685–2700. [CrossRef] [PubMed]

103. Heales, S.; Davies, S.; Bates, T.; Clark, J. Depletion of Brain Glutathione is Accompanied by Impaired Mitochondrial Function and Decreased *N*-acetyl Aspartate Concentration. *Neurochem. Res.* **1995**, *20*, 31–38. [CrossRef] [PubMed]

104. Merad-Saidoune, M.; Biotier, E.; Nicole, A.; Marsac, C.; Martinou, J.; Sola, B.; Sinet, P.; Ceballos-Picot, I. Overproduction of Cu/Zn-superoxide Dismutase or Bcl-2 Prevents the Brain Mitochondrial Respiratory Dysfunction Induced by Glutathione Depletion. *Exp. Neurol.* **1999**, *158*, 428–436. [CrossRef] [PubMed]

105. Jha, N.; Jurma, O.; Lalli, G.; Liu, Y.; Pettus, E.; Greenamyrel, J.; Liu, R.; Forman, H.; Anderson, J. Glutathione Depletion in PC12 Results in Selective Inhibition of Mitochondrial Complex I Activity: Implication for Parkinson's Disease. *J. Biol. Chem.* **2000**, *275*, 26096–26101. [CrossRef] [PubMed]

106. Hargreaves, I.; Sheena, Y.; Land, J.; Heales, S. Glutathione Deficiencey in Patients with Mitochondrial Disease: Implication for Pathogenesis and Treatment. *J. Inherit. Dis.* **2005**, *28*, 81–88. [CrossRef] [PubMed]

107. Jenner, P.; Dexter, D.; Sian, J.; Schapira, A.; Marsden, C. Oxidative stress as a Cause of Nigral Cell Death in Parkinson's Disease and Incidental Lewy Body Disease. The Royal Kings and Queens Parkinson's Disease Research Group. *Ann. Neurol.* **1992**, *32*, 82–87. [CrossRef]

108. Zeevalk, G.D.; Manzino, L.; Sonsalla, P.K.; Bernard, L.P. Characterization of Intracellular Elevation of Glutathione (GSH) with Glutathione Monoethyl Ester and GSH in Brain and Neuronal Cultures: Relevance to Parkinson's Disease. *Exp. Neurol.* **2007**, *203*, 512–520. [CrossRef] [PubMed]

109. Farr, S.; Poon, H.; Dogrukol-Ak, D.; Drake, J.; Banks, W.; Eyerman, E.; Butterfield, D.; Morley, J. The Antioxidants α-lipoic Acid and *N*-Acetylcysteine Reverse Memory Impairment and Brain Oxidative Stress in Aged SAMP8 Mice. *J. Neurochem.* **2003**, *84*, 1173–1183. [CrossRef] [PubMed]

110. Kerksick, C.; Willoughby, D. The Antioxidant Role of Glutathione and *N*-Acetyl-Cysteine Supplements and Exercise-Induced Oxidative Stress. *Int. Soc. Sports Nutr.* **2005**, *2*, 38–44. [CrossRef] [PubMed]

111. Munoz, M.; Rey, P.; Soto, R.; Guerra, M.; Labandeira, L. Systemic Administration of N Acetylcysteine Protects Dopaminergic Neurons Against 6 Hydroxydopamine Induced Degeneration. *J. Neurosci. Res.* **2004**, *76*, 551–562. [CrossRef] [PubMed]

112. Hargreaves, I.P. Coenzyme Q$_{10}$ as a Therapy for Mitochondrial Disease. *Int. J. Biochem. Cell Biol.* **2014**, *49*, 105–111. [CrossRef] [PubMed]

113. Duncan, A.; Heales, S.; Mills, K.; Eaton, S.; Land, J.; Hargreaves, I. Determination of Coenzyme Q$_{10}$ Status in Blood Mononuclear Cells, Skeletal Muscle, and Plasma by HPLC with Di-Propoxy-Coenzyme Q$_{10}$. *Clin. Chem.* **2005**, *51*, 2380–2382. [CrossRef] [PubMed]

114. Ogasahara, S.; Engel, A.; Frens, D.; Mack, D. Muscle Coenzymee Q Deficiency in Familial Mitocondrial Encephalomyopathy. *Proc. Natl. Acad. Sci. USA* **1989**, *86*, 2379–2382. [CrossRef] [PubMed]

115. Maldergem, L.; Trijbels, F.; DiMauro, S.; Sindelar, P.; Musumeci, O.; Janssen, A.; Delberghe, X.; Martin, J.; Gillerot, Y. Coenzyme Q-Responsive Leigh's Encephalopathy in Two Sisters. *Ann. Neurol.* **2002**, *52*, 750–754. [CrossRef] [PubMed]

116. Papadimitriou, A.; Hadjigeorgiou, G.; Divari, R.; Papagalanis, N.; Comi, G.; Bresolin, N. The Influence of Coenzyme Q$_{10}$ on Total Serum Calcium concentration in Two Patients with Kearns—Sayre syndrome and hypoparathyroidism. *Neuromuscul. Disord.* **1996**, *6*, 49–53. [CrossRef]

117. Quinzii, C.; DiMauro, S.; Hirano, M. Human Coenzyme Q$_{10}$ Deficiency. *Neurochem. Res.* **2007**, *32*, 723–727. [CrossRef] [PubMed]

118. Koene, S.; Smeitink, J. Metabolic Manipulators: A Well Founded Strategy to Combat Mitochondrial Dysfunction. *J. Inherit. Metab. Dis.* **2011**, *34*, 315–325. [CrossRef] [PubMed]

119. Napolitano, A.; Salyetti, S.; Vista, M.; Lombardi, V.; Siciliano, G.; Giraldi, C. Long-Term Treatment with Idebenone and Riboflavin in a Patient with MELAS. *Neurol. Sci.* **2000**, *21*, 981–982. [CrossRef]

120. Mancuso, M.; Orsucci, D.; Filosto, M.; Simoncini, C.; Siciliano, G. Drugs and Mitochondrial Diseases: 40 Queries and Answers. *Expert Opin. Pharmacother.* **2012**, *13*, 527–543. [CrossRef] [PubMed]

121. Jaber, S.; Polster, B. Idebenone and Neuroprotection: Antioxidant, Pro-Oxidant, or Electron Carrier? *J. Bioenergy Biomembr.* **2015**, *47*, 111–118. [CrossRef] [PubMed]

122. Shults, C.; Haas, R.; Passov, D.; Beal, M. Coenzyme Q$_{10}$ Levels Correlate with the Activities of Complex I and II/III in Mitochondria From Parkinsonian and Nonparkinsonian subjects. *Ann. Neurol.* **1997**, *42*, 261–264. [CrossRef] [PubMed]

123. Hargreaves, I.; Lane, A.; Sleiman, P. The Coenzyme Q$_{10}$ Status of the Brain Regions of Parkinson's Disease Patients. *Neurosci. Lett.* **2008**, *447*, 17–19. [CrossRef] [PubMed]

124. Sohmiya, M.; Tanaka, M.; Tak, N.W.; Yanagisawa, M.; Tanino, Y.; Suzuki, Y.; Okamoto, K.; Yamamoto, Y. Redox Status of Plasma Coenzyme Q$_{10}$ Indicates Elevated Systemic Oxidative Stress in Parkinson's Disease. *J. Neurol. Sci.* **2004**, *223*, 161–166. [CrossRef] [PubMed]

125. Gotz, M.; Gerstner, A.; Harth, R.; Dirr, A.; Janetzky, B.; Kuhn, W.; Riederer, P.; Gerlach, M. Altered Redox State of Platelet Coenzyme Q$_{10}$ in Parkinson's Disease. *J. Neural Transm. (Vienna)* **2000**, *107*, 41–48.

126. Gille, G.; Hung, S.; Reichmann, H.; Rausch, W. Oxidative Stress to Dopaminergic Neurons as Models of Parkinson's Disease. *Ann. N. Y. Acad. Sci.* **2004**, *1018*, 533–540. [CrossRef] [PubMed]

127. Kooncumchoo, P.; Sharma, S.; Porter, J.; Govitrapong, P.; Ebadi, M. Coenzyme Q (10) Provides Neuroprotection in Iron-Induced Apoptosis in Dopaminergic Neurons. *J. Mol. Neurosci.* **2006**, *28*, 125–141. [CrossRef]

128. Winkler-Stuck, K.; Wiedemann, F.; Wallesch, C.; Kunz, S. Effect of Coenzyme Q$_{10}$ on the Mitochondrial Function of Skin Fibroblasts from Parkinson Patients. *J. Neurol. Sci.* **2004**, *220*, 41–48. [CrossRef] [PubMed]

129. Moreira, P.I.; Zhu, X.; Wang, X.; Lee, Y.-G.; Nunomura, A.; Peterson, R.B.; Perry, G.; Smith, M.A. Mitochondria: A Therapeutic Target in Neurodegeneration. *Biochim. Biophys. Acta* **2010**, *1802*, 212–220. [CrossRef] [PubMed]

130. Shults, C.; Oakes, D.; Kieburtz, K.; Beal, M.; Haas, R.; Plumb, S.; Juncos, J.; Nutt, J.; Shoulson, I.; Carter, J.; et al. Parkinson Study Group. Effects of Coenzyme Q$_{10}$ in Early Parkinson Disease: Evidence of Slowing of the Functional Decline. *Arch. Neurol.* **2002**, *59*, 1541–1550. [CrossRef] [PubMed]

131. Josef, F. Treatment of Mitochondrial Disorders. *Eur. J. Pediatr. Neurol.* **2010**, *14*, 29–44.

132. Josef, F.; Bindu, P. Therapeutic Strategies for Mitochondrial Disorders. *Pediatr. Neurol.* **2015**, *52*, 302–313.

133. Jin, H.; Kanthasamy, A.; Ghosh, A.; Anantharam, V.; Kaylanaraman, B.; Kanthasamy, G. Mitochondria-targeted Antioxidants for Treatment of Parkinson's Disease: Preclinical and Clinical Outcomes. *Biochim. Biophys. Acta* **2014**, *1842*, 1282–1294. [CrossRef] [PubMed]

134. Teodoro, G.; Baraldi, G.; Sampaio, H.; Bomfim, H.; Queiroz, L.; Passos, A.; Carneiro, M.; Alberici, C.; Gomis, R.; Amaral, G.; et al. Melatonin Prevents Mitochondrial Dysfunction and Insulin Resistance in Rat Skeletal Muscle. *J. Pineal Res.* **2014**, *57*, 155–167. [CrossRef] [PubMed]

135. Smith, R.; Adlam, V.; Blaikie, F.; Manas, A.; Porteous, C.; James, A.; Ross, M.; Logan, A.; Cochemé, H.; Trnka, J.; et al. Mitochondria-Targeted Antioxidants in the Treatment of Disease. *Ann. N. Y. Acad. Sci.* **2008**, *1147*, 105–111. [CrossRef] [PubMed]

136. Henderson, C.B.; Filloux, F.M.; Alder, S.C.; Lyon, J.L.; Caplin, D.A. Efficacy of the Ketogenic Diet as a treatment Option for Epilepsy: Meta-Analysis. *J. Child. Neurol.* **2006**, *3*, 193–198.

137. Keene, D.L. A Systematic Review of the Use of the Ketogenic Diet in Childhood Epilepsy. *Pediatr. Neurol.* **2006**, *1*, 1–5. [CrossRef] [PubMed]

138. Neal, E.G.; Chaffe, H.; Schwartz, R.H.; Lawson, M.S.; Edwards, N.; Fitzsimmons, G.; Whitney, A.; Cross, J.H. A Randomized Trial of Classical and Medium-Chain Triglyceride Ketogenic Diets in the Treatment of Childhood Epilepsy. *Epilepsia* **2009**, *5*, 1109–1117. [CrossRef] [PubMed]

139. Bough, K.J.; Wetherington, J.; Hassel, B.; Pare, J.F.; Gawryluk, J.W.; Greene, J.G.; Shaw, R.; Smith, Y.; Geiger, J.D.; Dingledine, R.J. Mitochondrial Biogenesis in The Anticonvulsant Mechanism of the Ketogenic Diet. *Ann. Neurol.* **2006**, *2*, 223–235. [CrossRef] [PubMed]

140. Hughes, S.D.; Kanabus, M.; Anderson, G.; Hargreaves, I.P.; Rutherford, T.; O'Donnell, M.; Cross, J.H.; Rahman, S.; Eaton, S.; Heales, S.J. The Ketogenic Diet Component Decanoic Acid Increases Mitochondrial Citrate Synthase and Complex I Activity in Neuronal Cells. *J. Neurochem.* **2014**, *3*, 426–433. [CrossRef] [PubMed]

141. Kanabus, M.; Fassone, E.; Hughes, S.D.; Bilooei, S.F.; Rutherford, T.; O'Donnell, M.; Heales, S.J.R.; Rahman, S. The Pleiotropic Effects of Decanoic Acid Treatment on Mitochondrial Function in Fibroblasts from Patients with Complex I Deficient Leigh Syndrome. *J. Inherit. Metab. Dis.* **2016**, *39*, 415–426. [CrossRef] [PubMed]

Journal of
Clinical Medicine

MDPI

Review

Statins, Muscle Disease and Mitochondria

Radha Ramachandran [1,2,*] and Anthony S. Wierzbicki [1]

[1]   Departments of Chemical Pathology/Metabolic Medicine, Guys and St Thomas' Hospitals NHS Foundation Trust, London SE1 7EH, UK; anthony.wierzbicki@kcl.ac.uk
[2]   Adult Inherited Metabolic Diseases, Centre for Inherited Metabolic Diseases, Evelina, Guys and St Thomas' Hospitals NHS Foundation Trust, Lambeth Palace Road, London SE1 7EH, UK
*   Correspondence: radha.ramachandran@gstt.nhs.uk

Academic Editors: Iain P. Hargreaves and Jane Grant-Kels
Received: 6 May 2017; Accepted: 12 July 2017; Published: 25 July 2017

Abstract: Cardiovascular disease (CVD) accounts for >17 million deaths globally every year, and this figure is predicted to rise to >23 million by 2030. Numerous studies have explored the relationship between cholesterol and CVD and there is now consensus that dyslipidaemia is a causal factor in the pathogenesis of atherosclerosis. Statins have become the cornerstone of the management of dyslipidaemia. Statins have proved to have a very good safety profile. The risk of adverse events is small compared to the benefits. Nevertheless, the potential risk of an adverse event occurring must be considered when prescribing and monitoring statin therapy to individual patients. Statin-associated muscle disease (SAMS) is by far the most studied and the most common reason for discontinuation of therapy. The reported incidence varies greatly, ranging between 5% and 29%. Milder disease is common and the more serious form, rhabdomyolysis is far rarer with an incidence of approximately 1 in 10,000. The pathophysiology of, and mechanisms leading to SAMS, are yet to be fully understood. Literature points towards statin-induced mitochondrial dysfunction as the most likely cause of SAMS. However, the exact processes leading to mitochondrial dysfunction are not yet fully understood. This paper details some of the different aetiological hypotheses put forward, focussing particularly on those related to mitochondrial dysfunction.

Keywords: cardiovascular; statin; myopathy; muscle; mitochondria

1. Introduction

Cardiovascular disease (CVD) accounts for >17 million deaths globally every year, and this figure is predicted to rise to >23 million by 2030 [1]. Numerous studies have explored the relationship between cholesterol and CVD and there is a consensus that low density lipoprotein cholesterol (LDL-C) is a causal factor in the pathogenesis of atherosclerosis [2,3]. The epidemiological studies underlying this concept have been aggregated and meta-analysed by the Emerging Risk Factors Collaboration [4]. These results provided the impetus for discovery of cholesterol lowering drugs starting with the use of high dose niacin and then proceeding through bile acid sequestrants, fibrates and eventually statins [5,6]. Statins have now become the cornerstone of the management of dyslipidaemia [7].

The first step in cholesterol synthesis involves formation of 2-Hydroxymethylglutaryl-coenzyme A (HMG-CoA) by condensation of acetyl CoA and aceto-acetylCoA; HMG-CoA is then converted to Mevalonate by the enzyme HMG-CoA reductase (Figure 1). This is the rate-limiting step in cholesterol synthesis. HMG-CoA reductase was pursued as a viable target for cholesterol lowering drug development. This led to the development of HMG-CoA reductase inhibitors, known as "statins" [6]. Toxicity was limited as HMG-CoA, the immediate precursor before the block, is water soluble and can be metabolised via alternative metabolic pathways, thus preventing accumulation. Numerous attempts have been made to inhibit cholesterol synthesis at other points but these have been limited either by the knowledge of inherited errors of metabolism associated with defects at those sites or by toxicity of potential drug candidates, e.g., squalene synthase inhibition [8].

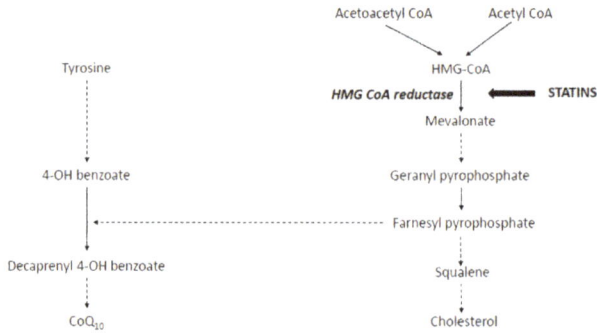

Figure 1. Schematic representation of cholesterol and CoQ10 synthetic pathway. Dotted arrows are used where some of the intermediate products in the pathway have been omitted in the diagram. Site of Statin action is shown. Statin inhibits enzymes HMG CoA reductase which is written in bold italics.

In September 1987, Lovastatin became the first statin to be given US Food and Drug Administration approval as a cholesterol lowering agent [8]. Two further semi-synthetic (pravastatin and simvastatin) and four synthetic statins (fluvastatin, pitavastatin, atorvastatin and rosuvastatin) of varying efficacy [9] have been successfully introduced into the market since but cerivastatin was withdrawn due to toxicity [10]. The reduction in plasma LDL-C caused by statins is due to upregulation of LDL receptor expression and not only from a decrease in cholesterol synthesis due to HMG-CoA reductase inhibition allied with decreased production of apolipoprotein B containing lipoproteins [11]. These drugs reduce LDL-C levels even in patients with heterozygous Familial Hypercholesterolemia (FH) due to LDL receptor mutations, but not in receptor null homozygous FH [11].

Statins were initially received with some scepticism due to uncertainties regarding benefit and anxieties concerning potential adverse effects [6]. These reservations were dispelled by the results of large long-term randomised controlled trials such as the Scandinavian Simvastatin Survival Study [12]. This study provided unequivocal evidence for reduction in all-cause mortality (30%, $p = 0.0003$), coronary artery related deaths (42%), major coronary events (34%) and revascularisation procedures (37%) with statin therapy. The Heart Protection Study provided further evidence for benefits in women and in patients with diabetes and previous history of cerebrovascular events [13]. Moreover these randomised controlled trials provided reassurance that there was no increase in adverse effects such as cataracts, previously observed in animal studies relating to an earlier cholesterol lowering drug candidate triparanol [14], clinical liver disease [15] or cancer [16], although some concerns continue to be raised [17]. A later patient-based meta-analysis of statin trials showed a 21% reduction in CVD events and an 11% reduction in CV mortality for each 1 mmol/LDL-C reduction [18]. Furthermore, maintaining a 2 mmol/L reduction in LDL cholesterol in 10,000 patients for 5 years prevented approximately 1000 major vascular events in patients with a high risk of coronary events [7,18]. Hence, with good reason, statins are now amongst the most widely prescribed medications across the globe. They are prescribed to roughly 30 million people, and had sales of $25 billion in 2005 [19].

Statins have proved to have a very good safety profile [7,20]. The risk of adverse events is small compared to benefits. Nevertheless, the potential risk of an adverse event occurring must be considered when prescribing and monitoring statin therapy to individual patients. Memory loss, impairment of liver/kidney function, new onset diabetes and muscle symptoms are some of the many adverse effects reported by patients taking statins. Of these, statin-associated muscle disease is by far the most studied [21] and the most common reason for discontinuation of therapy and hence will be the focus of the rest of this paper.

Presentations with statin intolerance often due to myopathy form up to 10% of the workload of clinical lipid services. Only recently has a classification of statin-related muscle symptoms been agreed [22]. Statin-induced muscle disease can be broadly classified into the rarer, more severe, often irreversible statin-induced necrotising inflammatory myopathy (SINIM) [23], and the relatively more common, reversible spectrum of muscle disease often referred to as statin-associated muscle symptoms (SAMS) [22,24]. There is no consensus about the optimal pathway for investigation of these cases. Investigations include exclusion of common autoimmune muscle diseases and vitamin D deficiency [25]. Vitamin D deficiency can exacerbate statin myopathy but there is no clear evidence for concurrent supplementation having benefits as trials are small [26] and usually not randomised. Some clinicians proceed to muscle biopsy and electron microscopy in severe cases. A case series of 279 biopsies from patients with statin myopathy show a 24% incidence (n = 67) of mitochondrial dysfunction on either histochemistry and/or electron microscopy [27]. Ten percent (n = 29) had abnormal respiratory chain enzyme activity [27].

2. Statin-Associated Muscle Symptoms

The reported incidence varies greatly, ranging between 5% and 29% with milder symptoms being common and, the rare, more serious form, rhabdomyolysis being far rarer with an incidence of approximately 1 in 10,000 [17,28–31]. The Statins on Skeletal Muscle Function and Performance (STOMP) study assessed the effect of 6 months of 80 mg of Atorvastatin on 420 statin-naïve healthy controls and found a significantly increased incidence of muscle-related symptoms in the statin versus placebo group (p < 0.05) but found no significant difference in exercise capacity or muscle strength in the statin versus placebo group [32].

Symptoms of SAMS are often non-specific and largely localised to proximal muscle groups such as thighs, buttocks and calves. Physical activity, female gender and Asian ethnicity have all been shown to be associated with an increased risk of SAMS [33,34]. Hypothyroidism, renal and liver impairment and diabetes are other risk factors [35,36]. Interactions due to co-administration of drugs sharing the same cytochrome P450 metabolic pathway may account for up to 60% of SAMS. Drugs such as glucocorticoids, gemfibrozil, protease inhibitors, antipsychotics such as risperidone and immunosuppressives such as cyclosporine, and common food-associated factors such as orange or cranberry juice and excess alcohol consumption have all been implicated [36]. Hence patients must be assessed for these pre-existing risk factors prior to being prescribed statins.

2.1. SAMS Diagnosis

Confirmation of SAMS remains a challenge in the absence of a specific and sensitive biomarker. The 2015 European Atherosclerosis Society Consensus statement suggested that diagnosis should be based on the triad of (i) temporal relationship of symptoms and/or CK elevation to initiation of statin therapy; (ii) disappearance of symptoms on withdrawal; and (iii) re-appearance on re-challenge with statin therapy [36]. Similar proposals were mooted by other expert groups including the Canadian Consensus Working Group [29].

2.2. SAMS Classification

SAMS can be further classified based on muscle symptoms, the presence and degree of CK elevation [22]. Muscle symptoms with no elevation in CK, often referred to as myalgia, is regarded as the mildest form. The term myositis is sometimes used to describe symptoms associated with significant CK elevation (>10 times upper limit of normal range). Rhabdomyolysis is the most severe form, and may result in myoglobinuria and renal impairment. CK levels in rhabdomyolysis may rise to >40 times upper limit of normal range.

The pathophysiology of and mechanisms leading to SAMS is yet to be fully understood. The rest of this paper will detail some of the different aetiological hypotheses put forward and, in particular, focus on the hypotheses related to mitochondrial dysfunction.

2.3. SAMS Pathobiology

2.3.1. Genetic Predisposition

Polymorphisms in a number of genes, including those coding for efflux ABC transporters (ABCB1 and ABCG2), influx transporter- organic anion–transporting polypeptide 1B1 (OATP1B1) (*SLCO1B1*) and Cytochrome P450 enzymes: CYP2D6, CYP3A4, and CYP3A5, have been associated with SAMS. However, thus far, only SNPs in the *SLCO1B1*, causing disruption in the hepatic uptake of simvastatin, have shown convincing associations [37]. Routine testing for this polymorphism is currently not recommended. In contrast, polymorphisms leading to reduced expression of *GATM* gene may offer a degree of protection against SAMS [38]. Whilst the exact mechanism is not known, it was proposed that reduced creatine synthesis, and hence reduced phosphocreatine stores, modify cellular energy stores and the AMPK signalling favourably. However, other studies have failed to replicate the results [39]. Furthermore, creatine deficiency related to a loss of function mutation in *GATM* can present with myopathy [38]. Statins precipitate myopathic symptoms in patients with genetic mutations for rare inherited metabolic disorders such as MELAS (Mitochondrial encephalomyopathy, lactic acidosis, and stroke-like episodes) syndrome [40–44]. A study of 110 patients with statin-induced myopathy reported a 4-fold ($p < 0.001$) increase in pathogenic mutant alleles for carnitine palmitoyltransferase II deficiency, McArdle disease and myoadenylate deaminase deficiency versus controls [43]. Despite these isolated reports in the literature, convincing evidence to recommend routine testing for genetic predispositions prior to initiation of treatment with statins remains elusive.

2.3.2. Mitochondrial Dysfunction

Mitochondrial dysfunction is defined as a decrease in the ability of the mitochondria to synthesise high energy compounds such as adenosine $5'$ triphosphate and a suboptimal electron transfer rate across the respiratory chain complex. Primary mitochondrial disease results in mitochondrial dysfunction due to mutations in genes coding for mitochondrial function. In addition, mitochondrial dysfunction related to oxidative damage is a recognised feature of aging and a number of chronic diseases including diabetes [44].

Mitochondrial dysfunction can also be precipitated by drugs and is now the most widely accepted and studied pathobiological mechanism for SAMS [45,46]. However, the exact nature/extent/type of mitochondrial dysfunction(s) causing SAMS remains unclear. Studies seeking to establish the aetiology of SAMS have demonstrated impairment of numerous mitochondrial processes. Results and hypotheses of some of these studies have been summarised below.

Coenzyme Q_{10} Deficiency

Statin-induced Coenzyme Q_{10} (CoQ_{10})/Ubiquinone deficiency is the most commonly mooted and extensively studied aetiological cause for SAMS. Ubiquinone comprises a quinone ring and a 10 isoprenoid unit side chain. Whilst some of it is derived from diet, a significant proportion, approximately half, is synthesised within the mitochondria (Figure 1) [47,48]. It is present within the inner mitochondrial membrane of eukaryotic cells and is involved in many different metabolic processes including electron transfer in the mitochondrial respiratory chain [49,50]. It is an essential cofactor for a number of dehydrogenases including those involved in fatty acid oxidation and pyrimidine synthesis. It is also an antioxidant and has a role in apoptosis.

Statins reduce synthesis of a number of other intermediary isoprenoid compounds downstream in the cholesterol biosynthesis pathway [51,52]. Low levels of CoQ_{10} secondary to inhibition of mevalonate synthesis by statins has been implicated in SAMS aetiology. However, many studies have produced conflicting/equivocal conclusions [50].

A number of intervention trials with CoQ_{10} have been performed in statin-associated myalgia following anecdotal case reports of benefit. A systematic review and meta-analysis assessed data from 6 randomised control trials with 8 placebo-controlled treatment arms which met the inclusion criteria: (i) placebo controlled study; and (ii) presented sufficient information about change in CoQ 10 levels following statin therapy [50]. This meta-analysis included a total of 240 participants and 210 controls, with duration of trial interval varying between 6 and 26 weeks. The trials included statin therapy with simvastatin (20–80 mg/day), atorvastatin (10–40 mg /day), rosuvastatin 40 mg/day and pravastatin 20 mg/day. The results showed a significant reduction in circulating CoQ_{10} levels (-0.44 umol/L; $p < 0.001$) in the statin versus placebo arms. The degree of reduction was independent of type or dose statin used and duration of study (>12 weeks versus <12 weeks). However, the clinical relevance of the reduced serum/plasma CoQ_{10} levels remains unclear. Most CoQ_{10} is found in LDL particles, so any decrease in measured circulating levels following statin therapy may have been related to lowering in LDL-C levels with little associated change in tissue levels [53]. Indeed, no correlation was found between plasma and muscle CoQ_{10} levels in the 48 patients studied in a randomised controlled trial assessing the effects of statins on cholesterol and CoQ_{10} metabolism [54]. Paiva et al. studied CoQ_{10} levels in skeletal muscle and demonstrated that CoQ_{10} levels were significantly lower in simvastatin-treated patients versus controls but confusingly no difference was observed in patients treated with Atorvastatin [54]. They also measured respiratory chain activity in six of the subjects and observed a concomitant decrease in respiratory chain activity, although the ratio between citrate synthase and complex activities remained unchanged [54]. They therefore hypothesised that lower muscle CoQ_{10} levels seen in simvastatin-treated patients were associated with mitochondrial volume loss, rather than a true decrease in CoQ_{10} levels within the mitochondria [54,55].

The reported lack of correlation between plasma and muscle CoQ_{10} levels show that, ideally, CoQ_{10} levels should be measured in the correct sub-compartment [56]. However, measuring CoQ_{10} levels in skeletal muscle involves a muscle biopsy, which is a technically challenging, requires admission, and is significantly more invasive than taking a simple blood sample. CoQ_{10} levels in peripheral blood mononuclear cells have been shown to correlate well with muscle CoQ_{10} levels and may be used as a good, less invasive surrogate marker for tissue CoQ_{10} levels in future studies [47,57]. Using this technique, Avis et al. were able to demonstrate a significant drop in CoQ_{10} levels in muscle and mononuclear cells of children with familial hypocholesterolaemia treated with Rosuvastatin [57]. But they were unable to demonstrate any associated drop in mitochondrial ATP synthesis and hypothesised that a decrease in mitochondrial ATP synthesis may only become apparent when CoQ_{10} levels fall below a certain minimal threshold [57]. The same group have previously reported reduced CoQ_{10} and complex IV levels in muscle biopsy samples from two patients presenting with simvastatin associated rhabdomyolysis [58]. It is of note that both of these patients had other predisposing factors that increase risk of myopathy. They were on medications (cyclosporine and itraconazole) that have been reported to result in increased circulating simvastatin levels on co-administration [59,60]. Furthermore, since the patients presented following rhabdomyolysis, an underlying mitochondrial muscle pathology predating statin therapy could not be confidently excluded [58].

There is little evidence to support routine CoQ_{10} measurement and supplementation for statin-related myalgia [29,36,61]. However, assessment of CoQ_{10} function prior to and after starting statin therapy using mononuclear cell levels may be suitable in a small group of patients with suspected inherited deficiencies of CoQ_{10} biosynthesis. The preferred treatment option in susceptible patients with these conditions would be to use alternative lipid lowering therapies such as PCSK9 inhibitors that have recently been approved for prescription in the UK.

Mitochondrial Depletion

A retrospective analysis of the muscle biopsy samples of 48 patients showed a significant decrease in mitochondrial DNA copy number versus nuclear DNA copy numbers (median -47%), thus suggesting mitochondrial depletion in the simvastatin-treated group [54], as did a study of muscle biopsies in 23 patients [62]. The degree of depletion varied, and was independent of clinical signs or symptoms. It was suggested that symptoms may only become apparent once a critical threshold of depletion is reached, similar to other inherited disorders of mitochondrial depletion [63]. A number of different mechanisms for mitochondrial depletion have been proposed. Several studies using rodent myocytes, and human and rodent cell cultures suggest a role for pathways including insulin-like growth factor 1(IGF-1)/Akt in mitochondrial damage and hence apoptosis [64]. Statins have been shown to induce apoptosis by increasing Atrogin-1 mRNA expression in rodent cardiomyocytes [65]. Other studies incriminate decreased mitochondrial biogenesis secondary to statin-induced downregulation of the transcriptional co-activator peroxisome proliferator activating receptor gamma co-activator 1 (PGC1α) as a cause of SAMS [66]. PGC1α upregulates mitochondrial biogenesis by activating nuclear respiratory factor 1, which in turn regulates transcription of transcription factor A. Some studies have reported a reduced PGC1α mRNA expression in human muscle cells on exposure to statins [66,67]. PGC1α also decreases atrophy gene expression, and thus muscle atrophy [68]. A reduced statin-related expression of PGC1α has therefore been proposed as a putative cause of increased atrophy gene expression and muscle atrophy. However, in one study, while muscle atrophy genes were upregulated in statin-exposed muscle, the upregulation was independent of changes in PGC1α or mitochondrial content [69]. Muscle tissue varies in its response to statin exposure. Rodent studies show differential effects on fast versus slow twitch muscle fibres [69–71]. Fast twitch fibres appear to be more susceptible to statin related damage. West Africans, who have more fast twitch muscle fibres [72], tend to have higher CK levels and show an increased rate of statin myopathy [36]. In contrast to skeletal muscles, statins appear to protect mitochondria in cardiac myocytes from oxidative stress [73] possibly by activating PGC1α via reactive oxygen species and inducing mitochondrial biogenesis. This has been proposed as a possible hypothesis for the protective effect of statins on the cardiac mitochondria. Studies on cardiac and skeletal muscles of patients on statins have demonstrated that statins generate low levels of ROS in the cardiac muscle of patients, thus promoting mitochondrial biogenesis [50]. This effect is complex, as large concentrations of ROS in the skeletal muscle have the opposite effect on mitochondrial biogenesis [66].

Inhibition of Mitochondrial Respiratory Chain Complexes

Direct inhibition of one or more complexes in the mitochondrial respiratory chain has been proposed as a possible cause of statin myopathy [27,51].

Adverse effects of statins on L6 rat myocyte cell lines and in rodent muscle biopsy specimens include impaired function of complexes I, III and IV of the respiratory chain [74] when exposed to >100 umol/L of cerivastatin, simvastatin, fluvastatin, atorvastatin. No significant toxicity was seen at concentrations of 1 umol/L. Secondary effects of statins included disturbances in mitochondrial membrane potential, fatty acid beta-oxidation, mitochondrial membrane permeability, DNA fragmentation and apoptosis. The effects were more pronounced in the lipophilic statins but were only seen at high concentrations with a hydrophilic statin. Importantly, pravastatin did not impair electron chain activity even at concentrations of 1 mmol/L [74]. This might suggest a relatively lower SAMS rate in pravastatin-treated patients, but this is not been validated in large studies [20].

Sirvent et al. conducted some elegant experiments to assess the effect of simvastatin on human myocytes. They were able to demonstrate respiratory inhibition of complexes I to IV, with the main effect being inhibition of complex I activity. ATP synthase activity (complex V) was not affected [75]. Both lipidic and glucidic pathways were equally effected, showing up to 20% reduction when treated with 50 uM Simvastatin [75].

Whilst these in vitro studies are convincing, it is important to note that the doses of simvastatin used were more than a thousand fold higher than levels achieved in patients treated with therapeutic doses of Simvastatin, where serum concentrations are typically 1–15 nmol/L, and concentrations in muscle are 30% of circulating concentrations [76]. A direct extrapolation of these effects to patient populations is therefore difficult to make. This could account for discordance between cell studies and relatively lower incidence of human adverse event reported by larger meta-analytical studies [20]. Nevertheless, case reports and case cohort studies have reported decreased respiratory chain activity, especially complex IV activity in patients on statin therapy.

Phillips et al. examined muscle biopsy specimens from four patients who gave history of muscle symptoms on statin therapy. Their symptoms resolved on discontinuation of statin. Biopsy specimens showed red ragged fibres, decreased staining for cytochrome oxidase, and increased lipid droplets. Repeat biopsy done on three patients 3–6 months after discontinuation of therapy showed complete resolution of symptoms. No other comorbidities were found in these patients and their creatine kinase levels remained within reference limits throughout [77]. Similar reports were published by Duncan et al., who showed decreased complex IV activity in addition to decreased CoQ_{10} levels in two patients who presented with rhabdomyolysis when simvastatin was co-administered with another medication (Cyclosporine and Itraconazole) [58]. Arenas et al. reported 60% reduction in cytochrome oxidase activity patient who presented with myoglobinuria on co-administration of cerivastatin and gemfibrozil [78]. They proposed that this was due to cerivastatin related depletion of isoprenoid farnesyl pyrophosphate (FPP) [78,79]. FPP serve as lipid moieties for a number of intracellular compounds and hence can affect a number of intracellular signal pathways and functions including complex IV activity [78,79]. In fact, the role of FPP depletion in a number of proposed beneficial and harmful effects of statins has been extensively studied [79]. It is important to note, however, that cerivastatin has been withdrawn from the market due to significant adverse effects [80]. Safety profile of currently available statins are significantly better that cerivastatin, and hence cerivastatin-mediated effects, cannot automatically be extrapolated to other statins. Furthermore, gemfibrozil is known to increase risk of SAMS when co-administered with statins [36].

Lactone Toxicity

Statins are converted to lactones from statin acids. Lactones are produced by uridine 5'-diphospho-glucuronosyltransferases, and may be responsible for the cytotoxic effects of statins [81]. Recently a comprehensive study investigated the effect of seven different statins on C2C12 myoblast cell lines [82]. Lactones were more powerful cytotoxic agents than their acid counterparts, and induced cytotoxity through apoptosis. Lactones also acutely decreased mitochondrial ATP generation through inhibition of complex III in the mitochondrial respiratory chain by binding with one of two binding sites involved in electron transfer from CoQ_{10} to cytochrome C. These findings were corroborated in muscle biopsy samples from 37 patients with history of statin-related myopathy. The decrease in complex III activity correlated with both symptoms and with muscle levels of statins. Higher levels of toxicity were observed for more lipophilic statins atorvastatin and simvastatin compared with the hydrophilic statins pravastatin and rosuvastatin [82].

Impaired Ca^{2+} Homeostasis

Rodent cell culture, rodent in vivo and human myocyte studies have shown impaired excitation-contraction coupling of skeletal muscle fibres in response increased cytosolic Ca^{2+} efflux secondary to altered mitochondrial function as a mechanism for statin-related myopathy [51,83,84]. Guis et al. conducted contraction tests and ^{31}P magnetic resonance spectroscopy studies of muscle biopsy specimens taken from nine patients with statin-related myopathy, which showed abnormal contraction in 7/9 patients and delayed proton efflux, suggestive of impaired calcium homeostasis [85]. A pharmacogenetic study of cerivastatin, a highly myopathic statin, noted an association of both increased and decreased adverse events with different ryanodine receptor polymorphisms (RYR2) in 185 patients with rhabdomyolysis [86]. Changes in calcium-release mechanism gene expression occur in patients without myalgia exposed to statins [87]. A 1.56 (95% confidence interval 1.20–2.10) fold increased risk of muscle symptoms or raised CK has been described in 332 patients with malignant hyperthermia and 3261 of their relatives from Sweden. This increases to 52 (22–123)-fold for drug-induced myopathy and a 30 (6–148) fold increase for hyperthermia when compared with 3320 controls and 30,728 relations [88]. Statins may therefore unmask disorders of calcium homeostasis such as malignant hyperthermia.

Substrate Overload

Impaired glucose oxidation secondary to statin-related induction of atrogin-1 mRNA expression causes glycogen accumulation in muscle due to inhibition of pyruvate dehydrogenase complex activity [89]. Statin-induced impairment of beta-oxidation leading to lipid accumulation in the muscle cells has also been reported [69]. Both these mechanisms can potentially lead to eventual development of insulin resistance and then muscle atrophy [90].

3. Statin-Induced Necrotising Inflammatory Myopathy

This is a rare autoimmune disease related to the presence of anti-HMG CoA reductase antibodies associated with a restricted HLA type (DRB1*11:01) [23,91,92]. Patients with previous statin exposure develop symmetrical proximal myopathy with grossly elevated creatine kinase (CK). Symptoms persist despite cessation of statin therapy. It is very rare, and has a reported incidence of less than 2 per million per year. Muscle biopsy reveals muscle fibre necrosis with minimal endomysial inflammatory infiltrates. Diagnosis is confirmed by the presence of anti-HMG CoA antibodies and characteristic findings on muscle MRI [23]. The mainstay of treatment is cessation of statin therapy and immunosuppression [92]. A detailed review of SINIM is beyond the scope of this paper [23].

4. Conclusions

Statins are one of the most widely prescribed therapeutic agents because of their proven track record in significantly reducing cardiovascular mortality. Whilst mostly well tolerated, SAMS is the most common cause of statin intolerance and discontinuation of therapy. A number of factors including genetic predispositions and drug interactions have been associated with an increased risk of SAMS. Whilst evidence in the literature points towards statin-induced mitochondrial dysfunction as the most likely cause of SAMS, the exact processes leading to mitochondrial dysfunction are not yet fully understood. Larger and more robust studies looking the plausible pathways are needed to enable a more thorough elucidation of SAMS pathology and to identify biomarkers of risk.

Author Contributions: RR wrote the first draft, ASW reviewed the draft and contributed to the manuscript.

Conflicts of Interest: The authors declare no conflict of interest.

Abbreviations

CVD	Cardiovascular disease
SAMS	Statin associated muscle disease
HMG-CoA	2-Hydroxymethylglutaryl-coenzyme A
LDL	Low density lipoprotein
FH	Familial Hypercholesterolemia
CV	Cardiovascular
CK	Creatine Kinase
SINIM	Statin induced necrotising inflammatory myopathy
CoQ10	Coenzyme Q_{10}
ATP	Adenosine triphosphate
PCSK9	proprotein convertase subtilisin–kexin type 9
IGF-1	insulin like growth factor 1
PGC1α	peroxisome proliferator activating receptor gamma co-activator 1
ROS	reactive oxygen species
FPP	farnesyl pyrophosphate
RYR2	ryanodine receptor polymorphisms

References

1. GBD 2015 Risk Factors Collaborators. Global, regional, and national comparative risk assessment of 79 behavioural, environmental and occupational, and metabolic risks or clusters of risks, 1990–2015: A systematic analysis for the Global Burden of Disease Study 2015. *Lancet* **2016**, *388*, 1659–1724.
2. Goldstein, J.L.; Brown, M.S. A century of cholesterol and coronaries: From plaques to genes to statins. *Cell* **2015**, *161*, 161–172. [CrossRef] [PubMed]
3. Ference, B.A.; Ginsberg, H.N.; Graham, I.; Ray, K.K.; Packard, C.J.; Bruckert, E.; Hegele, R.A.; Krauss, R.M.; Raal, F.J.; Schunkert, H.; et al. Low-density lipoproteins cause atherosclerotic cardiovascular disease. 1. Evidence from genetic, epidemiologic and clinical studies. A Consensus Statement from the European Atherosclerosis Society Consensus Panel. *Eur. Heart J.* 2017. [CrossRef]
4. Danesh, J.; Erqou, S.; Walker, M.; Thompson, S.G.; Tipping, R.; Ford, C.; Pressel, S.; Walldius, G.; Jungner, I.; Folsom, A.R.; et al. The emerging risk factors collaboration: Analysis of individual data on lipid, inflammatory and other markers in over 1.1 million participants in 104 prospective studies of cardiovascular diseases. *Eur. J. Epidemiol.* **2007**, *22*, 839–869. [PubMed]
5. Steinberg, D. The pathogenesis of atherosclerosis. An interpretive history of the cholesterol controversy, part IV: The 1984 coronary primary prevention trial ends it—Almost. *J. Lipid. Res.* **2006**, *47*, 1–14. [CrossRef] [PubMed]
6. Steinberg, D. Thematic review series: The pathogenesis of atherosclerosis. An interpretive history of the cholesterol controversy, part V: The discovery of the statins and the end of the controversy. *J. Lipid. Res.* **2006**, *47*, 1339–1351. [CrossRef] [PubMed]
7. Baigent, C.; Blackwell, L.; Emberson, J.; Holland, L.E.; Reith, C.; Bhala, N.; Peto, R.; Barnes, E.H.; Keech, A.; Simes, J.; et al. Efficacy and safety of more intensive lowering of LDL cholesterol: A meta-analysis of data from 170,000 participants in 26 randomised trials. *Lancet* **2010**, *376*, 1670–1681. [PubMed]
8. Stein, E.A.; Bays, H.; O'Brien, D.; Pedicano, J.; Piper, E.; Spezzi, A. Lapaquistat acetate: Development of a squalene synthase inhibitor for the treatment of hypercholesterolemia. *Circulation* **2011**, *123*, 1974–1985. [CrossRef] [PubMed]
9. Naci, H.; Brugts, J.J.; Fleurence, R.; Ades, A.E. Dose-comparative effects of different statins on serum lipid levels: A network meta-analysis of 256,827 individuals in 181 randomized controlled trials. *Eur. J. Prev. Cardiol.* **2013**, *20*, 658–670. [CrossRef] [PubMed]
10. Graham, D.J.; Staffa, J.A.; Shatin, D.; Andrade, S.E.; Schech, S.D.; La, G.L.; Gurwitz, J.H.; Chan, K.A.; Goodman, M.J.; Platt, R. Incidence of hospitalized rhabdomyolysis in patients treated with lipid-lowering drugs. *JAMA* **2004**, *292*, 2585–2590. [CrossRef] [PubMed]
11. Goldstein, J.L.; Brown, M.S. The LDL receptor. *Arterioscler. Thromb. Vasc. Biol.* **2009**, *29*, 431–438. [CrossRef] [PubMed]
12. The Scandinavian Simvastatin Survival Study (4S) investigators. Randomised trial of cholesterol lowering in 4444 patients with coronary heart disease: The scandinavian simvastatin survival study (4S). *Lancet* **1994**, *344*, 1383–1389.
13. MRC/BHF Heart Protection Study Investigators. MRC/BHF heart protection study of cholesterol lowering with simvastatin in 20,536 high-risk individuals: A randomised placebo-controlled trial. *Lancet* **2002**, *360*, 7–22.
14. Gehring, P.J. The cataractogenic activity of chemical agents. *CRC Crit. Rev. Toxicol.* **1971**, *1*, 93–118. [CrossRef] [PubMed]
15. Cash, J.; Callender, M.E.; McDougall, N.I.; Young, I.S.; Nicholls, D.P. Statin safety and chronic liver disease. *Int. J. Clin. Pract.* **2008**, *62*, 1831–1835. [CrossRef] [PubMed]
16. Emberson, J.R.; Kearney, P.M.; Blackwell, L.; Newman, C.; Reith, C.; Bhala, N.; Holland, L.; Peto, R.; Keech, A.; Collins, R.; et al. Lack of effect of lowering LDL cholesterol on cancer: Meta-analysis of individual data from 175,000 people in 27 randomised trials of statin therapy. *PLoS ONE* **2012**, *7*, e29849.
17. Collins, G.S.; Altman, D.G. Predicting the adverse risk of statin treatment: An independent and external validation of Qstatin risk scores in the UK. *Heart* **2012**, *98*, 1091–1097. [CrossRef] [PubMed]
18. Baigent, C.; Keech, A.; Kearney, P.M.; Blackwell, L.; Buck, G.; Pollicino, C.; Kirby, A.; Sourjina, T.; Peto, R.; Collins, R.; et al. Efficacy and safety of cholesterol-lowering treatment: Prospective meta-analysis of data from 90,056 participants in 14 randomised trials of statins. *Lancet* **2005**, *366*, 1267–1278. [PubMed]

19. Lessons from Lipitor and the broken blockbuster drug model. *Lancet* **2011**, *378*, 1976.
20. Naci, H.; Brugts, J.; Ades, T. Comparative tolerability and harms of individual statins: A study-level network meta-analysis of 246,955 participants from 135 randomized, controlled trials. *Circ. Cardiovasc. Qual. Outcomes* **2013**, *6*, 390–399. [CrossRef] [PubMed]
21. Magni, P.; Macchi, C.; Morlotti, B.; Sirtori, C.R.; Ruscica, M. Risk identification and possible countermeasures for muscle adverse effects during statin therapy. *Eur. J. Intern. Med.* **2015**, *26*, 82–88. [CrossRef] [PubMed]
22. Alfirevic, A.; Neely, D.; Armitage, J.; Chinoy, H.; Cooper, R.G.; Laaksonen, R.; Carr, D.F.; Bloch, K.M.; Fahy, J.; Hanson, A.; et al. Phenotype standardization for statin-induced myotoxicity. *Clin. Pharmacol. Ther.* **2014**, *96*, 470–476. [CrossRef] [PubMed]
23. Babu, S.; Li, Y. Statin induced necrotizing autoimmune myopathy. *J. Neurol. Sci.* **2015**, *351*, 13–17. [CrossRef] [PubMed]
24. Maghsoodi, N.; Wierzbicki, A.S. Statin myopathy: Over-rated and under-treated? *Curr. Opin. Cardiol.* **2016**, *31*, 417–425. [CrossRef] [PubMed]
25. Michalska-Kasiczak, M.; Sahebkar, A.; Mikhailidis, D.P.; Rysz, J.; Muntner, P.; Toth, P.P.; Jones, S.R.; Rizzo, M.; Kees Hovingh, G.; Farnier, M.; et al. Analysis of vitamin D levels in patients with and without statin-associated myalgia—A systematic review and meta-analysis of 7 studies with 2420 patients. *Int. J. Cardiol.* **2014**, *178*, 111–116. [CrossRef] [PubMed]
26. Khayznikov, M.; Hemachrandra, K.; Pandit, R.; Kumar, A.; Wang, P.; Glueck, C.J. Statin intolerance because of myalgia, myositis, myopathy, or myonecrosis can in most cases be safely resolved by vitamin D supplementation. *N. Am. J. Med. Sci.* **2015**, *7*, 86–93. [PubMed]
27. Hou, T.; Li, Y.; Chen, W.; Heffner, R.R.; Vladutiu, G.D. Histopathologic and biochemical evidence for mitochondrial disease among 279 patients with severe statin myopathy. *J. Neuromuscul. Dis.* **2017**, *4*, 77–87. [CrossRef] [PubMed]
28. Rosenson, R.S.; Baker, S.K.; Jacobson, T.A.; Kopecky, S.L.; Parker, B.A. An assessment by the statin muscle safety task force: 2014 update. *J. Clin. Lipidol.* **2014**, *8*, S58–S71. [CrossRef] [PubMed]
29. Mancini, G.B.; Tashakkor, A.Y.; Baker, S.; Bergeron, J.; Fitchett, D.; Frohlich, J.; Genest, J.; Gupta, M.; Hegele, R.A.; Ng, D.S. Diagnosis, prevention, and management of statin adverse effects and intolerance: Canadian Working Group Consensus update. *Can. J. Cardiol.* **2013**, *29*, 1553–1568. [CrossRef] [PubMed]
30. Bruckert, E.; Hayem, G.; Dejager, S.; Yau, C.; Begaud, B. Mild to moderate muscular symptoms with high-dosage statin therapy in hyperlipidemic patients—The PRIMO study. *Cardiovasc. Drugs Ther.* **2005**, *19*, 403–414. [CrossRef] [PubMed]
31. Zhang, H.; Plutzky, J.; Skentzos, S.; Morrison, F.; Mar, P.; Shubina, M.; Turchin, A. Discontinuation of statins in routine care settings: A cohort study. *Ann. Intern. Med.* **2013**, *158*, 526–534. [CrossRef] [PubMed]
32. Parker, B.A.; Capizzi, J.A.; Grimaldi, A.S.; Clarkson, P.M.; Cole, S.M.; Keadle, J.; Chipkin, S.; Pescatello, L.S.; Simpson, K.; White, C.M.; et al. Effect of statins on skeletal muscle function. *Circulation* **2013**, *127*, 96–103. [CrossRef] [PubMed]
33. Katz, D.H.; Intwala, S.S.; Stone, N.J. Addressing statin adverse effects in the clinic: The 5 Ms. *J. Cardiovasc. Pharmacol. Ther.* **2014**, *19*, 533–542. [CrossRef] [PubMed]
34. Cohen, J.D.; Brinton, E.A.; Ito, M.K.; Jacobson, T.A. Understanding Statin Use in America and Gaps in Patient Education (USAGE): An internet-based survey of 10,138 current and former statin users. *J. Clin. Lipidol.* **2012**, *6*, 208–215. [CrossRef] [PubMed]
35. Banach, M.; Rizzo, M.; Toth, P.P.; Farnier, M.; Davidson, M.H.; Al-Rasadi, K.; Aronow, W.S.; Athyros, V.; Djuric, D.M.; Ezhov, M.V.; et al. Statin intolerance—An attempt at a unified definition. Position paper from an International Lipid Expert Panel. *Expert Opin. Drug Saf.* **2015**, *14*, 935–955. [CrossRef] [PubMed]
36. Stroes, E.S.; Thompson, P.D.; Corsini, A.; Vladutiu, G.D.; Raal, F.J.; Ray, K.K.; Roden, M.; Stein, E.; Tokgozoglu, L.; Nordestgaard, B.G.; et al. Statin-associated muscle symptoms: Impact on statin therapy-European Atherosclerosis Society Consensus Panel Statement on Assessment, Aetiology and Management. *Eur. Heart J.* **2015**, *36*, 1012–1022. [CrossRef] [PubMed]
37. Canestaro, W.J.; Austin, M.A.; Thummel, K.E. Genetic factors affecting statin concentrations and subsequent myopathy: A HuGENet systematic review. *Genet. Med.* **2014**, *16*, 810–819. [CrossRef] [PubMed]
38. Mangravite, L.M.; Engelhardt, B.E.; Medina, M.W.; Smith, J.D.; Brown, C.D.; Chasman, D.I.; Mecham, B.H.; Howie, B.; Shim, H.; Naidoo, D.; et al. A statin-dependent QTL for GATM expression is associated with statin-induced myopathy. *Nature* **2013**, *502*, 377–380. [CrossRef] [PubMed]

39. Luzum, J.A.; Kitzmiller, J.P.; Isackson, P.J.; Ma, C.; Medina, M.W.; Dauki, A.M.; Mikulik, E.B.; Ochs-Balcom, H.M.; Vladutiu, G.D. GATM polymorphism associated with the risk for statin-induced myopathy does not replicate in case-control analysis of 715 dyslipidemic individuals. *Cell Metab.* **2015**, *21*, 622–627. [CrossRef] [PubMed]

40. Tay, S.K.; DiMauro, S.; Pang, A.Y.; Lai, P.S.; Yap, H.K. Myotoxicity of lipid-lowering agents in a teenager with MELAS mutation. *Pediatr. Neurol.* **2008**, *39*, 426–428. [CrossRef] [PubMed]

41. Chariot, P.; Abadia, R.; Agnus, D.; Danan, C.; Charpentier, C.; Gherardi, R.K. Simvastatin-induced rhabdomyolysis followed by a MELAS syndrome. *Am. J. Med.* **1993**, *94*, 109–110. [CrossRef]

42. Thomas, J.E.; Lee, N.; Thompson, P.D. Statins provoking MELAS syndrome. A case report. *Eur. Neurol.* **2007**, *57*, 232–235. [CrossRef] [PubMed]

43. Vladutiu, G.D.; Simmons, Z.; Isackson, P.J.; Tarnopolsky, M.; Peltier, W.L.; Barboi, A.C.; Sripathi, N.; Wortmann, R.L.; Phillips, P.S. Genetic risk factors associated with lipid-lowering drug-induced myopathies. *Muscle Nerve* **2006**, *34*, 153–162. [CrossRef] [PubMed]

44. Nicolson, G.L. Mitochondrial dysfunction and chronic disease: Treatment with natural supplements. *Integr. Med.* **2014**, *13*, 35–43.

45. Du Souich, P.; Roederer, G.; Dufour, R. Myotoxicity of statins: Mechanism of action. *Pharmacol. Ther.* **2017**, *175*, 1–16. [CrossRef] [PubMed]

46. Apostolopoulou, M.; Corsini, A.; Roden, M. The role of mitochondria in statin-induced myopathy. *Eur. J. Clin. Investig.* **2015**, *45*, 745–754. [CrossRef] [PubMed]

47. Duncan, A.J.; Heales, S.J.; Mills, K.; Eaton, S.; Land, J.M.; Hargreaves, I.P. Determination of coenzyme Q10 status in blood mononuclear cells, skeletal muscle, and plasma by HPLC with di-propoxy-coenzyme Q10 as an internal standard. *Clin. Chem.* **2005**, *51*, 2380–2382. [CrossRef] [PubMed]

48. Ericsson, J.; Dallner, G. Distribution, biosynthesis, and function of mevalonate pathway lipids. *Subcell. Biochem.* **1993**, *21*, 229–272. [PubMed]

49. Desbats, M.A.; Lunardi, G.; Doimo, M.; Trevisson, E.; Salviati, L. Genetic bases and clinical manifestations of coenzyme Q10 (CoQ 10) deficiency. *J. Inherit. Metab. Dis.* **2015**, *38*, 145–156. [CrossRef] [PubMed]

50. Banach, M.; Serban, C.; Sahebkar, A.; Ursoniu, S.; Rysz, J.; Muntner, P.; Toth, P.P.; Jones, S.R.; Rizzo, M.; Glasser, S.P.; et al. Effects of coenzyme Q10 on statin-induced myopathy: A meta-analysis of randomized controlled trials. *Mayo Clin. Proc.* **2015**, *90*, 24–34. [CrossRef] [PubMed]

51. Sirvent, P.; Mercier, J.; Lacampagne, A. New insights into mechanisms of statin-associated myotoxicity. *Curr. Opin. Pharmacol.* **2008**, *8*, 333–338. [CrossRef] [PubMed]

52. Wierzbicki, A.S.; Poston, R.; Ferro, A. The lipid and non-lipid effects of statins. *Pharmacol. Ther.* **2003**, *99*, 95–112. [CrossRef]

53. Littarru, G.P.; Langsjoen, P. Coenzyme Q10 and statins: Biochemical and clinical implications. *Mitochondrion* **2007**, *7*, S168–S174. [CrossRef] [PubMed]

54. Paiva, H.; Thelen, K.M.; Van, C.R.; Smet, J.; De, P.B.; Mattila, K.M.; Laakso, J.; Lehtimaki, T.; von, B.K.; Lutjohann, D.; et al. High-dose statins and skeletal muscle metabolism in humans: A randomized, controlled trial. *Clin. Pharmacol. Ther.* **2005**, *78*, 60–68. [CrossRef] [PubMed]

55. Larsen, S.; Stride, N.; Hey-Mogensen, M.; Hansen, C.N.; Bang, L.E.; Bundgaard, H.; Nielsen, L.B.; Helge, J.W.; Dela, F. Simvastatin effects on skeletal muscle: Relation to decreased mitochondrial function and glucose intolerance. *J. Am. Coll. Cardiol.* **2013**, *61*, 44–53. [CrossRef] [PubMed]

56. Hargreaves, I.P.; Duncan, A.J.; Heales, S.J.; Land, J.M. The effect of HMG-CoA reductase inhibitors on coenzyme Q10, Possible biochemical/clinical implications. *Drug Saf.* **2005**, *28*, 659–676. [CrossRef] [PubMed]

57. Avis, H.J.; Hargreaves, I.P.; Ruiter, J.P.; Land, J.M.; Wanders, R.J.; Wijburg, F.A. Rosuvastatin lowers coenzyme Q10 levels, but not mitochondrial adenosine triphosphate synthesis, in children with familial hypercholesterolemia. *J. Pediatr.* **2011**, *158*, 458–462. [CrossRef] [PubMed]

58. Duncan, A.J.; Hargreaves, I.P.; Damian, M.S.; Land, J.M.; Heales, S.J. Decreased ubiquinone availability and impaired mitochondrial cytochrome oxidase activity associated with statin treatment. *Toxicol. Mech. Methods* **2009**, *19*, 44–50. [CrossRef] [PubMed]

59. Arnadottir, M.; Eriksson, L.O.; Thysell, H.; Karkas, J.D. Plasma concentration profiles of simvastatin 3-hydroxy-3-methyl-glutaryl-coenzyme A reductase inhibitory activity in kidney transplant recipients with and without ciclosporin. *Nephron* **1993**, *65*, 410–413. [CrossRef] [PubMed]

60. Neuvonen, P.J.; Kantola, T.; Kivisto, K.T. Simvastatin but not pravastatin is very susceptible to interaction with the CYP3A4 inhibitor itraconazole. *Clin. Pharmacol. Ther.* **1998**, *63*, 332–341. [CrossRef]

61. Rabar, S.; Harker, M.; O'Flynn, N.; Wierzbicki, A.S. Lipid modification and cardiovascular risk assessment for the primary and secondary prevention of cardiovascular disease: Summary of updated NICE guidance. *BMJ* **2014**, *349*, G4356. [CrossRef] [PubMed]

62. Stringer, H.A.; Sohi, G.K.; Maguire, J.A.; Cote, H.C. Decreased skeletal muscle mitochondrial DNA in patients with statin-induced myopathy. *J. Neurol. Sci.* **2013**, *325*, 142–147. [CrossRef] [PubMed]

63. Schick, B.A.; Laaksonen, R.; Frohlich, J.J.; Paiva, H.; Lehtimaki, T.; Humphries, K.H.; Cote, H.C. Decreased skeletal muscle mitochondrial DNA in patients treated with high-dose simvastatin. *Clin. Pharmacol. Ther.* **2007**, *81*, 650–653. [CrossRef] [PubMed]

64. Mullen, P.J.; Zahno, A.; Lindinger, P.; Maseneni, S.; Felser, A.; Krahenbuhl, S.; Brecht, K. Susceptibility to simvastatin-induced toxicity is partly determined by mitochondrial respiration and phosphorylation state of Akt. *Biochim. Biophys. Acta* **2011**, *1813*, 2079–2087. [CrossRef] [PubMed]

65. Bonifacio, A.; Mullen, P.J.; Mityko, I.S.; Navegantes, L.C.; Bouitbir, J.; Krahenbuhl, S. Simvastatin induces mitochondrial dysfunction and increased atrogin-1 expression in H9c2 cardiomyocytes and mice in vivo. *Arch. Toxicol.* **2016**, *90*, 203–215. [CrossRef] [PubMed]

66. Bouitbir, J.; Charles, A.L.; Echaniz-Laguna, A.; Kindo, M.; Daussin, F.; Auwerx, J.; Piquard, F.; Geny, B.; Zoll, J. Opposite effects of statins on mitochondria of cardiac and skeletal muscles: A 'mitohormesis' mechanism involving reactive oxygen species and PGC-1. *Eur. Heart J.* **2012**, *33*, 1397–1407. [CrossRef] [PubMed]

67. Vaughan, R.A.; Garcia-Smith, R.; Bisoffi, M.; Conn, C.A.; Trujillo, K.A. Ubiquinol rescues simvastatin-suppression of mitochondrial content, function and metabolism: Implications for statin-induced rhabdomyolysis. *Eur. J. Pharmacol.* **2013**, *711*, 1–9. [CrossRef] [PubMed]

68. Sandri, M.; Lin, J.; Handschin, C.; Yang, W.; Arany, Z.P.; Lecker, S.H.; Goldberg, A.L.; Spiegelman, B.M. PGC-1alpha protects skeletal muscle from atrophy by suppressing FoxO3 action and atrophy-specific gene transcription. *Proc. Natl. Acad. Sci. USA* **2006**, *103*, 16260–16265. [CrossRef] [PubMed]

69. Goodman, C.A.; Pol, D.; Zacharewicz, E.; Lee-Young, R.S.; Snow, R.J.; Russell, A.P.; McConell, G.K. Statin-induced increases in atrophy gene expression occur independently of changes in PGC1alpha protein and mitochondrial content. *PLoS ONE* **2015**, *10*, e0128398. [CrossRef] [PubMed]

70. Westwood, F.R.; Bigley, A.; Randall, K.; Marsden, A.M.; Scott, R.C. Statin-induced muscle necrosis in the rat: Distribution, development, and fibre selectivity. *Toxicol. Pathol.* **2005**, *33*, 246–257. [CrossRef] [PubMed]

71. Westwood, F.R.; Scott, R.C.; Marsden, A.M.; Bigley, A.; Randall, K. Rosuvastatin: Characterization of induced myopathy in the rat. *Toxicol. Pathol.* **2008**, *36*, 345–352. [CrossRef] [PubMed]

72. Holden, C. Peering under the hood of Africa's runners. *Science* **2004**, *305*, 637–639. [CrossRef] [PubMed]

73. Jones, S.P.; Teshima, Y.; Akao, M.; Marban, E. Simvastatin attenuates oxidant-induced mitochondrial dysfunction in cardiac myocytes. *Circ. Res.* **2003**, *93*, 697–699. [CrossRef] [PubMed]

74. Kaufmann, P.; Torok, M.; Zahno, A.; Waldhauser, K.M.; Brecht, K.; Krahenbuhl, S. Toxicity of statins on rat skeletal muscle mitochondria. *Cell. Mol. Life Sci.* **2006**, *63*, 2415–2425. [CrossRef] [PubMed]

75. Sirvent, P.; Bordenave, S.; Vermaelen, M.; Roels, B.; Vassort, G.; Mercier, J.; Raynaud, E.; Lacampagne, A. Simvastatin induces impairment in skeletal muscle while heart is protected. *Biochem. Biophys. Res. Commun.* **2005**, *338*, 1426–1434. [CrossRef] [PubMed]

76. Bjorkhem-Bergman, L.; Lindh, J.D.; Bergman, P. What is a relevant statin concentration in cell experiments claiming pleiotropic effects? *Br. J. Clin. Pharmacol.* **2011**, *72*, 164–165. [CrossRef] [PubMed]

77. Phillips, P.S.; Haas, R.H.; Bannykh, S.; Hathaway, S.; Gray, N.L.; Kimura, B.J.; Vladutiu, G.D.; England, J.D. Statin-associated myopathy with normal creatine kinase levels. *Ann. Intern. Med.* **2002**, *137*, 581–585. [CrossRef] [PubMed]

78. Arenas, J.; Fernandez-Moreno, M.A.; Molina, J.A.; Fernandez, V.; del Hoyo, P.; Campos, Y.; Calvo, P.; Martin, M.A.; Garcia, A.; Moreno, T.; et al. Myoglobinuria and COX deficiency in a patient taking cerivastatin and gemfibrozil. *Neurology* **2003**, *60*, 124–126. [CrossRef] [PubMed]

79. Mans, R.A.; McMahon, L.L.; Li, L. Simvastatin-mediated enhancement of long-term potentiation is driven by farnesyl-pyrophosphate depletion and inhibition of farnesylation. *Neuroscience* **2012**, *202*, 1–9. [CrossRef] [PubMed]

80. Kalaria, D.; Wassenaar, W. Rhabdomyolysis and cerivastatin: Was it a problem of dose? *CMAJ* **2002**, *167*, 737. [PubMed]

81. Subramanian, R.; Fang, X.; Prueksaritanont, T. Structural characterization of in vivo rat glutathione adducts and a hydroxylated metabolite of simvastatin hydroxy acid. *Drug Metab. Dispos.* **2002**, *30*, 225–230. [CrossRef] [PubMed]

82. Schirris, T.J.; Renkema, G.H.; Ritschel, T.; Voermans, N.C.; Bilos, A.; van Engelen, B.G.; Brandt, U.; Koopman, W.J.; Beyrath, J.D.; Rodenburg, R.J.; et al. Statin-induced myopathy is associated with mitochondrial complex III inhibition. *Cell Metab.* **2015**, *22*, 399–407. [CrossRef] [PubMed]

83. Nakahara, K.; Yada, T.; Kuriyama, M.; Osame, M. Cytosolic Ca^{2+} increase and cell damage in L6 rat myoblasts by HMG-CoA reductase inhibitors. *Biochem. Biophys. Res. Commun.* **1994**, *202*, 1579–1585. [CrossRef] [PubMed]

84. Inoue, R.; Tanabe, M.; Kono, K.; Maruyama, K.; Ikemoto, T.; Endo, M. Ca^{2+}-releasing effect of cerivastatin on the sarcoplasmic reticulum of mouse and rat skeletal muscle fibers. *J. Pharmacol. Sci.* **2003**, *93*, 279–288. [CrossRef] [PubMed]

85. Guis, S.; Figarella-Branger, D.; Mattei, J.P.; Nicoli, F.; Le Fur, Y.; Kozak-Ribbens, G.; Pellissier, J.F.; Cozzone, P.J.; Amabile, N.; Bendahan, D. In vivo and in vitro characterization of skeletal muscle metabolism in patients with statin-induced adverse effects. *Arthrit. Rheum.* **2006**, *55*, 551–557. [CrossRef] [PubMed]

86. Marciante, K.D.; Durda, J.P.; Heckbert, S.R.; Lumley, T.; Rice, K.; McKnight, B.; Totah, R.A.; Tamraz, B.; Kroetz, D.L.; Fukushima, H.; et al. Cerivastatin, genetic variants, and the risk of rhabdomyolysis. *Pharmacogenet. Genom.* **2011**, *21*, 280–288. [CrossRef] [PubMed]

87. Draeger, A.; Sanchez-Freire, V.; Monastyrskaya, K.; Hoppeler, H.; Mueller, M.; Breil, F.; Mohaupt, M.G.; Babiychuk, E.B. Statin therapy and the expression of genes that regulate calcium homeostasis and membrane repair in skeletal muscle. *Am. J. Pathol.* **2010**, *177*, 291–299. [CrossRef] [PubMed]

88. Hedenmalm, K.; Granberg, A.G.; Dahl, M.L. Statin-induced muscle toxicity and susceptibility to malignant hyperthermia and other muscle diseases: A population-based case-control study including 1st and 2nd degree relatives. *Eur. J. Clin. Pharmacol.* **2015**, *71*, 117–124. [CrossRef] [PubMed]

89. Mallinson, J.E.; Constantin-Teodosiu, D.; Glaves, P.D.; Martin, E.A.; Davies, W.J.; Westwood, F.R.; Sidaway, J.E.; Greenhaff, P.L. Pharmacological activation of the pyruvate dehydrogenase complex reduces statin-mediated upregulation of FOXO gene targets and protects against statin myopathy in rodents. *J. Physiol.* **2012**, *590*, 6389–6402. [CrossRef] [PubMed]

90. Hafizi Abu Bakar, M.; Kian Kai, C.; Wan Hassan, W.N.; Sarmidi, M.R.; Yaakob, H.; Zaman, H.H. Mitochondrial dysfunction as a central event for mechanisms underlying insulin resistance: The roles of long chain fatty acids. *Diabetes Metab. Res. Rev.* **2015**, *31*, 453–475. [CrossRef] [PubMed]

91. Needham, M.; Fabian, V.; Knezevic, W.; Panegyres, P.; Zilko, P.; Mastaglia, F.L. Progressive myopathy with up-regulation of MHC-I associated with statin therapy. *Neuromuscul. Disord.* **2007**, *17*, 194–200. [CrossRef] [PubMed]

92. Grable-Esposito, P.; Katzberg, H.D.; Greenberg, S.A.; Srinivasan, J.; Katz, J.; Amato, A.A. Immune-mediated necrotizing myopathy associated with statins. *Muscle Nerve* **2010**, *41*, 185–190. [CrossRef] [PubMed]

Journal of
Clinical Medicine

MDPI

Review

Evidence of Oxidative Stress and Secondary Mitochondrial Dysfunction in Metabolic and Non-Metabolic Disorders

Karolina M. Stepien [1,*], Robert Heaton [2], Scott Rankin [2], Alex Murphy [2], James Bentley [2], Darren Sexton [2] and Iain P. Hargreaves [2,*]

[1] The Mark Holland Metabolic Unit Salford Royal NHS Foundation Trust Stott Lane, Salford M6 8HD, UK
[2] School of Pharmacy, Liverpool John Moore University, Byrom Street, Liverpool L3 3AF, UK;
r.heaton@2013.ljmu.ac.uk (R.H.); s.rankin@2014.ljmu.ac.uk (S.R.); a.murphy3@2013.ljmu.ac.uk (A.M.);
j.bentley@2014.ljmu.ac.uk (J.B.); d.w.sexton@ljmu.ac.uk (D.S.)
* Correspondence: kstepien@doctors.org.uk (K.M.S.); i.p.hargreaves@ljmu.ac.uk (I.P.H.);
Tel.: +44-161-206-3645 (K.M.S.); +44-151-231-2711 (I.P.H.)

Received: 7 June 2017; Accepted: 14 July 2017; Published: 19 July 2017

Abstract: Mitochondrial dysfunction and oxidative stress have been implicated in the pathogenesis of a number of diseases and conditions. Oxidative stress occurs once the antioxidant defenses of the body become overwhelmed and are no longer able to detoxify reactive oxygen species (ROS). The ROS can then go unchallenged and are able to cause oxidative damage to cellular lipids, DNA and proteins, which will eventually result in cellular and organ dysfunction. Although not always the primary cause of disease, mitochondrial dysfunction as a secondary consequence disease of pathophysiology can result in increased ROS generation together with an impairment in cellular energy status. Mitochondrial dysfunction may result from either free radical-induced oxidative damage or direct impairment by the toxic metabolites which accumulate in certain metabolic diseases. In view of the importance of cellular antioxidant status, a number of therapeutic strategies have been employed in disorders associated with oxidative stress with a view to neutralising the ROS and reactive nitrogen species implicated in disease pathophysiology. Although successful in some cases, these adjunct therapies have yet to be incorporated into the clinical management of patients. The purpose of this review is to highlight the emerging evidence of oxidative stress, secondary mitochondrial dysfunction and antioxidant treatment efficacy in metabolic and non-metabolic diseases in which there is a current interest in these parameters.

Keywords: mitochondria; electron transport chain; reactive oxygen species; reactive nitrogen species; oxidative stress; phenylketonuria; methylmalonic acidemia; methylmalonic acid; peroxisome; glutathione; catalase; superoxide dismutase; coenzyme Q_{10}; sepsis; nitrosative stress; nitric oxide synthase

---

1. Introduction

Oxidative stress has been implicated as a major contributory factor to the pathophysiology of a number of diseases and conditions including cancer [1], sepsis [2] and metabolic diseases [3–8]. The origin of oxidative stress in disease is generally multifactorial and can rarely be attributed to one mechanism [9]. Although, impairment of mitochondrial function as a secondary consequence of disease pathophysiology is thought to make a major contribution to reactive oxygen species (ROS) generation in a number of disorders [9]. Factors responsible for this mitochondrial dysfunction include toxic metabolites which accumulate in metabolic disorders [10,11] as well as ROS and reactive nitrogen species (RNS) generated as part of the pathogenesis of other diseases [2,12]. These factors are then able to directly impair the electron transport chain (ETC) which is the site of mitochondrial ROS generation [13,14].

The cell has several means available to tackle free radical generation including antioxidants and antioxidant enzymes; however, as soon as pro-oxidants exceed the antioxidant capacity of the cell, free radicals accumulate and oxidative stress occurs with the resultant damage to proteins, lipids and DNA causing cellular and consequently organ dysfunction [9]. In view of the detrimental effects of oxidative stress, a number of studies have investigated the utility of antioxidant interventions in disease and have shown evidence of therapeutic efficacy in some cases [15,16].

It is the purpose of this review to highlight evidence of oxidative stress and secondary mitochondrial dysfunction in disease, highlighting putative mechanisms and therapeutic strategies in disorders in which there is a growing interest in the association between these parameters. Although this review will primarily focus upon oxidative stress, evidence of nitrosative stress as the result of RNS accumulation will also be outlined in the metabolic and non-metabolic diseases discussed in this review.

2. Phenyloketonuria (PKU)

PKU is an autosomal recessive inherited metabolic disorder of amino acid metabolism which is caused by mutations in the gene encoding the enzyme, phenylalanine hydroxylase (EC1.14.16.1) [17]. Phenylalanine (Phe) is an essential amino acid obtained exclusively from the diet or by proteolysis. It is crucial for protein synthesis, as well as for the synthesis of tyrosine and its derivatives, such as dopamine, norepinephrine and melanin [18,19]. However, a deficiency of phenylalanine hydroxylase leads to accumulation of Phe in the blood and other tissues of affected patients [20–22]. Phe concentrations in plasma may reach very high levels (mmol/L) and, as a result, some of the accumulated Phe can then be metabolized by alternative pathways making phenylketones such as phenylpyruvate, phenyllactate and phenylacetate [20].

Untreated PKU patients present with severe mental retardation, microcephaly, developmental delay, epilepsy, behavioral alterations, cerebral white matter abnormalities and progressive supranuclear motor disturbances [17,23,24]. Newborn screening for PKU has enabled early diagnosis and treatment of this condition [25]. This will help prevent the possibility of mental retardation, although slightly reduced neurophysiological outcomes may occur, in particular in combination with poor compliance to PKU diet [26]. The main findings presented by PKU patients are severe neurological damage, including corpus callosum, striatum, and cortical alterations and hypomyelination, that result in intellectual deficit and neurodegeneration [27–30]. However, the pathophysiology underlying the brain damage has yet to be fully elucidated, although oxidative stress may play an important role [15]. In PKU, oxidative stress appears to be already present at the time of diagnosis and persists even in the presence of dietary compliance [31,32]. Evidence of oxidative stress in PKU patients has been indicated by increased levels of plasma thiobarbituric acid-reactive species (TBAR), an indicator of lipid peroxidation [33], malondialdehyde (a lipid peroxidation marker) [31] and 8-hydroxy-2-deoxygyanosine (marker of DNA oxidation) [34]. The oxidative stress associated with PKU may result from the effect of the restricted diet of patients as well as the elevated levels of Phe or its metabolites upon cellular antioxidant defenses [15]. Historically, a deficiency in the status of the trace metal, selenium (Se), was considered to be an important contributory factor to the oxidative stress associated with PKU [35]. Se is required for the biological activity of selenoproteins, one of which is the antioxidant enzyme, glutathione peroxidase (GSH-Px; EC: 1.11.1.9), and therefore, a deficiency in Se status may compromise the activity of this enzyme [36]. However, evidence of decreased GSH-Px activity has been reported in PKU patients with plasma Se levels within the reference range suggesting that other factors may be responsible for the deficit in enzyme activity [33]. One of these factors may be the low level of methionine present in the diet of PKU patients, which may result in impaired GSH-Px synthesis [5]. Phe itself may directly inhibit the activity of GSH-Px [33]. In addition, animal studies have reported the potential for hyperphenylalaninemia to directly suppress the production of

GSH-Px as well as enhance its degradation [37]. A decreased level of the cellular antioxidant, reduced glutathione (GSH), has also been reported in PKU, although it was uncertain whether this was caused by oxidative stress or the restricted diet [38]. However, a subsequent study in rat astrocytoma cells reported evidence of decreased GSH status in conjunction with increased oxidative stress in cells exposed to Phe at levels commonly detected in PKU patients (1000–1500 µmol/L) [39]. This study indicated the vulnerability of neural cells to Phe-induced oxidative stress which may be an important contributory factor to the neurological dysfunction associated with PKU. Kienzle-Hagen and colleagues (2002) reported a significant ($p < 0.01$) inhibitory effect of the hyperphenylalaninemia on the cerebral catalase activity of rat [37]; however, studies in PKU patients have found no evidence of an inhibition of this enzyme in peripheral tissue [31]. Indeed, a number of studies have reported an increase in the activity of this enzyme in patients [40].

In addition to oxidative stress, one study has reported evidence of nitrosative stress in PKU patients by measurement of serum NOx (nitrite/nitrate), the stable breakdown products of nitric oxide (NO), which was found to be significantly increased compared to control levels [33]. However, NOx tended to be lower in patients with plasma Phe levels > 900 µM. This study suggested an impairment in the regulation of NO metabolism in PKU with the increase in serum NOx < 900 µM Phe thought to reflect the increased oxidative stress. The decrease in serum NOx at Phe > 900 µM originates from the oxidative stress-induced transcriptional suppression of the nitric oxide synthase (NOS) gene, or as a result of structural changes in the NOS enzyme [33].

The mevalonate pathway enzymes, 3-hydroxy-3-methylglutaryl-CoA (HMG-CoA; EC1.1.1.98) reductase, and mevalonate 5-pyrophosphate decarboxylase (EC4.1.1.33) have been reported to be inhibited by Phe and its metabolite, phenylacetate; however, only Phe-induced inhibition within its physiological range ($\geq$250 µmol/L) [41]. Since HMG-CoA reductase is the major regulatory enzyme in the synthesis of both cholesterol and the lipid soluble antioxidant, coenzyme Q$_{10}$ (CoQ$_{10}$), since they share a common pathway, it is therefore unsurprising that perturbations in the synthesis of both of these isoprenoids have been associated with PKU [6,42]. The availability of tyrosine is also essential for the synthesis of CoQ$_{10}$; however, in PKU, no association has been observed between the plasma level of tyrosine and that of CoQ$_{10}$, although this relationship was not investigated in tissues [6]. The results of cellular CoQ$_{10}$ status in PKU has been contradictory with a study by Colome et al. (2002) finding evidence of a deficit in this isoprenoid in the lymphocytes from well-controlled PKU patients [43]. In contrast, a study by Hargreaves et al. (2002) found no evidence of a CoQ$_{10}$ deficiency in blood mononuclear cells from an older group of PKU patients [44].

The reported ability of hyperphenylalaninaemia to impair the activity of the mitochondrial electron transport chain (ETC) [45] may also contribute to the oxidative stress associated with PKU, since ETC dysfunction has been associated with reactive oxygen species (ROS) generation [13]. In the study by Rech et al. (2002), ETC complex I–III (NADH cytochrome c reductase; EC1.3.5.1 + EC1.10.2.2) activity was found to be reduced following chemically induced hyperphenylalaninemia in rat brain cortex [45]. ETC complex II (succinate: ubiquinone reductase; EC1.3.5.1) and complex IV (cytochrome c oxidase; EC1.9.3.1) were unaffected. It was surmised that the impairment of ETC complex I–III activity was the result of Phe competing with NADH for the active site of complex I (NADH ubiquinone reductase; EC: 1.6.5.3). Subsequent studies in human astrocytoma cells [46] and blood mononuclear cells [44] have found no evidence of inhibition of either ETC complex I or ETC complex II–III (succinate:cytochrome reductase; EC1.3.5.1 + EC1.10.2.2) activities, respectively under conditions of hyperphenylalaninemia. However, since no studies have as yet directly assessed the effect of hyperphenylalaninemia on ETC complex III (ubiquinol: cytochrome c reductase; EC1.10.2.2) activity, the possibility that this enzyme is susceptible to Phe-induced toxicity cannot be discounted. In addition, the suggested ability of hyperphenylalaninemia to induce a CoQ$_{10}$ deficiency in some studies may also result in secondary ETC dysfunction in some PKU patients [6,43].

The effect of hyperphenylalaninaemia on the mitochondrial oxidative metabolism was investigated by the authors by determining the lactate concentration of cell culture medium derived from immortalised HEPG2 liver cells that had been exposed to 900 and 1200 µmol/L Phe, respectively, for 72 h. Following 72 h of culture, the lactate concentration in the cell culture media was determined by the method outlined in the study by Kyprianou et al. (2009) and no significant difference was found between the control and Phe-treated HEPG2 cell groups following Student's *t*-test analysis ($p < 0.05$ was considered statistically significant, Figure 1) [46], which suggests no evidence of Phe-induced ETC impairment in the immortalised human liver cells.

The putative mechanisms that have been implicated for ETC dysfunction and oxidative stress in PKU are outlined in Figure 2.

Treatments for PKU patients consist of restriction of Phe intake, through natural-protein-restricted diet supplemented with Phe-free amino acid mixtures enriched with trace elements, vitamins and minerals [47–49]. Strict low-protein diet, however, causes some micronutrient and antioxidant deficiencies including zinc, copper, Se, magnesium and iron (Fe) deficiencies [50–53]. A deficiency in Fe may also result in a secondary diminution in the level of carnitine, since Fe is required for the synthesis of this compound [54]. In view of the antioxidant properties of carnitine, which is able to act as an ROS scavenger, a deficit in the status of this compound which has been reported in some PKU patients may comprise antioxidant status [38,55]. Indeed, supplementation of PKU patients with Se and carnitine has been recommended as a means to ameliorate the oxidative stress associated with this condition [35]. At present however, there is no overall consensus on the use of antioxidant supplementation in the treatment of PKU, although this adjunct therapy may offer some protection against the neurological dysfunction associated with this condition [56].

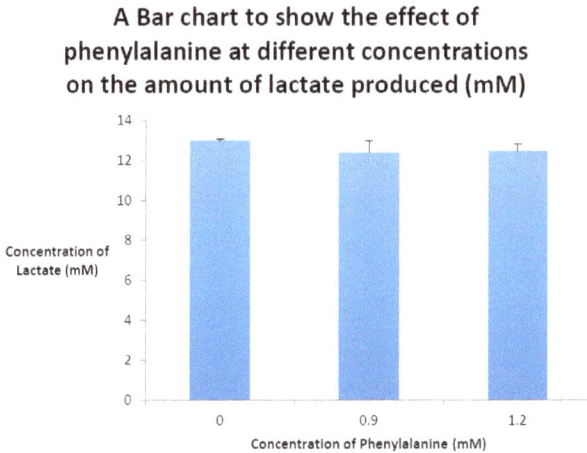

A Bar chart to show the effect of phenylalanine at different concentrations on the amount of lactate produced (mM)

Figure 1. Bar chart displaying the mean cell culture lactate concentrations determined following culture of human HEPG2 liver cells for 72 h with 0.9 and 1.2 mM phenylalanine, respectively. Results are expressed as the mean and standard deviation of four determinations.

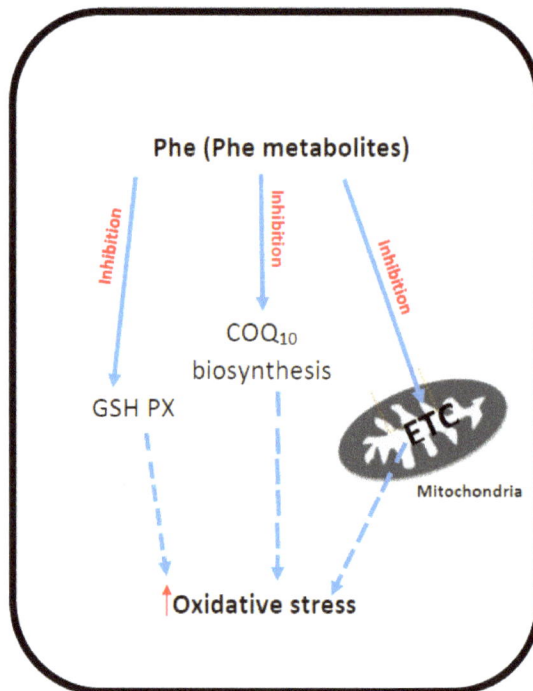

Figure 2. Putative mechanisms of oxidative stress generation and mitochondrial dysfunction in PKU. PKU: Phenylketonuria; Phe: Phenylalanine; ETC: Mitochondrial electron transport chain; CoQ$_{10}$: Coenzyme Q$_{10}$; GSH-PX: Glutathione peroxidase.

3. Methylmalonic Acidemia

Methylmalonic acidemia is one of the organic acidemias, which is primarily caused by severe deficiency of the enzyme, L-methylmalonyl-CoA mutase (MCM; EC: 5.4.99.2), or by defects in the synthesis of 5-deoxyadenosyl cobalamin, the active form of vitamin B12 and an essential cofactor required for the activation of MCM [57]. This condition leads to an increase in the level of methylmalonyl-CoA, which is spontaneously converted to methylmalonic acid (MMA) [58]. Biochemically, the condition is characterized by tissue accumulation of MMA. The levels of MMA in the blood and cerebrospinal fluid are usually around 2.5 mmol/L during acute metabolic crises [58,59] but may be even higher in the brain [60].

Clinical features of this condition include lethargy, coma, vomiting, failure to thrive, muscular hypotonia, progressive neurological deterioration and kidney failure [61].

The mechanisms responsible for the neurological and renal dysfunction in this organic acidemia have so far not been fully elucidated, although ETC dysfunction and oxidative stress are thought to contribute to the pathophysiology of this disorder [62,63].

Evidence of ETC dysfunction in methylmalonic acidemia was first suggested by the unexplained lactic acidosis in patients with this condition [64]. This was later confirmed in the study Hayasaka et al. (1982), which reported evidence of ETC complex IV deficiency in post-mortem liver of a single patient [62]. A number of subsequent studies have demonstrated evidence of ETC dysfunction in association with methylmalonic acidemia, with evidence of both single [65,66] and multiple ETC enzyme deficiencies [67–70] being reported in patient and animal studies. In addition, animal and patient studies have also reported morphological abnormalities in mitochondria as the result of methylmalonic acidemia. Proteinuria, renal tubular injury, dilated tubuli and mitochondrial swelling and disorganization of cristae in the tubulum epithelium was observed in an experimental study on rats exposed chronically to MMA [71]. Cell autonomous ETC complex IV deficiency was demonstrated in megamitochondria from renal tubules in a patient with MMA [72], confirming the observations from the previous animal study [71]. Brain imaging and histopathological investigations have revealed a symmetric degeneration of the basal ganglia, particularly the globus pallidus, as well as a mild spongiosis of the subthalamic nucleus, mammillary bodies, and internal capsule [73–75]. Interestingly, symmetrical lesions in the basal ganglia are also found in patients with inherited ETC complex II deficiencies [76].

An increase in lactate concentration together with a reduction in N-acetyl aspartate were observed in the globus pallidus of patients with methylmalonic acidemia which in conjunction with an elevation in cerebrospinal fluid (CSF) lactate levels indicated a possible perturbation in mitochondrial oxidative metabolism [77]. The pathological changes in methylmalonic acidemia are thought to result from the accumulation of toxic organic acids during decompensation [78], and this toxicity has been ascribed to MMA and its metabolites, methylcitrate and malonate [10,79,80]. However, it has been suggested that the mitochondrial dysfunction observed in methylmalonic acidemia is the result of inhibition of the ETC by methylcitrate and malonate rather than by MMA, which has been reported not to inhibit ETC enzyme activity [10]. Although, results from other studies have suggested the propensity for MMA to inhibit ETC activity [66,68,79,81–84]. The ETC dysfunction associated with methylmalonic acidemia may therefore be the result of synergistic inhibition of the ETC by MMA, methylcitrate and malonate [59]. Evidence of oxidative stress in methylmalonic acidemia has been reported in a number of studies both in patients [56,85] and animal models [68,86–88]. ETC dysfunction is thought to be the major cause of oxidative stress in methylmalonic acidemia [86]; however, increased expression of the mitochondrial enzyme, glycerophosphate dehydrogenase, may also contribute to the ROS generation in this condition [63]. The effect of methylmalonic acidemia on cellular antioxidant status has been documented in a number of studies. In 1996, Treacy et al. reported a blood GSH deficiency in a seven-year-old child with this condition [59]. The patient was treated with high-dose ascorbic acid therapy and showed some clinical improvement which the authors suggested may have resulted as a consequence of the vitamin supplementation eliciting a replenishment of cellular antioxidant capacity. Evidence of a decrease in GSH status was also reported in the liver of a mouse model of methylmalonic acidemia [69]. In this study, a decrease in the level of GSSG (the oxidised form of GSH) was also reported, indicating that an impairment in cellular ATP generation may also have contributed to the loss of total glutathione (GSH + GSSG) status. Since glutathione synthesis is ATP-dependent [89], the ETC deficiencies also reported in the liver tissue of the animal model may have been sufficient to compromise oxidative phosphorylation [69]. Decreased plasma [90] and monocyte levels of GSH [85] have also been reported in patients with methylmalonic acidemia, which in both studies accompanied evidence of increased oxidative stress. In view of the number of toxic organic acids which have been implicated in the pathogenesis of methylmalonic acidemia [10,79,80], the authors investigated the propensity of MMA to induce a deficit in the level of neuronal cell GSH status. In this human neuroblastoma, SHS-5Y cells were incubated with MMA at concentrations reported in the plasma of patients with methylmalonic acidemia (2 and 5 mmol/L) [69]. Cellular GSH levels were determined by the HPLC electrochemical method outlined in the study by Hargreaves et al. (2005) following 6 and 10 days in culture, respectively (Figure 3) [89]. Although no evidence of decrease of GSH status was detected after 6 days of culture, evidence of a significant ($p < 0.05$) decrease in SHS-5Y cell GSH status following 10 days of culture with 5 mmol/L MMA was determined following Student's *t* test analysis of the data.

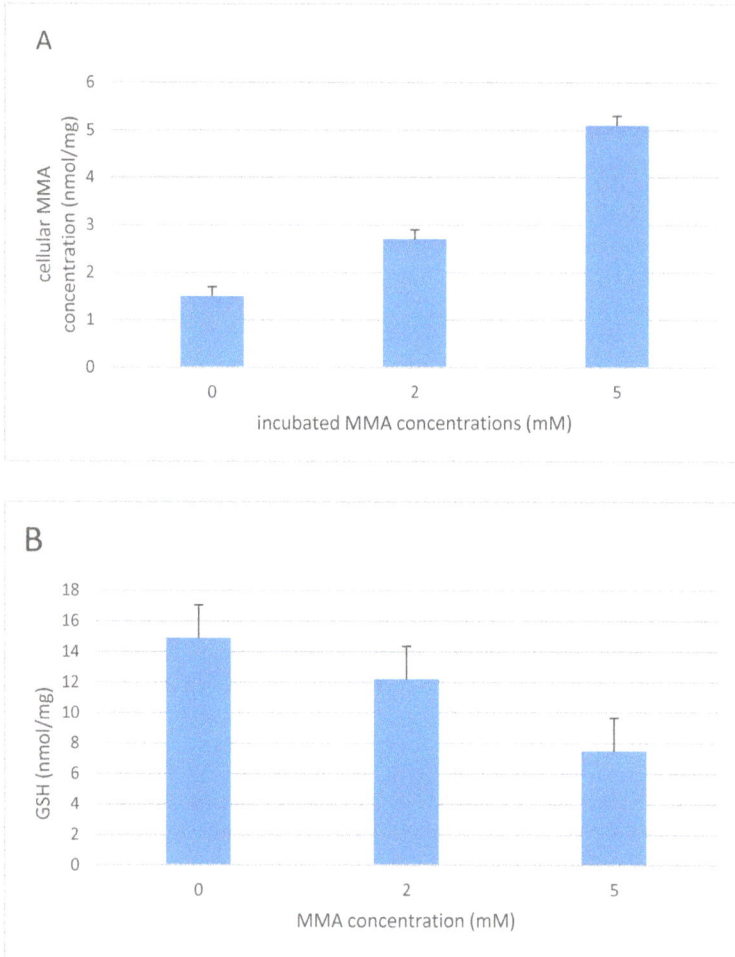

Figure 3. The concentration of cellular MMA (**A**) and GSH (**B**) in human neuroblastoma SHS-5Y cells following 10 days of incubation with MMA (0, 2 and 5 mM). Results are expressed as the mean and standard deviation of five determinations. MMA: Methylmalonic acid; GSH: Reduced glutathione. Previously unpublished data obtained by the authors of this paper with permission given for its publication.

The CoQ$_{10}$ status of fibroblasts from patients with MMA as the result of either L-methylmalonyl-CoA mutase deficiency or by defects in the synthesis of 5-deoxyadenosyl cobalamin were found to be significantly ($p < 0.05$) decreased compared with aged-matched controls [91]. Furthermore, a decreased level of CoQ$_{10}$ was also reported in a mouse model of this condition [88]. However, the level of Coenzyme Q$_9$, which is the predominant ubiquinone species in mice [92], was comparable to control levels discounting the possibility of impairment in ubiquinone biosynthesis [92]. The putative mechanisms that have been implicated in ETC dysfunction and oxidative stress in methylmalonic acidemia are outlined in Figure 4.

Antioxidant have been recommended as an adjunct therapy to treatment regime of methylmalonic acidemia patients; however, few studies have evaluated their potential therapeutic efficacy [93]. CoQ$_{10}$ treatment in conjunction with vitamin E was reported to improve visual acuity in a 15-year-old methylmalonic acidemia patient with optic neuropathy [94]. Although this report contrasts with a previous case study by Williams et al. (2009), which failed to demonstrate any evidence of visual improvement following CoQ$_{10}$ therapy, vitamin E was not included in the treatment regime of the latter patient [95]. A significant improvement in glomerular filtration rate was also reported in a mouse model of methylmalonic acidemia following co-treatment with CoQ$_{10}$ and vitamin E, suggesting that the beneficial effects of this therapy may not be restricted to the nervous system [88]. In light of evidence demonstrating a deficit in GSH status in methylmalonic acidemia [69,85,90], therapeutic strategies aimed at replenishing this tripeptide may prove beneficial to patients with this condition, although as far as the authors are aware, no such studies have been undertaken.

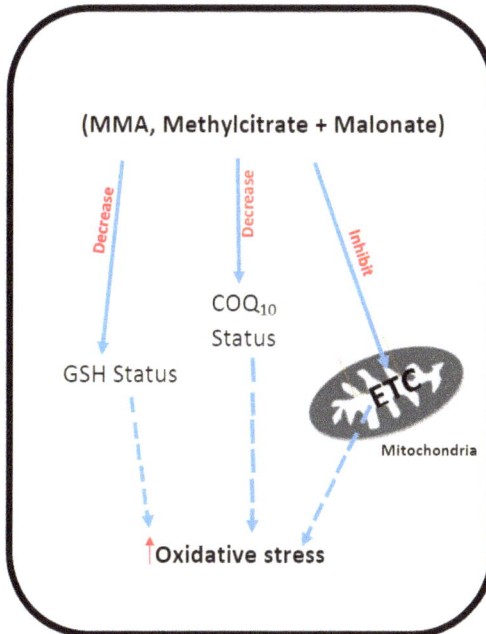

Figure 4. Putative mechanisms of oxidative stress generation and mitochondrial dysfunction in Methylmalonic acidemia. MMA: Methylmalonic acid; ETC: Mitochondrial electron transport chain; CoQ$_{10}$: Coenzyme Q$_{10}$; GSH: Reduced glutathione; ROS: Reactive oxygen species.

4. Peroxisomal Disorders

Peroxisome disorders are a heterogeneous group of rare metabolic diseases that can result from either a single peroxisomal enzyme deficiency (Refsum disease and X-linked adrenoleucodystrophy; X-ALD) [96] or as the result of a perturbation in the biogenesis of the organelle (Zellweger Syndrome spectrum disorders and rhizomelic chondrodysplasia punctate: RCDP) [97].

Zellweger Syndrome, neonatal adrenoleucodystrophy (ALD) and infantile Refsum disease all belong to the Zellweger spectrum of peroxisome biogenesis disorders and result from mutations in the PEX genes which encode superperoxins, proteins required for the import of protein into peroxisome, as well as the assembly of the organelle [97]. Patients with Zellweger Syndrome spectrum disorders lack functional peroxisomes and, as a result, have matrix proteins from the organelle mislocalized in the cytosol [97].

The disparity in the biochemical and clinical phenotypes of patients with Zellweger Syndrome spectrum peroxisomal disorders suggests that a large set of PEX mutations is likely to contribute to their pathogenesis, possibly via additional molecular mechanisms independent of their role in peroxisome biogenesis [98]. Clinical manifestation of Zellweger Syndrome group of disorders varies and includes liver disease, variable neurodevelopmental delay, retinopathy and sensorineural deafness. Patients with RCDP disorders present with skeletal dysplasia including proximal shortening of the limbs (rhizomelia) and punctate calcifications in cartilage present at birth, profound growth deficiency, cataracts and severe psychomotor defects [99]. ALD is the most frequent inherited leukodystrophy and peroxisomal disorder, characterized by an inflammatory cerebral demyelination, or a progressive axonopathy in the spinal cord, causing spastic paraparesis [100–102].

Peroxisomes have multiple biosynthetic functions and play a role in the β-oxidation of very-long-chain fatty acids (VLCFA) [103], prostaglandins, dicarboxylic acids, xenobiotic fatty acids and hydroxylated 5-β-cholestanoic acids [104]. In peroxisomal β-oxidation, the electrons liberated during the degradation of very-long-chain acyl-CoAs (VLCAC) are transferred directly to oxygen to generate hydrogen peroxide (H_2O_2) [105]. In addition, peroxisomes also contain a number of other ROS-generating enzymes such as Xanthine oxidase, which liberates H_2O_2 and superoxide during the catabolism of purines, and therefore these organelles are a major site of ROS generation within the cell [106]. In order to compensate for the abundance of ROS generated, the peroxisomes are well equipped with antioxidant defense systems, most notable of these being the catalase enzyme which converts H_2O_2 to oxygen and water [107]. Therefore, not unsurprisingly, peroxisomal disorders have been associated with oxidative stress, which is thought to contribute to disease progression [108,109]. The origin of oxidative stress in this disorder is thought to result from either an impairment of the peroxisomal antioxidant defense system and/or an accumulation of VLCFAs as well as VLCACs from the β-oxidation system of this organelle [110]. Peroxisomes also contain the inducible form of NOS, iNOS which catalyses the oxidation of *L*-arginine to citrulline and NO [111]. However, in peroxisomes this enzyme is thought to exist in its inactive monomeric form, whilst the cytosol contains both the monomeric and active homodimer forms of the enzyme [111]. Although, it has been speculated that under the circumstances peroxisomal iNOS may produce NO and this may be an explanation for the significant ($p = 0.022$) increase in NOx (marker of NO production) reported in the serum of patients with peroxisomal biogenesis disorders in the study by El-bassyouni et al. (2012) [109].

Peroxisome biogenesis defects resulting from PEX gene mutations may impair the import of matrix proteins such as catalase [112], impairing the antioxidant capacity of the organelle and rendering the cell more susceptible to free radical-induced oxidative damage [113]. This is also observed in aging cells where catalase is also mislocalized to the cytosol, resulting in an accumulation of cellular ROS with associated damage to protein, lipids and DNA [114]. In addition, peroxisomal biogenesis defects will also cause an impairment in the synthesis of the phospholipid antioxidant species, plasmalogens, which will compromise the ability of the cell to detoxify ROS [115,116].

Studies in fibroblasts from patients with X-ALD have revealed that hexacosanoic acid (C26:0), the VLCFA which accumulates in this disorder, causes a direct impairment of oxidative phosphorylation resulting in an increase in ROS generation and, consequently, the oxidation of mitochondrial DNA and proteins [117]. The mechanism by which C26:0 impairs oxidative phosphorylation in X-ALD is as yet uncertain, but may result from the ability of C26:0 to disrupt the physicochemical properties of the mitochondrial membrane [118]. The accumulated VLCFAs and VLCACs resulting from peroxisomal dysfunction may directly impair ETC function causing an increase in ROS generation from the chain as illustrated by the ability of phytanic acid, the C20 branch fatty acid that accumulates in Refsum disease to inhibit ETC complex I activity whilst concomitantly causing mitochondrial oxidative stress [11]. Since a number of studies have reported evidence of impaired oxidative phosphorylation in peroxisomal disorders [108,119–124], and the ETC is a major source of ROS generation [13], it does appear judicious to suggest that mitochondrial dysfunction may be a major contributor to the oxidative stress detected in these diseases [125–127]. The putative mechanisms that have been implicated in ETC dysfunction and oxidative stress in peroxisomal disorders are outlined in Figure 5.

In an animal model of X-ALD, oxidative damage, metabolic failure and axonal degeneration were reversed following treatment with the antioxidants, *n*-acetyl cysteine (NAC), α-lipoic acid, and α-tocopherol, providing proof of concept on the pivotal contribution of oxidative damage to disease pathogenesis in addition to illustrating the efficacy of antioxidant interventions [128–130]. A subsequent human study in X-ALD documented the ability of NAC treatment to replenish plasma GSH levels and ameliorate oxidative damage to proteins under in vitro conditions [131]. Evidence of decreased plasma CoQ_{10} status was reported in patients with a defect in peroxisome β-oxidation enzyme, D-bifunctional protein, which was associated with markers of increased oxidative stress [132]. It has been suggested that, based on the integral involvement of oxidative stress in the pathogenesis of peroxisomal disorders, the administration of antioxidants should be considered as a potential adjunct therapy for patients with these diseases [109,131,132].

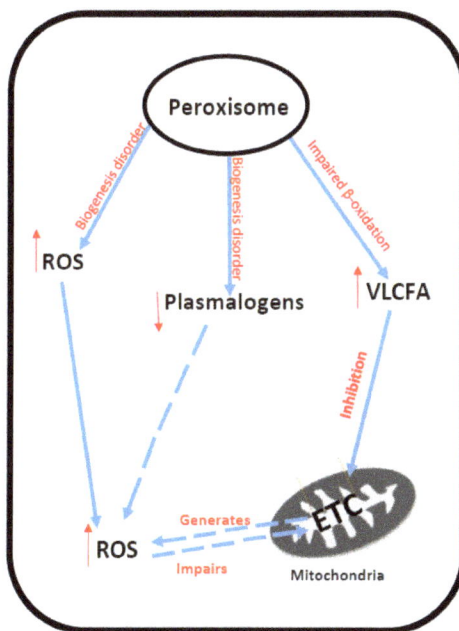

Figure 5. Putative mechanisms of oxidative stress generation and mitochondrial dysfunction in peroxisome disorders. VLCFA: Very-long-chain fatty acid; ROS: Reactive oxygen species; ETC: Mitochondrial electron transport chain.

5. Xeroderma Pigmentosum

Xeroderma pigmentosum (XP) is a rare condition characterized by an extreme sensitivity to ultraviolet (UV) rays from sunlight often causing skin burn. It affects the eyes and areas of skin exposed to the sun and is associated with an increased risk of skin cancer of lips, eyelids as well as brain tumors [133]. Patients with XP may present with neurological complications such as cerebellar ataxia, chorea, hearing loss, poor coordination, difficulty walking, movement problems, loss of intellectual function, difficulty swallowing and talking, and seizures [134]. Mutations in eight genes have been associated with XP.

XP is caused by autosomal recessive mutations in genes encoding for proteins that play a role in the nucleotide excision repair system (NER) [135]. There are eight complementation groups of XP, seven correspond to dysfunctional NER complex components, XP-A to -G, and one which affects DNA polymerase-η involved in translation synthesis and post-replication repair: XP-Variant (XP-V) [136]. XP cells lack a functional NER mechanism and so UV-induced bulky DNA lesions resulting from exposure to UV rays cannot be corrected. Unrepaired lesions occur in many genes, including those that encode cell growth and proliferation factors leading to a high rate of mutagenesis during DNA replication [137]. As well as the role of NER in UV-induced DNA damage repair, there is increasing support for the involvement of NER proteins in the repair of oxidative DNA damage [138,139]. Evidence of oxidative DNA damage in the form of free radical-induced DNA lesions such as 8-hydroxy-2-deoxygyanosine and cyclodeoxypurines have been detected in tumours and autopsied brains of neurological XP patients and animal models [140–142]. In XP-A, no evidence of DNA repair was reported in a study by Hayahi et al. (2008) and lesions were found to accumulate in patient cells [143]. The accumulation of such unrepaired DNA may be the source of internal carcinogenesis [144] and neuronal cell death, explaining the progressive neurodegeneration in XP [139,142].

Studies have been undertaken to elucidate the origin of oxidative stress in XP-C, the commonest form of this condition in Caucasians [145], and have indicated that the activation of the cytosolic enzyme, NADPH oxidase (NOx), may be a major contributor to ROS generation in this disease [146,147]. Furthermore, the NOx activation-induced ROS production has been suggested as a possible cause of the mitochondrial dysfunction detected in XP-C and possibly other forms of XP [146]. However, a study by Fang et al. (2014) suggested that impaired mitophagy may also contribute to the mitochondrial dysfunction observed in XP-A [148]. Interestingly, impaired mitophagy has also been associated with increased cellular ROS generation [149].

Evidence of mitochondrial dysfunction in XP has been indicated by mitochondrial DNA (mtDNA) deletions [150,151], ETC enzyme dysfunction [147,152] and morphological abnormalities [153,154]. Interestingly, studies have suggested that mitochondria are the major source of ROS generation in human XP-C cells and that mtDNA is the primary target for damage accumulation [152]. Since mtDNA lacks an NER, with repair being elicited through other mechanisms [155], this does suggest that mitochondrial abnormalities reported in XP are a secondary consequence of abnormalities in the nuclear DNA repair system.

Decreased activities of the antioxidant enzymes, catalase [156], SOD (superoxide dismutase) [143] and GSH-PX [152] have been reported in patient tissue and cell models of XP. In addition, decreased plasma CoQ_{10} levels were reported in patients with XP, with improvements in their daily activity being documented in a subset of these patients following CoQ_{10} supplementation [157]. The putative mechanisms that have been implicated in ETC dysfunction and oxidative stress in XP are outlined in Figure 6.

The authors are aware of no studies as yet to evaluate the therapeutic potential of antioxidants in the treatment of XP, although genetic strategies to ameliorate ROS generation are being considered [158].

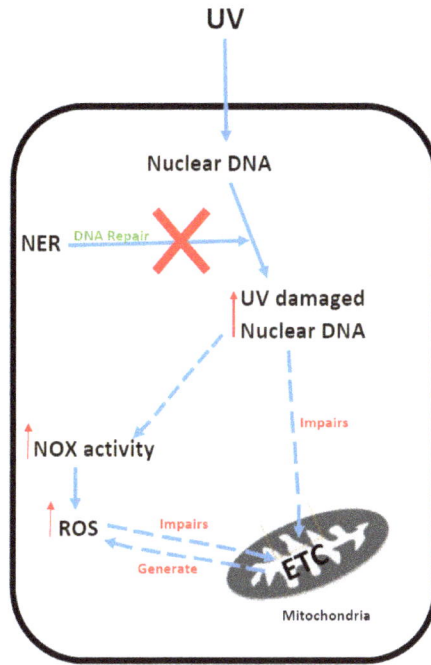

Figure 6. Putative mechanisms of oxidative stress generation and mitochondrial dysfunction in Xeroderma Pigmentation. UV: Ultraviolet radiation; NER: Nucleotide excision repair system; NOx: NADPH oxidase; ETC: Mitochondrial electron transport chain; ROS: Reactive oxygen species.

6. Sepsis

Sepsis is a chain of pathophysiological and metabolic reactions in response to infection, also identified as the systemic inflammatory response syndrome (SIRS) [16,159]. Clinically, sepsis may present in different forms depending on severity and include SIRS, septic shock and, in severe cases, multiple organ dysfunction syndrome including septic shock. The mortality rate is significantly increased (up to 34%) in patients with acute kidney injury versus 7% in patients without acute kidney injury [160]. Sepsis, together with hypoperfusion, is responsible for half of all cases of acute kidney injury in Intensive Care Units [161–163].

The precise pathophysiologic mechanisms underlying the development of multi-organ failure remain elusive [164]. However, the main causes of sepsis have been identified and include infection by gram-positive and gram-negative bacteria, fungi, or both. Concomitant factors, such as diabetes, transplantation, surgical intervention, chronic obstructive pulmonary disease, congestive heart failure, and renal disease increase a person's susceptibility to sepsis or aggravate their clinical score [16]. Additionally, an excessive degree of inflammation in response to the infectious insult triggers an activation of multiple downstream pathways. As a result, activated leukocytes release inflammatory cytokines such as tumour necrosis factor (TNF)-a, IL-1a, IL-1b, and IL-6, and chemokines such as IL-8 and KC that also impact upon the severity of sepsis [16]. Sepsis-related organ failure is associated with a significant morbidity and mortality [165,166] with long-term physical and neurocognitive problems affecting many survivors of critical illness [167,168].

It has been suggested for many years that both oxidative and nitrosative stress play a central role in the pathogenesis of sepsis and that ETC dysfunction may be an important causative factor in the multi-organ dysfunction associated with this condition [16]. Within the confines of this review, it would not be possible to outline all the mechanisms that have been proposed to account for the generation of free radical species or ETC dysfunction reported in sepsis. However, a paradigm will be offered based on the results of studies from the literature.

The inflammatory cytokines released by activated leukocytes following exposure to exo- and endo-toxins (most notably lipopolysaccharides; LPS) produced by gram-positive and -negative bacteria, respectively cause the overproduction of the RNS, NO, by the induction of iNOS activity in a number of vital organs including the heart and kidney as well as skeletal muscle [169–171]. LPS treatment has also been reported to induce NOx expression in renal cells resulting in a concomitant increase in ROS production [172].

The over-production of ROS and RNS by the cell may then result in the impairment of ETC function [2,12]. NO can combine with the ROS species, superoxide, to form the highly RNS species peroxynitrite, which can cause irreversible inhibition of the ETC [173]. Multiple ETC enzyme deficiencies have been reported in patients and animal models of sepsis [174,175]. As a consequence of ETC dysfunction, the mitochondria may also become a source of cellular ROS generation in sepsis, which can further exacerbate oxidative phosphorylation [172]. Decreased tissue ATP levels associated with ETC dysfunction have been linked to both organ failure and an increased mortality rate in sepsis [2]. The putative mechanisms that have been implicated in ETC dysfunction and oxidative stress in sepsis are outlined in Figure 7.

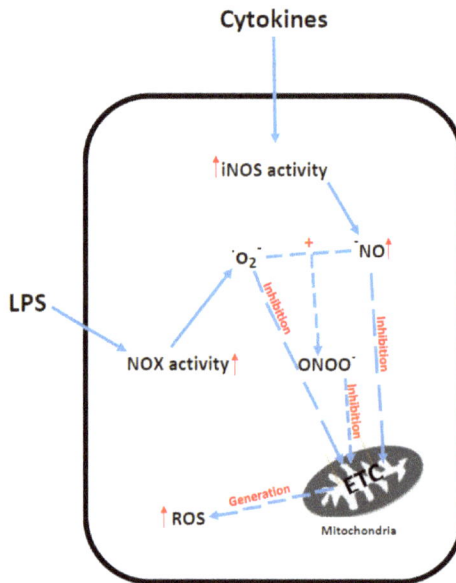

Figure 7. Putative mechanisms of oxidative stress and mitochondrial dysfunction in sepsis. iNos: Inducible nitric oxide synthase; NOx: NADPH oxidase; O$_2^-$: Superoxide; NO: Nitric oxide; ROS: Reactive oxygen species; ETC: Mitochondrial electron transport chain; ONOO$^-$: Peroxynitrite; LPS: Lipopolysaccharides.

In view of the ability of the ROS and RNS generated in sepsis to overwhelm cellular antioxidant defenses [2] and inhibit ETC function, a number of therapeutic strategies aimed at replenishing cellular antioxidant status have been investigated in patients and animal models of the disease [16]. The ability to replenish tissue GSH levels which have been found to be deficient in sepsis patients has been associated with clinical and biochemical improvement in animal models [176,177]. In addition, Se supplementation has been associated with increased GSH-PX activity [178] and a decreased mortality rate in septic patients [179]. It has been suggested, however, that mitochondrial-targeted antioxidants using compounds such as MitoQ or mitoVit E may offer novel therapeutic avenues to explore in the future [180]. Although, ubiquinol, the reduced form of CoQ_{10} has been reported to reduce peroxynitrite levels and attenuate the damage of the ETC associated with this RNS [181].

7. Conclusions

Mitochondrial dysfunction and oxidative stress are inextricably linked to the pathophysiology of a number of diseases as indicated by the disorders referred to in this review (Table S1). Within the mitochondria, the ETC is particularly vulnerable to ROS- and RNS-induced impairment either as the result of oxidative damage to the enzyme complexes, mtDNA or the mitochondrial membrane phospholipids [182,183]. Once impaired, the ETC then becomes a major source of ROS generation, resulting in further ETC dysfunction and compounding cellular oxidative stress [13,184]. The cell possesses a number of antioxidant defense systems to combat ROS and RNS; however, during pathological condition these defenses become overwhelmed, causing oxidative damage to the biomolecules of the cell and resulting in cellular and, consequently, organ dysfunction [9]. The use of appropriate antioxidants as an adjunct therapy may be particularly important in the treatment of diseases associated with oxidative stress, although treatment protocols have yet to be standardized or indeed instigated in some clinical centres. Since the mitochondria can make a major contribution to cellular oxidative stress in the disease state, antioxidant strategies which target this organelle may offer particular therapeutic potential [180]. Evidence of oxidative stress can be detected in patients through non-invasive means such as by assessing plasma antioxidant status or the stable end products of lipid, DNA or protein oxidation as alluded to in this review. For this reason, it is particularly important to engender some consensus with an aim to establishing a unified approach to monitoring evidence of oxidative stress in patients with diseases associated with this parameter together with the development of appropriate therapeutic strategies. It is also essential to take into account the possibility of nitrosative as well as oxidative stress in patients, the former being implicated as a major contributory factor to a number of chronic diseases and conditions [185]. In diseases which have been associated with oxidative stress and/or nitrosative stress, it is important to firstly, select a reliable and sensitive marker of this/these parameter(s) and, secondly, to choose an appropriate surrogate tissue for monitoring purposes. In the clinical studies outlined in this review, a number of different end-point markers were used to monitor evidence of oxidative stress and the sensitivity and specificity of these markers may vary [31,33,34,108,109]. In addition, in view of the sophistication and/or laboriousness of a number of these methods, it may be difficult to translate them into a clinical laboratory setting. Therefore, the ability to assess a number of markers of both oxidative and nitrosative stress together in a large-scale panel by either Liquid chromatography-mass spectrometry and/or ELISA as suggested by Frijhoff et al. (2015) [186] may overcome the problems of sensitivity/specificity as well as decrease the assay time for these determinations. The surrogate which is generally used to measure levels of ROS, RNS as well as antioxidant status in clinical studies [33,34,132,178,179] is plasma/serum; however, it is uncertain whether levels of these parameters in plasma/serum reflect those of tissue. This is certainly the case for CoQ_{10}, and plasma CoQ_{10} status has been reported not to reflect that of muscle [187]. Blood mononuclear cells or lymphocytes have been suggested as appropriate alternative surrogates for this determination in patients [43,44]. Furthermore, lymphocytes have also been suggested as an appropriate surrogate to assess intracellular GSH status in patients [188]. Therefore, the assessment of ROS, RNS or antioxidant status in white blood cells rather than plasma/serum in future clinical studies may give a better indicator of these parameters in tissue. A compound to consider for future treatment strategies in diseases associated with mitochondrial dysfunction and oxidative and nitrosative stress is the synthetic quinone, EPI-743 [188]. This compound has demonstrated some therapeutic efficacy in the treatment of patients with primary mitochondrial disorders by its ability to replenish cellular GSH status as well as its proposed capacity to interact with the transcription factor, nuclear factor E2-related factor 2 (Nrf2) which regulates both the expression of antioxidant proteins as well as cellular energy metabolism [189,190]. However, one reason why antioxidants in general have been relatively ineffective in treating either acute or chronic diseases is that they are only targeting oxidative stress and do not take into account nitrosative stress, which can make a major contribution to disease pathophysiology in a number of disorders [2,12]. Therefore, antioxidant treatments that target both oxidative as well as nitrosative stress are important considerations for future therapeutic strategies.

Acknowledgments: All authors would like to thank all patients for their contribution into research.

Author Contributions: K.M.S. and I.P.H. have completed literature search and wrote the manuscript. R.H., S.R., A.M., J.B., D.S. helped with the literature search and drew figures.

Conflicts of Interest: The authors declare no conflict of interest.

Abbreviations

ALD	adrenoleuklodystrophy
CSF	cerebrospinal fluid
ETC	electron transport chain
CoQ_{10}	coenzyme C_{10}
Fe	iron
GSH	glutathione
GSH-PX	glutathione peroxidase
GSSG	the oxidised form of GSH
HMG-CoA	3-hydroxy-3-methylglutaryl-CoA
H_2O_2	hydrogen peroxide
LPS	lipopolysaccharides
Inos	induction of nitric oxide synthase
MCM	L-methylmalonyl-CoA mutase
MMA	methylmalonic acid
mtDNA	mitochondrial DNA
NADPH	nicotinamide adenine dinucleotide phosphate-oxidase
NAC	*N*-acetyl-cysteine cysteine
NER	nucleotide excision repair system
NOx	NADPH oxidase
Phe	phenylalanine
PKU	Phenyloketonuria
RCDP	rhizomeric chondrodysplasia punctate
SOD	superoxide dismutase
SIRS	systemic inflammatory response syndrome
ROS	reactive oxygen species
RNS	reactive nitrogen species
Se	selenium
VLCFA	very-long-chain fatty acids
VLCAC	very-long-chain acyl-CoAs
TBAR	thiobarbituric acid-reactive species
XP	Xeroderma pigmentosum

References

1. Sosa, V.; Moline, T.; Somoza, R.; Paciucci, R.; Kondoh, H.; Leonard, L. Oxidative stress and cancer: An overview. *Aging Res. Rev.* **2013**, *12*, 376–390. [CrossRef] [PubMed]
2. Brealey, D.; Brand, M.; Hargreaves, I.; Heales, S.; Land, J.; Smolenski, R.; Davis, N.A.; Cooper, C.E.; Singer, M. Association between mitochondrial dysfunction and severity and outcome of septic shock. *Lancet* **2002**, *360*, 219–223. [CrossRef]

3. Heales, S.J.; Bolanos, J.P.; Brand, M.P.; Clark, J.B.; Land, J.M. Mitochondrial damage: An important feature in a number of inborn errors of metabolism? *J. Inherit. Metab. Dis.* **1996**, *19*, 140–142. [CrossRef] [PubMed]

4. Moyano, D.; Vilaseca, M.A.; Pineda, M.; Campistol, J.; Vernet, A.; Poo, P.; Artuch, R.; Sierra, C. Tocopherol in inborn errors of intermediary metabolism. *Clin. Chim. Acta* **1997**, *263*, 147–155. [CrossRef]

5. Sierra, C.; Vilaseca, M.A.; Moyano, D.; Brandi, N.; Campistol, J.; Lambruschini, N.; Cambra, F.J.; Deulofeu, R.; Mira, A. Antioxidant status in hyperphenylalaninemia. *Clin. Chim. Acta* **1998**, *276*, 1–9. [CrossRef]

6. Artuch, R.; Vilaseca, M.A.; Moreno, J.; Lambruschini, N.; Cambra, F.J.; Campistol, J. Decreased serum ubiquinone-10 concentrations in phenylketonuria. *Am. J. Clin. Nutr.* **1999**, *70*, 892–895. [PubMed]

7. Fisberg, R.M.; Silva-Femandes, M.E.; Fisberg, M.; Schmidt, B.J. Plasma zinc, copper, and erythrocyte superoxide dismutase in children with phenylketonuria. *Nutrition* **1999**, *15*, 449–452. [CrossRef]

8. Van Bakel, M.M.E.; Printzen, G.; Wermuth, B.; Wiesmann, U.N. Antioxidant and thyroid hormone status in selenium-deficient phenylketonuric and hyperphenylalaninemic patients. *Am. J. Clin. Nutr.* **2000**, *72*, 976–981. [PubMed]

9. Rani, V.; Deep, G.; Singh, R.K. Oxidative stress and metabolic disorders: Pathogenesis and therapeutic strategies. *Life Sci.* **2016**, *148*, 183–193. [CrossRef] [PubMed]

10. Kolker, S.; Schwab, M.; Horster, F.; Sauer, S.; Hinz, A.; Wolf, N.I.; Mayatepek, E.; Hoffmann, G.F.; Smeitink, J.A.M.; Okun, J.G. Methylmalonic acid, a biochemical hallmark of methylmalonic acidurias but no inhibitor of mitochondrial respiratory chain. *J. Biol. Chem.* **2003**, *278*, 47388–47393. [CrossRef] [PubMed]

11. Schonfeld, P.; Reiser, G. Rotenone-like action of the branch chain phytanic acid induces oxidative stress in mitochondria. *J. Biol. Chem.* **2006**, *281*, 7136–7142. [CrossRef] [PubMed]

12. Zapelini, P.H.; Retin, G.T.; Cardoso, M.R.; Ritter, C.; Ritter, C.; Klamt, F.; Moreira, J.C.; Streck, E.L.; Dal-Pizzol, F. Antioxidant treatment reverses mitochondrial dysfunction in a sepsis animal model. *Mitochondrion* **2008**, *8*, 211–218. [CrossRef] [PubMed]

13. Paradies, G.; Ruggiero, F.M.; Petrosillo, G.; Quagliariello, E. Peroxidative damage to cardiac mitochondria: Cytochrome c oxidase and cardiolipin alterations. *FEBS Lett.* **1998**, *424*, 155–158. [CrossRef]

14. Murphy, M.P. How mitochondria produce reactive oxygen species. *Biochem. J.* **2009**, *417*, 1–13. [CrossRef] [PubMed]

15. Rocha, C.R.; Martins, M.J. Oxidative stress in phenylketonuria: Future directions. *J. Inherit. Metab. Dis.* **2012**, *35*, 381–398. [CrossRef] [PubMed]

16. Prauchner, C.A. Oxidative stress in sepsis: Pathophysiological implications justifying antioxidant co-therapy. *Burns* **2016**, *43*, 471–485. [CrossRef] [PubMed]

17. Scriver, C.R.; Kaufman, S.; Eisensmith, R.C.; Woo, S.L.C. The phenylalaninemias. In *The Metabolic and Molecular Bases of Inherited Disease*, 7th ed.; Scriver, C.R., Beaudet, A.L., Valle, D., Sly, W.S., Eds.; McGraw-Hill: New York, NY, USA, 1995; pp. 1015–1075.

18. Williams, R.A.; Mamotte, C.D.; Burnett, J.R. Phenylketonuria: An inborn error of phenylalanine metabolism. *Clin. Biochem. Rev.* **2008**, *29*, 31–41. [PubMed]

19. Velema, M.; Boot, E.; Engelen, M.; Hollak, C. Parkinsonism in phenylketonuria: A consequence of dopamine depletion? *JIMD Rep.* **2015**, *20*, 35–38. [PubMed]

20. Krause, W.; Halminski, M.; McDonald, L.; Demure, P.; Salvo, R.; Friedes, S.R.; Elsas, L. Biochemical and neuropsychological effects of elevated plasma phenylalanine in patients with treated phenylketonuria. *J. Clin. Investig.* **1985**, *75*, 40–48. [CrossRef] [PubMed]

21. Ushakova, G.A.; Gubkina, H.A.; Kachur, V.A.; Lepekhin, E.A. Effect of experimental hyperphenylalaninemia on the postnatal rat brain. *Int. J. Dev. Neurosci.* **1997**, *15*, 29–36. [CrossRef]

22. Ercal, N.; Aykin-Burns, N.; Gurer-Orhan, H.; McDonald, J.D. Oxidative stress in a phenylketonuria animal model. *Free Radic. Biol. Med.* **2002**, *32*, 906–911. [CrossRef]

23. Pietz, J. Neurological aspects of adult phenylketonuria. *Curr. Opin. Neurol.* **1998**, *11*, 679–688. [CrossRef] [PubMed]

24. Smith, I.; Knowles, J. Behaviour in early treated phenylketonuria: A systematic review. *Eur. J. Pediatr.* **2000**, *159* (Suppl. 2), S89–S93. [CrossRef] [PubMed]

25. Berry, S.A.; Brown, C.; Grant, M.; Greene, C.L.; Jurecki, E.; Koch, J.; Moseley, K.; Suter, R.; van Calcar, S.C.; Wiles, J.; et al. Newborn screening 50 years later: Access issues faced by adults with PKU. *Genet. Med.* **2013**, *15*, 591–599. [CrossRef] [PubMed]

26. Weglage, J.; Pietsch, M.; Funders, B.; Koch, H.G.; Ullrich, K. Neurological findings in early treated phenylketonuria. *Acta Paediatr.* **1995**, *84*, 411–415. [CrossRef] [PubMed]

27. Dyer, C.A. Comments on the neuropathology of phenylketonuria. *Eur. J. Pediatr.* **2000**, *159* (Suppl. 2), S107–S108. [CrossRef] [PubMed]

28. Huttenlocher, P.R. The neuropathology of phenylketonuria: Human and animal studies. *Eur. J. Pediatr.* **2000**, *159* (Suppl. 2), S102–S106. [CrossRef] [PubMed]

29. Rocha, J.C.; Martel, F. Large neutral amino acids supplementation in phenylketonuric patients. *J. Inherit. Metab. Dis.* **2009**, *32*, 472–480. [CrossRef] [PubMed]

30. Duarte, J.M.; Schuck, P.F.; Wenk, G.L.; Ferreira, G.C. Metabolic disturbances in diseases with neurological involvement. *Aging Dis.* **2013**, *5*, 238–255. [PubMed]

31. Sirtori, L.R.; Dutra-Filho, C.S.; Fitarelli, D.; Sitta, A.; Haeser, A.; Barschak, A.G.; Wajner, M.; Coelho, D.M.; Llesuy, S.; Bello-Klein, A.; et al. Oxidative stressvin patients with phenylketonuria. *Biochim. Biophys.* **2005**, *1740*, 68–73. [CrossRef] [PubMed]

32. Sitta, A.; Barschak, A.G.; Deon, M.; Barden, A.T.; Biancini, G.B.; Vargas, P.R.; de Souza, C.F.; Netto, C.; Wajner, M.; Vargas, C.R. Effect of short- and long-term exposition to high phenylalanine blood levels on oxidative damage in phenylketonuric patients. *Int. J. Dev. Neurosci.* **2009**, *27*, 243–247. [CrossRef] [PubMed]

33. Sanayama, Y.; Nagasaka, H.; Takayanagi, M.; Ohura, T.; Sakamoto, O.; Ito, T.; Ishige-Wada, M.; Usui, H.; Yoshino, M.; Ohtake, A.; et al. Experimental evidence that phenylalanine is strongly associated to oxidative stress in adolescents and adults with phenylketonuria. *Mol. Genet. Metab.* **2011**, *103*, 220–225. [CrossRef] [PubMed]

34. Schulpis, K.H.; Tsakiris, S.; Traeger-Synodinos, J.; Papassotiriou, I. Low total antioxidant status is implicated with high 8-hydroxy-2-deoxyguanosine serum concentrations in phenylketonuria. *Clin. Biochem.* **2005**, *38*, 239–242. [CrossRef] [PubMed]

35. Sitta, A.; Vanzin, C.S.; Biancini, G.B.; Manfredoni, V.; De Oliviera, A.B.; Wayhs, C.A.Y.; Ribas, G.O.S.; Giuliani, L.; Schwartz, I.V.D.; Bohrer, D.; et al. Evidence that L-carnitine and selenium supplementation reduces oxidative stress in phenylketonuric patients. *Cell Mol. Neurobiol.* **2011**, *31*, 429–436. [CrossRef] [PubMed]

36. Hatanaka, N.; Nakaden, H.; Yamamoto, Y.; Matsuo, S.; Fujikawa, T.; Matsusue, S. Selenium Kinetics and Changes in Glutathione Peroxidase Activities in Patients Receiving Long-Term Parenteral Nutrition and Effects of Supplementation With Selenite. *Nutrition* **2000**, *16*, 22–26. [CrossRef]

37. Kienzle-Hagen, M.E.; Pederzolli, C.D.; Sgaravatti, A.M.; Bridi, R.; Wajner, R.; Wannmacher, C.M.; Wyse, A.T.; Dutra-Filho, C.S. Experimental hyperphenylalaninemia provokes oxidative stress in rat brain. *Biochim. Biophys. Acta* **2000**, *1586*, 344–352. [CrossRef]

38. Sitta, A.; Barschak, A.G.; Deon, M.; de Mari, J.F.; Barden, A.T.; Vanzin, C.S.; Biancini, G.B.; Schwartz, I.V.; Wajner, M.; Vargas, C.R. L-Carnitine blood levels and oxidative stress in treated phenylketonuria patients. *Cell Mol. Neurobiol.* **2009**, *29*, 211–218. [CrossRef] [PubMed]

39. Preissler, T.; Bristot, I.J.; Costa, B.M.L.; Fernandes, E.K.; Rieger, E.; Bortoluzzi, V.T.; de Franceschi, I.D.; Dudra-Filho, C.S.; Moreira, J.C.F.; Wannmacher, C.M.D. Phenylalanine induces oxidative stress and decreases the viability of rat astrocytes: Possible relevance for the pathophysiology of neurodegeneration in phenylketonuria. *Metab. Brain Dis.* **2016**, *31*, 529–537. [CrossRef] [PubMed]

40. Schuck, P.F.; Malgarin, F.; Cararo, J.H.; Cardoso, F.; Streck, E.L.; Ferreira, G.C. Phenylketonuria pathophysiology: On the role of the metabolic alterations. *Aging Dis.* **2015**, *6*, 390–399. [CrossRef]

41. Castillo, M.; Martinez-Cayuela, M.; Zafra, M.F.; Garcia-Peregrin, E. Effect of phenylalanine derivatives on the main regulatory enzymes of hepatic cholesterogenesis. *Mol. Cell Biochem.* **1991**, *105*, 21–25. [CrossRef] [PubMed]

42. Shefer, S.; Tint, G.S.; Jean-Guillaume, D.; Daikhin, E.; Kendler, A.; Nguyen, L.B.; Yudkoff, M.; Dyer, C.A. Is there a relationship between 3-hydroxy-3-methylglutaryl coenzyme A reductase activity and forebrain pathology in the PKU mouse? *J. Neurosci. Res.* **2000**, *61*, 549–563. [CrossRef]

43. Colome, C.; Artuch, R.; Vilaseca, M.A.; Sierra, C.; Brandi, N.; Cambra, F.J.; Lambruschini, N.; Campistol, J. Ubiquinone-10 content in lymphocytes of phenylketonuric patients. *Clin. Biochem.* **2002**, *35*, 81–84. [CrossRef]

44. Hargreaves, I.P.; Heales, S.J.; Briddon, A.; Land, J.M.; Lee, P.J. Mononuclear cell coenzyme Q (coq) Concentration and mitochondrial respiratory chain succinate cytochrome C reductase (complex li-iii) activity in phenyloketonuric patiens. *J. Inher. Metab. Dis.* **2002**, *25*, 18. [CrossRef]

45. Rech, V.C.; Feksa, L.R.; Dudra-Filho, C.S.; Wyse, A.T.D.S.; Wajner, M.; Wannmacher, C.M.D. Inhibition of the mitochondrial respiratory chain by phenylalanine in rat cerebral; cortex. *Neurochem. Res.* **2002**, *27*, 353–357. [CrossRef] [PubMed]

46. Kyprianou, N.; Murphy, E.; Lee, P.; Hargreaves, I. Assessment of mitochondrial respiratory chain function in hyperphenylalaninemia. *J. Inherit. Metab.* **2009**, *32*, 289–296. [CrossRef] [PubMed]

47. Przyrembel, H.; Bremer, H.J. Nutrition, physical growth, and bone density in treated phenylketonuria. *Eur. J. Pediatr.* **2000**, *159* (Suppl. 2), S129–S135. [CrossRef] [PubMed]

48. Giovannini, M.; Verduci, E.; Salvatici, E.; Fiori, L.; Riva, E. Phenylketonuria: Dietary and therapeutic challenges. *J. Inherit. Metab. Dis.* **2007**, *30*, 145–152. [CrossRef] [PubMed]

49. Poustie, V.J.; Wildgoose, J. Dietary interventions for phenylketonuria. *Cochrane Libr.* 2010. [CrossRef]

50. McMurry, M.P.; Chan, G.M.; Leonard, C.O.; Ernst, S.L. Bone mineral status in children with phenylketonuria—Relationship to nutritional intake and phenylalanine control. *Am. J. Clin. Nutr.* **1992**, *55*, 997–1004. [PubMed]

51. Wilke, B.C.; Vidailhet, M.; Richard, M.J.; Ducros, V.; Arnaud, J.; Favier, A. Trace elements balance in treated phenylketonuria children. Consequences of selenium deficiency on lipid peroxidation. *Arch. Latinoam. Nutr.* **1993**, *43*, 119–122. [PubMed]

52. Ragsdale, S. Metal-carbon bonds in enzymes and cofactors. *Coord. Chem. Rev.* **2010**, *254*, 1948–1949. [CrossRef] [PubMed]

53. Robert, M.; Rocha, J.C.; van Rijn, M.; Ahring, K.; Bélanger-Quintana, A.; MacDonald, A.; Dokoupil, K.; Ozel, H.G.; Lammardo, A.M.; Goyens, P.; et al. Micronutrient status in phenylketonuria. *Mol. Genet. Metab.* **2013**, *110*, S6–S17. [CrossRef] [PubMed]

54. Bohler, H.; Ulrich, K.; Endres, W.; Behbehani, A.W.; Wendel, U. Inadequate iron availability as a possible cause of low serum carnitine concentrations in patients with phenylketonuria. *Eur. J. Pediatr.* **1991**, *15*, 425–428. [CrossRef]

55. Gullcin, I. Antioxidant and antiradical activities of L-carnitine. *Life Sci.* **2006**, *78*, 803–811. [CrossRef] [PubMed]

56. Ribas, G.S.; Sitta, A.; Wajner, M.; Vargas, C.R. Oxidative stress in phenylketonuria: What is the evidence? *Cell Mol. Neurobiol.* **2011**, *31*, 653–662. [CrossRef] [PubMed]

57. Dobson, C.M.; Wai, T.; Leclerc, D.; Wilson, A.; Wu, X.; Dore, C.; Hudson, T.; Rosenblatt, D.S.; Gravel, R.A. Identification of the gene responsible for the cblA complementation group of vitamin B12responsive methylmalonic acidemia based on analysis of prokaryotic gene arrangements. *Proc. Natl. Acad. Sci. USA* **2002**, *99*, 15554–15559. [CrossRef] [PubMed]

58. Fenton, W.A.; Gravel, R.A.A.; Rosenblatt, D.S. Disorders of propionate and methylmalonate metabolism. In *The Metabolic and Molecular Bases of Inherited Disease*; Scriver, C.R., Beaudet, A.L., Sky, W.S., Valle, D., Eds.; McGraw-Hill: New York, NY, USA, 2011; pp. 2165–2193.

59. Treacy, E.; Arbour, L.; Chessex, P.; Graham, G.; Kasprzak, L.; Casey, K.; Bell, L.; Mamer, O.; Scriver, C.R. Glutathione deficiency as a complication of methylmalonic acidemia: Response to high doses of ascorbate. *J. Pediatr.* **1996**, *129*, 445–448. [CrossRef]

60. Hoffmann, G.F.; Meier-Augenstein, W.; Stocker, S.; Surtees, R.; Rating, D.; Nyhan, W. Physiology and pathophysiology of organic acids in cerebrospinal fluid. *J. Inherit. Metab. Dis.* **1993**, *16*, 648–669. [CrossRef] [PubMed]

61. Manoli, I.; Sloan, J.L.; Venditti, C.P. Isolated methylmalonic acidemia. In *Genereviews [Internet]*; Pagon, R.A., Adam, M.P., Ardinger, H.H., Wallace, S.E., Amemiya, A., Bean, L.J.H., Bird, T.D., Ledbetter, N., Mefford, H.C., Smith, R.J.H., et al., Eds.; University of Washington: Seattle, WA, USA, 2016; pp. 1993–2017.

62. Hayasaka, K.; Metoki, K.; Satoh, T.; Narisawa, K.; Tada, K.; Kawakami, T.; Matsuo, N.; Aoki, T. Comparison of cytosolic and mitochondrial enzyme alterations in the livers of propionic or methylmalonic acidemia: A reduction of cytochrome oxidase activity. *Tohoku J. Exp. Med.* **1982**, *137*, 329–333. [CrossRef] [PubMed]

63. Richard, E.; Alvarez-Barrientos, A.; Perez, B.; Desviat, L.R.; Ugarte, M. Methylmalonic acidaemia leads to increased production reactive oxygen species and induction of apoptosis through the mitocondrial/caspase pathway. *J. Pathol.* **2007**, *213*, 453–461. [CrossRef] [PubMed]

64. Lindblad, B.; Lindblad, B.S.; Olin, P.; Svanberg, B.; Zetterstrom, R. Methylmalonic academia. A disorder associated with acidosis, hyperlycaemia, and hyperlactatemia. *Acta Paediatr. Scand.* **1968**, *57*, 417–424. [PubMed]

65. Okun, J.C.; Horster, F.; Farkas, L.; Feyh, P.; Hinz, A.; Sauer, S.; Hoffman, G.F.; Unisicker, K.; Mayatepek, E.; Kolker, S. Neurodegeneration in Methylmalonic Aciduria Involves Inhibition of Complex II and the Tricarboxylic Acid Cycle, and Synergistically Acting Excitotoxicity. *J. Biol. Chem.* **2002**, *277*, 14674–14680. [CrossRef] [PubMed]

66. Marisco, P.C.; Ribeiro, M.C.; Bonini, J.S.; Lima, T.T.; Mann, K.C.; Brenner, G.M.; Dutra-Filho, C.S.; Mello, C.F. Ammonia potentiates methylmalonic acid-induced convulsions and TBARS production. *Exp. Neurol.* **2003**, *182*, 455–460. [CrossRef]

67. Krahenbuhl, S.; Chang, M.; Brass, E.P.; Hoppel, C.L. Decreased activities of ubiquinol; ferricytochrome c oxidoreductase (complex III) and ferrocytochrome c: Oxygen oxidoreductase (complex IV) in liver mitochondria from rats with hydroxycobalamin[C-lactam]-induced methylmalonic aciduria. *J. Biol. Chem.* **1991**, *266*, 20998–21003. [PubMed]

68. Pettenuzzo, L.F.; Ferreira, G.D.C.; Schmidt, A.L.; Dutra-Filho, C.S.; Wyse, A.T.S.; Wajner, M. Differential inhibitory effects of methylmalonic acid on respiratory chain complex activities in rat tissues. *Int. J. Dev. Neurosci.* **2006**, *24*, 45–52. [CrossRef] [PubMed]

69. Chandler, R.J.; Zerfas, P.M.; Shanske, S.; Sloan, J.; Hoffman, V.; DiMauro, S.; Venditti, C.P. Mitochondrial dysfunction in mut methylmalonic academia. *FASEB J.* **2009**, *23*, 1252–1261. [CrossRef] [PubMed]

70. De Keyzer, Y.; Valayannopoulos, V.; Benoist, J.F.; Batteux, F.; Lacaille, F.; Hubert, L.; Chretien, D.; Chadefeaux-Vekemans, B.; Niaudet, P.; Touati, G.; et al. Multiple OXPHOS deficiency in the liver, kidney, heart, and skeletal muscle of patients with methylmalonic aciduria and propionic aciduria. *Pediatr. Res.* **2009**, *66*, 91–95. [CrossRef] [PubMed]

71. Kashtan, C.E.; Abousedira, M.; Rozen, S.; Manivel, J.C.; McCann, M.; Tuchman, M. Chronic administration of methylmaloic acid (MMA) to rats causes proteinuria and renal tubular injury (abstract). *Pediatr. Res.* **1998**, *43*, 309. [CrossRef]

72. Zsengeller, Z.K.; Alkinovic, N.; Teot, L.A.; Korson, M.; Rodig, N.; Sloan, J.L.; Venditti, C.P.; Berry, G.T.; Rosen, S. Methylmalonic academia: A megamitochondrial disorder affecting the kidney. *Pediatr. Nephrol.* **2014**, *29*, 2139–2146. [CrossRef] [PubMed]

73. De Souza, C.; Piesowicz, A.T.; Brett, E.M.; Leonard, J.V. Focal changes in the globi pallidi associated with neurological dysfunction in methylmalonic academia. *Neuropediatrics* **1989**, *20*, 199–201. [CrossRef] [PubMed]

74. Brismar, J.; Ozand, P.T. CT and MR of the brain in disorders of propionate and methylmalonate metabolism. *Am. J. Neuroradiol.* **1994**, *15*, 1459–1473. [PubMed]

75. Larnaout, A.; Mongalgi, M.A.; Kaabachi, N.; Khiari, D.; Debbabi, A.; Mebazza, A.; Ben Hamida, M.; Hentati, F. Methylmalonic acidaemia with bilateral globus pallidus involvement: A neuropathological study. *J. Inherit. Metab. Dis.* **1998**, *21*, 639–644. [CrossRef] [PubMed]

76. Martin, J.J.; van de Vyver, F.L.; Scholte, H.R.; Roodhooft, A.M.; Martin, C.C.; Luyt-Houwen, I.E.J. Defect in succinate oxidation by isolated muscle mitochondria in a patient with symmetrical lesions in the bassel ganglia. *J. Neurol. Sci.* **1988**, *84*, 189–200.

77. Trinh, B.C.; Melhem, E.R.; Barker, P.B. Multi-slice proton MR spectroscopy and diffusion-weighted imaging in methylmalonic acidemia: Report of two cases and review of the literature. *Am. J. Neuroradiol.* **2001**, *22*, 831–833. [PubMed]

78. Heidenreich, R.; Natowicz, M.; Hainline, B.E.; Berman, P.; Kelley, R.I.; Hillman, R.E.; Berry, G.T. Acute extrapyramidal syndrome in methylmalonic acidemia: "Metabolic stroke" involving the globus pallidus. *J. Pediatr.* **1988**, *113*, 1022–1027. [CrossRef]

79. Brusque, A.M.; Borba Rosa, R.; Schuck, P.F.; Dalcin, K.B.; Ribeiro, C.A.J.; Silva, C.G.; Wannmacher, C.M.D.; Dutra-Filho, C.S.; Wyse, A.T.S.; Briones, P.; et al. Inhibition of the mitochondrial respiratory chain complex activities in rat cerebral cortex by methylmalonic acid. *Neurochem. Int.* **2002**, *40*, 593–601. [CrossRef]

80. Kolker, S.; Okun, J.G. Methylmalonic acid—An endogenous toxin? *Cell Mol. Life Sci.* **2005**, *62*, 621–624. [CrossRef] [PubMed]

81. McLaughlin, B.A.; Nelson, D.; Silver, I.A.; Erecinska, M.; Chesselet, M.F. Methylmalonate toxicity in primary neuronal cultures. *Neuroscience* **1998**, *86*, 279–290. [CrossRef]

82. Calabresi, P.; Gubellini, P.; Picconi, B.; Centonze, D.; Pisani, A.; Bonsi, P.; Greengard, P.; Hipskind, R.A.; Borrelli, E.; Bernardi, G. Inhibition of mitochondrial complex II induces a long-term potentiation of NMDA-mediated synaptic excitation in the striatum requiring endogenous dopamine. *J. Neurosci.* **2001**, *21*, 5110–5120. [PubMed]

83. Royes, L.F.; Fighera, M.R.; Furian, A.F.; Oliveira, M.S.; da Silva, L.G.; Malfatti, C.R.; Schneider, P.H.; Braga, A.L.; Wajner, M.; Mello, C.F. Creatine protects against the convulsive behavior and lactate production elicited by the intrastriatal injection of methylmalonate. *Neuroscience* **2003**, *118*, 1079–1090.

84. Fleck, J.; Ribeiro, M.C.; Schneider, C.M.; Sinhorin, V.D.; Rubin, M.A.; Mello, C.F. Intrastriatal malonate administration induces convulsive behaviour in rats. *J. Inherit. Metab. Dis.* **2004**, *27*, 211–219. [CrossRef] [PubMed]

85. Atkuri, K.R.; Cowan, T.M.; Kwan, T.; Ng, A.; Herzenberg, L.A.; Herzenberg, LA.; Enns, G.M. Inherited disorders affecting mitochondrial function are associated with glutathione deficiency and hypocitrullinemia. *Proc. Natl. Acad. Sci. USA* **2009**, *106*, 3941–3944. [CrossRef] [PubMed]

86. Fontella, F.; Pulronick, V.; Gassen, E.; Wannmacher, C.M.D.; Klein, A.B.; Wajner, M.; Dutra-Filho, C.S. Propionic and *l*-methylmalonic acids induce oxidative stress in brain of young rats. *Neuroreport* **2000**, *28*, 541–544. [CrossRef]

87. Manoli, I.; Myles, J.G.; Sloan, J.L.; Carrillo-Carrasco, N.; Morava, E.; Strauss, K.A.; Morton, H.; Venditti, C.P. A critical reappraisal of dietary practices in methylmalonic academia raises concerns about the safety of medical foods. Part 2: Cobalamin C deficiency. *Genet. Med.* **2016**, *18*, 396–404. [PubMed]

88. Manoli, I.; Sysol, J.R.; Li, L.; Houillier, P.; Garone, C.; Wang, C.; Zerfas, P.M.; Cusmano-Ozog, K.; Young, S.; Trivedi, N.S.; et al. Targeting proximal tubule mitochondrial dysfunction attenuates the renal disease of methylmalonic academia. *Proc. Natl. Acad. Sci. USA* **2013**, *110*, 13552–13557. [CrossRef] [PubMed]

89. Hargreaves, I.P.; Sheena, Y.; Land, J.M.; Heales, S.J. Glutathione deficiency in patients with mitochondrial disease: Implications for pathogenesis and treatment. *J. Inherit. Metab. Dis.* **2005**, *28*, 1–88. [CrossRef] [PubMed]

90. Salmi, H.; Leonard, J.; Lapatto, R. Patients with organic acidaemias have an alteredthiol status. *Acta Paediatr.* **2012**, *101*, e505–e508. [CrossRef] [PubMed]

91. Haas, D.; Niklowitz, P.; Horster, F.; Baumgartner, E.R.; Prasad, C.; Rodenburg, R.J.; Hoffmann, T.; Menke, T.; Okun, J.G. Coenzyme Q$_{10}$ is decreased in fibroblasts of patients with methylmalonic aciduria but not in mevalonic aciduria. *J. Inherit. Metab. Dis.* **2009**, *4*, 570–575. [CrossRef] [PubMed]

92. Hargreaves, I.P. Ubiquinone: Cholesterol's reclusive cousin. *Ann. Clin. Biochem.* **2003**, *40*, 207–218. [CrossRef] [PubMed]

93. Baumgarther, M.R.; Horster, F.; Dionisi-Vici, C.; Haliloglu, G.; Karall, D.; Chapman, K.A.; Huemer, M.; Hochuli, M.; Assoun, M.; Ballhausen, D.; et al. Proposed guidelines for the diagnosis and management of methylmalonic and propionic acidemia. *Orphanet J. Rare Dis.* **2014**, *9*, 130. [CrossRef] [PubMed]

94. Pinar-Sueiro, S.; Martinez-Fernondez, R.; Lage-Medines, S.; Aldamiz-Echevarria, L.; Vecino, E. Optic neuropathy in methylmalonic acidemia: The role of neuroprotection. *J. Inherit. Metab. Dis.* **2010**, *3*, S199–S203. [CrossRef] [PubMed]

95. Williams, Z.R.; Hurley, P.E.; Altipamak, V.E.; Feldon, S.E.; Arnold, G.L.; Eggenberger, E.; Mejico, L.J. Late onset optic neuropathy in methylmalonic and propionic academia. *Am. J. Ophthalmol.* **2009**, *147*, 929–933. [CrossRef] [PubMed]

96. Leipnitz, G.; Amaral, A.U.; Fernandes, C.G.; Seminotti, B.; Zanatta, A.; Knebel, L.A.; Vargas, C.R.; Wajner, M. Pristanic acid promotes oxidative stress in brain damage in peroxisomal disorders. *Brain Res.* **2011**, *1382*, 259–265. [CrossRef] [PubMed]

97. Weller, S.; Gould, S.J.; Valle, D. Peroxisomes biogenesis disorders. *Ann. Rev. Genom. Hum. Genet.* **2003**, *4*, 165–211. [CrossRef] [PubMed]

98. Lee, Y.M.; Sumpter, R.; Zou, Z.; Sirasanagandia, S.; Wei, Y.; Mishra, P.; Rosewich, H.; Crane, D.I.; Levine, B. Peroxisomal protein PEX13 functions in selective autophagy. *EMBO Rep.* **2017**, *18*, 48–60. [CrossRef] [PubMed]

99. White, A.L.; Modaff, P.; Holland-Morris, P.; Pauli, P.M. Natural history of rhizomelic chondrodysplasia punctate. *Am. J. Med. Genet.* **2003**, *118*, 332–342. [CrossRef] [PubMed]

100. Ferrer, I.; Aubourg, P.; Pujol, A. General aspects and neuropathology of X-linked adrenoleukodystrophy. *Brain Pathol.* **2010**, *20*, 817–830. [CrossRef] [PubMed]

101. Moser, H.; Smith, K.D.; Watkins, P.A.; Powers, J.; Moser, A.B. X-linked adrenoleukodystrophy. In *The Metabolic and Molecular Bases of Inherited Disease*, 8th ed.; Scriver, C., Ed.; McGraw-Hill: New-York, NY, USA, 2001; Volume 2, pp. 3257–3301.

102. Powers, J.M.; DeCiero, D.P.; Ito, M.; Moser, A.B.; Moser, H.W. Adrenomyeloneuropathy: A neuropathologic review featuring its noninflammatory myelopathy. *J. Neuropathol. Exp. Neurol.* **2000**, *59*, 89–102. [CrossRef] [PubMed]

103. Wanders, R.J.A.; van Roermund, W.T.; Shutgens, R.B.H.; Barth, P.G.; Heymans, H.S.A.; van den Bosch, H.; Tager, J.M. The inborn errors of peroxisomal beta-oxidation. A review. *J. Inher. Metab. Dis.* **1990**, *13*, 4–36. [CrossRef] [PubMed]

104. Schulz, H. Beta oxidation of fatty acids. *Biochim. Biophys. Acta* **1991**, *1081*, 109–120. [CrossRef]

105. Poirier, Y.; Antonenkov, V.D.; Glumoff, T.; Hiltunen, J.K. Peroxisomal beta-oxidation- a metabolic pathway with multiple functions. *Biochim. Biophys. Acta* **2006**, *1763*, 1413–1426. [CrossRef] [PubMed]

106. Angermuller, S.; Bruder, G.; Volkl, A.; Wesch, H.; Fahimi, H.D. Localization of xanthine oxidase in crystalline cores of peroxisomes. A cytochemical and biochemical study. *Eur. J. Cell Biol.* **1987**, *45*, 137–144. [PubMed]

107. Lismont, C.; Nordgen, M.; van Veldhoven, P.P.; Fransen, M. Redox interplay between mitochondria and peroxisomes. *Front. Cell Dev. Biol.* **2015**, *3*, 1–19. [CrossRef] [PubMed]

108. Vargas, C.R.; Wajner, M.; Sirtori, L.R.; Goulart, L.; Chiochetta, M.; Coelho, D.; Latini, A.; Llesuy, S.; Bello-Klein, A.; Giugliani, R.; et al. Evidence that oxidative stress is increased in patients with X-linked adrenoleukodystrophy. *Biochim. Biophys. Acta* **2004**, *1688*, 26–32. [CrossRef] [PubMed]

109. El-Bassyouni, H.T.; Abel Maksoud, S.A.; Salem, F.A.; Badr El-Deen, R.; Abdel Aziz, H.; Thomas, M.M. Evidence of oxidative stress in peroxisomal disorders. *Singap. Med. J.* **2012**, *53*, 608.

110. Schrader, M.; Fahimi, H.D. Mammalian peroxisomes and reactive oxygen species. *Histochem. Cell Biol.* **2004**, *122*, 383–393. [CrossRef] [PubMed]

111. Fransen, M.; Nordgren, M.; Wang, B.; Apanasets, O. Role of peroxisomes in ROS/RNS-metabolism: Implications for human disease. *Biochim. Biophys. Acta* **2012**, *1822*, 1363–1373. [CrossRef] [PubMed]

112. Fujiwara, C.; Imamure, A.; Hashiguchi, V.; Shimozawa, N.; Suzuki, Y.; Kondo, N.; Imanaka, T.; Tsukamoto, T.; Osumi, T. Catalase-less Peroxisomes: Implication in the milder forms of peroxisome biogenesis disorder. *J. Biol. Chem.* **2000**, *275*, 37271–37277. [CrossRef] [PubMed]

113. Baumgart, E.; Vanhorebeek, I.; Grabenbauer, M.; Borgers, M.; Declercq, P.E.; Fahimi, H.D.; Baes, M. Mitochondrial alterations caused by defective peroxisomal biogenesis in a mouse model of Zellweger syndrome (PEX5 knock out mouse). *Am. J. Pathol.* **2001**, *159*, 1477–1494. [CrossRef]

114. Schrakamp, G.; Schalkwijk, C.G.; Schutgens, R.B.; Wanders, R.J.; Tager, J.M.; van den Bosch, H. Plasmalogen biosynthesis in peroxisomal disorders: Fatty alcohol versus alkylglycerol precursors. *J. Lipid Res.* **1988**, *29*, 325–334. [PubMed]

115. Wood, C.S.; Koepke, J.I.; Teng, H.; Boucher, K.K.; Katz, S.; Chang, P.; Terlecky, L.J.; Papanayotou, I.; Walton, P.A.; Terlecky, S.R. Hypocatalasemic fibroblasts accumulate hydrogen peroxide and display age-associated pathologies. *Traffic* **2006**, *7*, 97–107. [CrossRef] [PubMed]

116. Wanders, R.J.; Schutgens, R.B.; Barth, P.G. Peroxisomal disorders: A review. *J. Neuropathol. Exp. Neurol.* **1995**, *54*, 726–739. [CrossRef] [PubMed]

117. Lopez-Erauskin, J.; Galino, J.; Ruiz, M.; Cuezva, J.M.; Fabregat, I.; Cacabelos, D.; Boada, J.; Martinez, J.; Ferrer, I.; Pamplona, R.; et al. Impaired mitochondrial oxidative phosphorylation in the peroxisomal disease X-linked adrenoleukodystrophy. *Hum. Mol. Genet.* **2013**, *22*, 3296–3305. [CrossRef] [PubMed]

118. Ho, J.K.; Moser, H.; Kishimoto, Y.; Hamilton, J.A. Interactions of a very long chain fatty acid with model membranes and serum albumin. Implications for the pathogenesis of adrenoleukodystrophy. *J. Clin. Investig.* **1995**, *96*, 1455–1463. [CrossRef] [PubMed]

119. Sarnat, H.B.; Machin, G.; Darwish, H.Z.; Rubin, S.Z. Mitochondrial myopathy of cerebrohepato-renal (Zellweger) syndrome. *Can. J. Neurol. Sci.* **1983**, *10*, 170–177. [CrossRef] [PubMed]

120. Muller-Hocker, J.; Walther, J.R.; Bise, K.; Pongratz, D.; Hubner, G. Mitochondrial myopathy with loosely coupled oxidative phosphorylation in a case of Zellweger syndrome. *Virchows Arch. B Cell Pathol. Zell-Pathol.* **1984**, *45*, 125–138. [CrossRef]

121. Wolff, J.; Nyhanf, W.L.; Powell, H.; Takahashi, D.; Hutzler, J.; Hajra, A.K.; Datta, N.S.; Singh, I.; Moser, H.W. Myopathy in an infant with a fatal peroxisomal disorder. *Pediatr. Neurol.* **1986**, *2*, 141. [CrossRef]

122. Powers, J.M.; Pei, Z.; Heinzer, A.K.; Deering, R.; Moser, A.B.; Moser, H.W.; Watkins, P.A.; Smith, K.D. Adreno-leukodystrophy: Oxidative stress of mice and men. *J. Neuropathol. Exp. Neurol.* **2005**, *64*, 1067–1079. [CrossRef] [PubMed]

123. Fourcade, S.; Ruiz, M.; Guilera, C.; Hahnen, E.; Brichta, L.; Naudi, A.; Portero-Otin, M.; Dacremont, G.; Cartier, N.; Wanders, R.; et al. Valproic acid induces antioxidant effects in X-linked adrenoleukodystrophy. *Hum. Mol. Genet.* **2010**, *19*, 2005–2014. [CrossRef] [PubMed]

124. Salpietro, V.; Phadke, R.; Saggar, A.; Hargreaves, I.P.; Yates, R.; Fokoloros, C.; Mankad, K.; Hertecant, J.; Ruggieri, M.; McCormick, D.; et al. Zellweger syndrome and secondary mitochondrial myopathy. *Eur. J. Pediatr.* **2015**, *174*, 557–563. [CrossRef] [PubMed]

125. Fourcade, S.; Lopez-Erauskin, J.; Galino, J.; Duval, C.; Naudi, A.; Jove, M.; Kemp, S.; Villarroya, F.; Ferrer, I.; Pamplona, R.; et al. Early oxidative damage underlying neurodegeneration in X-adrenoleukodystrophy. *Hum. Mol. Genet.* **2008**, *17*, 1762–1773. [CrossRef] [PubMed]

126. Lopez-Erauskin, J.; Galino, J.; Bianchi, P.; Fourcade, S.; Andreu, A.L.; Ferrer, I.; Munoz-Pinedo, C.; Pujol, A. Oxidative stress modulates mitochondrial failure and cyclophilin D function in X-linked adrenoleukodystrophy. *Brain* **2012**, *135*, 3584–3598. [CrossRef] [PubMed]

127. Singh, I.; Pujol, A. Pathomechanisms underlying X-adrenoleukodystrophy: A three-hit hypothesis. *Brain Pathol.* **2010**, *20*, 838–844. [CrossRef] [PubMed]

128. Galea, E.; Launay, N.; Portero-Otin, M.; Ruiz, M.; Pamplona, R.; Aubourg, P.; Ferrer, I.; Pujol, A. Oxidative stress underlying axonal degeneration in adrenoleukodystrophy: A paradigm for multifactorial neurodegenerative diseases? *Biochim. Biophys. Acta* **2012**, *9*, 1475–1488. [CrossRef] [PubMed]

129. Galino, J.; Ruiz, M.; Fourcade, S.; Schluter, A.; Lopez-Erauskin, J.; Guilera, C.; Jove, M.; Naudi, A.; Garcia-Arumi, E.; Andreu, A.L.; et al. Oxidative damage compromises energy metabolism in the axonal degeneration mouse model of X-adrenoleukodystrophy. *Antioxid. Redox Signal.* **2011**, *15*, 2095–2107. [CrossRef] [PubMed]

130. Lopez-Erauskin, J.; Fourcade, S.; Galino, J.; Ruiz, M.; Schluter, A.; Naudi, A.; Jove, M.; Portero-Otin, M.; Pamplona, R.; Ferrer, I.; et al. Antioxidants halt axonal degeneration in a mouse model of X-adrenoleukodystrophy. *Ann. Neurol.* **2011**, *70*, 84–92. [CrossRef] [PubMed]

131. Marchetti, D.P.; Donida, B.; Deon, M.; Jacques, C.E.; Jardim, L.B.; Vargas, C.R. In vitro effects of *N*-acetyl-L-cysteine on glutathione and sulfhryl levels in X-linked adrenoleukodystrophy patients. *Clin. Biomed. Res.* **2017**, *37*, 33–37. [CrossRef]

132. Ferdinandusse, S.; Finckh, B.; de Hingh, Y.C.; Stroomer, L.E.; Denis, S.; Kohlschutter, A.; Wanders, R.J. Evidence for increased oxidative stress in peroxisomal D-bifunctional protein deficiency. *Mol. Genet. Metab.* **2003**, *79*, 281–287. [CrossRef]

133. Huang, J.; Liu, X.; Tang, L.L.; Long, J.T.; Zhu, J.; Hua, R.X.; Li, J. XPG gene polymorphisms and cancer susceptibility: Evidence from 47 studies. *Oncotarget* 2017. [CrossRef] [PubMed]

134. Carre, G.; Marelli, C.; Anheim, M.; Geny, C.; Renaud, M.; Rezvani, H.R.; Koenig, M.; Guissart, C.; Tranchant, C. Xeroderma pigmentosum complementation group F: A rare cause of cerebellar ataxia with chorea. *J. Neurol. Sci.* **2017**, *376*, 198–201.

135. Niedernhofer, L.J.; Bohr, V.A.; Sander, M.; Kraemer, K.H. Xeroderma pigmentosum and other diseases of human premature aging and DNA repair: Molecules to patients. *Mech. Ageing Dev.* **2011**, *132*, 340–347. [CrossRef] [PubMed]

136. DiGiovanna, J.J.; Kraemer, K.H. Shining a light on xeroderma pigmentosum. *J. Investig. Dermatol.* **2012**, *132*, 785–796. [CrossRef] [PubMed]

137. Copeland, N.E.; Hanke, C.W.; Michalak, J.A. The molecular basis of xeroderma pigmentosum. *Dermatol. Surg.* **1997**, *23*, 447–455. [CrossRef] [PubMed]

138. Wang, H.T.; Choi, B.; Tang, M.S. Melanocytes are deficient in repair of oxidative DNA damage and UV-induced photoproducts. *Proc. Natl. Acad. Sci. USA* **2010**, *107*, 12180–12185. [CrossRef] [PubMed]

139. Pascucci, B.; D'errico, M.; Parlanti, E.; Giovannini, S.; Dogliotti, E. Role of nucleotide excision repair proteins in oxidative DNA damage repair: An updating. *Biochemistry* **2011**, *76*, 4–15. [CrossRef] [PubMed]

140. Hayashi, M.; Itoh, M.; Araki, S.; Kumada, S.; Shioda, K.; Tamagawa, K.; Mizutani, T.; Morimatsu, Y.; Minagawa, M.; Oda, M. Oxidative stress and disturbed glutamate transport in hereditary nucleotide repair disorders. *J. Neuropathol. Exp. Neurol.* **2001**, *60*, 350–356. [CrossRef] [PubMed]

141. Murai, M.; Enokido, Y.; Inamura, N.; Yoshino, M.; Nakatsu, Y.; van der Horst, G.T.; Hoeijmakers, J.H.; Tanaka, K.; Hatanaka, H. Early postnatal ataxia and abnormal cerebellar development in mice lacking Xeroderma pigmentosum Group A and Cockayne syndrome Group B DNA repair genes. *Proc. Natl. Acad. Sci. USA* **2001**, *98*, 13379–13384. [CrossRef] [PubMed]

142. Brooks, P.J. The 8,5′-cyclopurine-2′-deoxynucleosides: Candidate neurodegenerative DNA lesions in xeroderma pigmentosum, and unique probes of transcription and nucleotide excision repair. *DNA Repair* **2008**, *7*, 1168–1179. [CrossRef] [PubMed]

143. Hayashi, M.; Ahmad, S.I.; Hanaoka, F. Roles of oxidative stress in xeroderma pigmentosum. *Adv. Exp. Med. Biol.* **2008**, *637*, 120–127. [PubMed]

144. Melis, J.P.; van Steeg, H.; Luijten, M. Oxidative DNA damage and nucleotide excision repair. *Antioxid. Redox Signal.* **2013**, *18*, 2409–2419. [CrossRef] [PubMed]

145. Li, L.; Bales, E.S.; Peterson, C.A.; Legerski, R.J. Characterization of molecular defects in xeroderma pigmentosum group C. *Nat. Genet.* **1993**, *5*, 413–417. [CrossRef] [PubMed]

146. Rezvani, H.R.; Rossignol, R.; Ali, N.; Bernard, G.; Tang, X.; Yang, H.S.; Jouary, T.; de Verneuil, H.; Taieb, A.; Kim, A.L.; et al. XPC silencing in normal human keratinocytes triggers metabolic alterations through NOX-1 activation-mediated reactive oxygen species. *Biochim. Biophys. Acta Bioenerg.* **2011**, *1807*, 609–619. [CrossRef] [PubMed]

147. Hosseini, M.; Mahfouf, W.; Serrano-Sanchez, M.; Raad, H.; Harfouche, G.; Bonneu, M.; Claverol, S.; Mazurier, F.; Rossignol, R.; Taieb, A.; et al. Premature Skin Aging Features Rescued by Inhibition of NADPH Oxidase Activity in XPC-Deficient Mice. *J. Investig. Dermatol.* **2015**, *135*, 1108–1118. [CrossRef] [PubMed]

148. Fang, E.F.; Scheibye-Knudsen, M.; Brace, L.E.; Kassahun, H.; Sengupta, T.; Nilsen, H.; Mitchell, J.R.; Croteau, D.L.; Bohr, V.A. Defective Mitophagy in XPA via PARP-1 Hyperactivation and NAD(+)/SIRT1 Reduction. *Cell* **2014**, *157*, 882–896. [CrossRef] [PubMed]

149. Osellame, L.D.; Rahim, A.A.; Hargreaves, I.P.; Gegg, M.E.; Richard-Londt, A.; Brandner, S.; Waddington, S.N.; Schapira, A.H.; Duchen, M.R. Mitochondria and quality control defects in a mouse model of Gaucher disease-links to Parkinson's disease. *Cell Metab.* **2013**, *17*, 941–953. [CrossRef] [PubMed]

150. Reza Rezvani, H.; Taieb, A. Le xeroderma pigmentosum. *M/S Med. Sci.* **2011**, *27*, 467.

151. Liu, J.; Fang, H.; Chi, Z.; Wu, Z.; Wei, D.; Mo, D.; Niu, K.; Balajee, A.S.; Hei, T.K.; Nie, L.; et al. XPD localizes in mitochondria and protects the mitochondrial genome from oxidative DNA damage. *Nucl. Acids Res.* **2015**, *43*, 5476–5488. [CrossRef] [PubMed]

152. Mori, M.P.; Costa, R.P.; Soltys, D.T.; Freire, T.D.S.; Rossato, F.A.; Amigo, I.; Kowaltowski, A.J.; Vercesi, A.E.; de Souza-Pinto, N.C. Lack of XPC leads to a a shift between respiratory complexes I and II but sensitizes cells to mitochondrial stress. *Sci. Rep.* **2017**, *7*, 155. [CrossRef] [PubMed]

153. Rothe, M.; Werner, D.; Thielmann, H.W. Enhanced expression of mitochondrial genes in xeroderma pigmentosum fibroblast strains from various complementation groups. *J. Cancer Res. Clin. Oncol.* **1993**, *119*, 675–684. [CrossRef] [PubMed]

154. Berg, D.; Otley, C.C. Skin cancer in organ transplant recipients: Epidemiology, pathogenesis and management. *J. Am. Acad. Dermatol.* **2002**, *47*, 1–17. [CrossRef] [PubMed]

155. Boesch, P.; Weber-Lotfi, F.; Ibrahim, N.; Tarasenko, V.; Cosset, A.; Paulus, F.; Lightowlers, R.N.; Dietrich, A. DNA repair in organelles: Pathways, organisation, regulation, relevance in disease and aging. *Biochim. Biophys. Acta* **2011**, *1813*, 186–200. [CrossRef] [PubMed]

156. Quillet, X.; Chevallier-Lagente, O.; Zeng, L.; Calvayrac, R.; Mezzina, M.; Sarasin, A.; Vuillaume, M. Retroviral-mediated correction of DNA repair defect in xeroderma pigmentosa cells is associated with recovery of catalase activity. *Mutat. Res. DNA Repair* **1997**, *385*, 235–242. [CrossRef]

157. Taneka, J.; Nagai, T.; Okada, S. Serum concentration of coenzyme Q in xeroderma pigmentosum. *Rinsho Shinkeigaku Clin. Neurol.* **1998**, *38*, 57–59.

158. Goncalves-Maia, M.; Magnaldo, T. Genetic therapy of xeroderma pigmentosa, analysis of strategies and translation. *Exp. Opin. Orphan Drugs* **2017**, *5*, 5–17. [CrossRef]

159. Bone, R.C.; Balk, R.A.; Cerra, F.B.; Dellinger, R.P.; Fein, A.M.; Knaus, W.A.; Schein, R.M.; Sibbald, W.J. Definitions for sepsis and organ failure and guidelines for the use of innovative therapies in sepsis. *Chest* **1992**, *101*, 1644–1655. [CrossRef] [PubMed]

160. Chertow, G.M.; Burdick, E.; Honour, M.; Bonventre, J.V.; Bates, D.W. Acute kidney injury, mortality, length of stay, and costs in hospitalized patients. *J. Am. Soc. Nephr.* **2005**, *16*, 3365–3370. [CrossRef] [PubMed]

161. Uchino, S.; Kellum, J.A.; Bellomo, R.; Doig, G.S.; Morimatsu, H.; Morgera, S.; Achetz, M.; Tan, I.; Bouman, C.; Macedo, E.; et al. Acute renal failure in critically ill patients: A multinational, multicenter study. *JAMA* **2005**, *294*, 813–818. [CrossRef] [PubMed]

162. Bagshaw, S.M.; Uchino, S.; Bellomo, R.; Morimatsu, H.; Morgera, S.; Schetz, M.; Tan, I.; Bouman, C.; Macedo, E.; Gibney, N.; et al. Septic acute kidney injury in critically ill patients: Clinical characteristics and outcomes. *Clin. J. Am. Soc. Nephrol.* **2007**, *2*, 431–439. [CrossRef] [PubMed]

163. Keir, I.; Kellum, J.A. Acute kidney injury in severe sepsis: Pathophysiology, diagnosis, and treatment recommendations. *J. Veter. Emerg. Crit. Care* **2015**, *25*, 200–209. [CrossRef] [PubMed]

164. Abraham, E.; Singer, M. Mechanisms of sepsis-induced organ dysfunction. *Crit. Care Med.* **2007**, *35*, 2408–2416. [CrossRef] [PubMed]

165. Martin, G.S.; Mannino, D.M.; Eaton, S.; Moss, M. The epidemiology of sepsis in the United States from 1979 through 2000. *N. Engl. J. Med.* **2000**, *348*, 1546–1554. [CrossRef] [PubMed]

166. Weycker, D.; Akhras, K.S.; Edelsberg, J.; Angus, D.C.; Oster, G. Long-term mortality and medical care charges in patients with severe sepsis. *Crit. Care Med.* **2003**, *31*, 2316–2323. [CrossRef] [PubMed]

167. Jackson, J.C.; Girard, T.D.; Gordon, S.M.; Thompson, J.L.; Shintani, A.K.; Thomason, J.W.W.; Pun, B.T.; Canonico, A.E.; Dunn, J.G.; Bernard, G.R.; et al. Long-term cognitive and psychological outcomes in the awakening and breathing controlled trial. *Am. J. Respir. Crit. Care Med.* **2010**, *182*, 183–191. [CrossRef] [PubMed]

168. Iwashyna, T.J.; Ely, E.W.; Smith, D.M.; Langa, K.M. Long-term cognitive impairment and functional disability among survivors of severe sepsis. *JAMA* **2010**, *304*, 1787–1794. [CrossRef] [PubMed]

169. Trumbeckaite, S.; Opalka, J.R.; Neuhof, C.; Zierz, S.; Gellerich, F.N. Different sensitivity of a rabbit heart and skeletal muscle to endotoxin-induced impairment of mitochondrial function. *Eur. J. Biochem.* **2001**, *268*, 1422–1429. [CrossRef] [PubMed]

170. Belcher, E.; Mitchell, J.; Evans, T. Myocardial dysfunction in sepsis: No role for NO. *Heart* **2000**, *87*, 507–509. [CrossRef]

171. Singer, M. The role of mitochondrial dysfunction in sepsis-induced multi-organ failure. *Landes Biosci.* **2014**, *5*, 66–72. [CrossRef] [PubMed]

172. Quoilin, C.; Mouithys-Mickalad, A.; Lecart, S.; Fontaine-Aupart, M.P.; Hoebeke, M. Evidence of oxidative stress and mitochondrial respiratory chain dysfunction in an in vitro model of sepsis-induced kidney injury. *Biochim. Biophys. Acta Bioenerg.* **2014**, *1837*, 1790–1800. [CrossRef] [PubMed]

173. Moncada, S.; Erusalimsky, J.D. Does nitric oxide modulate mitochondrial energy generation and apoptosis? *Nat. Rev. Mol. Cell Biol.* **2002**, *3*, 214–220. [CrossRef] [PubMed]

174. Adrie, C.; Bachelet, M.; Vayssier-Taussat, M.; Russo-Marie, F.; Bouchaert, I.; Adib-Conquy, M.; Cavaillon, J.M.; Pinski, M.R.; Dhainaut, J.F.; Polla, B.S. Mitochondrial membrane potential and apoptosis in peripheral blood monocytes in severe human sepsis. *Am. J. Respir. Crit. Care Med.* **2001**, *164*, 389–395. [CrossRef] [PubMed]

175. Gellerich, F.N.; Trumbeckaite, S.; Opalka, J.R.; Gellerich, J.F.; Chen, Y.; Zierz, S.; Werdan, K.; Neuhof, C.; Redl, H. Mitochondrial dysfunction in sepsis: Evidence from bacteraemic baboons and endotoxaemic rabbits. *Biosci. Rep.* **2002**, *22*, 99–113. [CrossRef] [PubMed]

176. Sener, G.; Toklu, H.; Ercan, F.; Erkanli, G. Protective effect of β-glucan against oxidative organ injury in a rat model of sepsis. *Int. Immunopharmacol.* **2005**, *5*, 1387–1396. [CrossRef] [PubMed]

177. Kim, J.-Y.; Lee, S.-M. Vitamins C and E protect hepatic cytochrome P450 dysfunction induced by polymicrobial sepsis. *Eur. J. Pharmacol.* **2006**, *534*, 202–209. [CrossRef] [PubMed]

178. Mishra, V.; Baines, M.; Perry, S.E.; McLaughlin, P.J.; Carson, J.; Wenstone, R.; Shenkin, A. Effect of selenium supplementation on biochemical markers and outcome in critically ill patients. *Clin. Nutr.* **2007**, *26*, 41–50. [CrossRef] [PubMed]

179. Angstwurm, M.W.; Engelman, L.; Zimmermann, T.; Lehmann, C.; Spes, C.H.; Abel, P.; Strau, R.; Meier-Hellmann, A.; Insel, R.; Radke, J.; et al. Selenium in intensive care (SIC): Results of a prospective randomized, placebo-controlled, multiple-centre study in patients with severe systemic inflammatory response syndrome, sepsis and septic shock. *Crit. Care Med.* **2007**, *35*, 118–126. [CrossRef] [PubMed]

180. Galley, H.F. Oxidative stress and mitochondrial dysfunction in sepsis. *Br. J. Anaesth.* **2011**, *107*, 57–64. [CrossRef] [PubMed]

181. Radi, R.; Cassina, A.; Hodara, R.; Quijano, C.; Castro, L. Peroxynitrite reactions and formation in mitochondria. *Free Radic. Biol. Med.* **2002**, *33*, 1451–1464. [CrossRef]

182. Laganiere, S.; Yu, B.P. Modulation of membrane phospholipid fatty acid composition and food restriction. *Gerontology* **1993**, *39*, 7–18. [CrossRef] [PubMed]

183. Kang, D.; Hamasaki, N. Mitochondrial oxidative stress and mitochondrial DNA. *Clin. Chem. Lab. Med.* **2003**, *41*, 1281–1288. [CrossRef] [PubMed]

184. Hargreaves, I.P.; Duncan, A.J.; Wu, L.; Agrawal, A.; Land, J.M.; Heales, S.J. Inhibition of mitochondrial complex IV leads to secondary loss complex II–III activity: Implications for the pathogenesis and treatment of mitochondrial encephalomyopathies. *Mitochondrion* **2007**, *7*, 284–287. [CrossRef] [PubMed]

185. Pacher, P.; Beckman, J.S.; Liaudet, L. Nitric oxide and peroxynitrite in health and disease. *Physiol. Rev.* **2007**, *87*, 315–424. [CrossRef] [PubMed]

186. Frijhoff, J.; Winyard, P.G.; Zarkovic, N.; Davies, S.S.; Stocker, R.; Cheng, D.; Gasparovic, A.C. Clinical Relevance of Biomarkers of Oxidative Stress. *Antioxid. Redox Signal.* **2015**, *23*, 1144–1170. [CrossRef] [PubMed]

187. Duncan, A.J.; Heales, S.J.; Mills, K.; Eaton, S.; Land, J.M.; Hargreaves, I.P. Determination of coenzyme Q_{10} status in blood mononuclear cells, skeletal muscle and plasma by HPLC with a di-propoxy-coenzyme Q_{10} as an internal standard. *Clin. Chem.* **2005**, *51*, 2380–2382. [CrossRef] [PubMed]

188. Enns, G.M.; Cowan, T.M. Glutathione as a Redox Biomarker in Mitochondrial Disease-Implications for Therapy. *J. Clin. Med.* **2017**, *6*, 50. [CrossRef]

189. Enns, G.M.; Kinsma, S.L.; Perlman, S.L.; Spicer, K.M.; Abdenur, J.E.; Cohen, B.H.; Thoolen, M. Initial experience in the treatment of inherited mitochondrial disease with EPI-743. *Mol. Genet. Metab.* **2012**, *105*, 91–102. [CrossRef] [PubMed]

190. Homstrom, K.H.; Baird, L.; Zhang, Y.; Hargreaves, I.; Chalasani, A.; Land, J.M.; Abramov, A.Y. NrF$_2$ impacts cellular bioenergetics by controlling substrate availability for mitochondrial respiration. *Biol. Open* **2013**, *2*, 761–770. [CrossRef] [PubMed]

Journal of
Clinical Medicine

MDPI

Review

Myopathology of Adult and Paediatric Mitochondrial Diseases

Rahul Phadke [1,2]

[1] Division of Neuropathology, UCL Institute of Neurology, National Hospital for Neurology and
    Neurosurgery, UCLH NHS Foundation Trust, London WC1N 3BG, UK; r.phadke@ucl.ac.uk;
    Tel.: +44-020-344-84393
[2] Dubowitz Neuromuscular Centre, Great Ormond Street Hospital for Children NHS Foundation Trust,
    London WC1N 3JH, UK

Academic Editor: Jane Grant-Kels
Received: 12 June 2017; Accepted: 28 June 2017; Published: 4 July 2017

Abstract: Mitochondria are dynamic organelles ubiquitously present in nucleated eukaryotic
cells, subserving multiple metabolic functions, including cellular ATP generation by oxidative
phosphorylation (OXPHOS). The OXPHOS machinery comprises five transmembrane respiratory
chain enzyme complexes (RC). Defective OXPHOS gives rise to mitochondrial diseases (mtD). The
incredible phenotypic and genetic diversity of mtD can be attributed at least in part to the RC
dual genetic control (nuclear DNA (nDNA) and mitochondrial DNA (mtDNA)) and the complex
interaction between the two genomes. Despite the increasing use of next-generation-sequencing
(NGS) and various omics platforms in unravelling novel mtD genes and pathomechanisms, current
clinical practice for investigating mtD essentially involves a multipronged approach including clinical
assessment, metabolic screening, imaging, pathological, biochemical and functional testing to guide
molecular genetic analysis. This review addresses the broad muscle pathology landscape including
genotype–phenotype correlations in adult and paediatric mtD, the role of immunodiagnostics in
understanding some of the pathomechanisms underpinning the canonical features of mtD, and recent
diagnostic advances in the field.

Keywords: mitochondrial; muscle biopsy; ragged red; COX-negative; subsarcolemmal;
immunohistochemistry

---

1. Introduction

The diagnosis of mtD is challenging due to the incredible phenotypic and genetic diversity
associated with these diseases. This partly stems from the dual genetic control (nDNA and
mtDNA) of the RC, the complexity of intergenomic signalling and its functional consequences. mtD
can be inherited in an autosomal dominant, autosomal recessive, X-linked or mitochondrial (i.e.,
maternal) fashion. The circular mtDNA encodes 13 RC subunits, 22 mitochondrial tRNAs and two
ribosomal RNAs. Additionally, the mitoproteome requires over 1300 nuclear encoded proteins to
produce, assemble and support the five multimeric OXPHOS RC (I–V), and ancillary mitochondrial
processes [1–3]. Tissues and organs affected in mtD are often those with high-energy requirements.
Clinical symptoms can manifest at any age, and can affect a single organ or be multisystemic [4].
Typically, the more severe phenotypes present early, and milder phenotypes present later in life [5].
There are classic clinical syndromes with stereotypic features such as Leigh syndrome (subacute
necrotising encephalomyopathy), MELAS (mitochondrial myopathy, encephalopathy, lactic acidosis
and stroke-like episodes) and Alpers disease (epilepsy and liver failure). Point mutations and
large-scale mtDNA deletions represent the two most common causes of primary mtDNA disease, the
former usually being maternally inherited, and the latter typically arising de novo during embryonic

development [1]. Exemplary genotype–phenotype associations are recognised, often in a syndromic context, e.g., mitochondrial protein synthesis *tRNA* gene point mutations are associated with MELAS, myoclonic epilepsy with ragged red fibres (MERRF) and syndromic forms of maternally transmitted diabetes; mutations in mitochondrial RC protein coding genes are associated with Leber hereditary optic neuropathy (LHON), neuropathy, ataxia, retinitis pigmentosa (NARP), and maternally inherited Leigh syndrome (MILS); single mtDNA deletions are associated with Pearson syndrome, chronic progressive external ophthalmoplegia (CPEO) and Kearns-Sayre syndrome (KSS); multiple mtDNA deletions are associated with mitochondrial neurogastrointestinal encephalomyopathy (MNGIE), autosomal dominant or autosomal recessive progressive external ophthalmoplegia (AD/AR-PEO), and sensory ataxic neuropathy, ataxia, ophthalmoplegia (SANDO); mtDNA depletion are associated with early-onset myopathic and hepatocerebral forms; Leigh syndrome is associated with a variable genotypic background including nDNA and mtDNA mutations; and so on [6,7]. However, many patients display non-specific features of developmental delay or regression, further hindering accurate diagnosis [8]. The onset of symptoms, phenotypic variability, and variable penetrance of mtD are influenced by the peculiarities of mitochondrial genetics including the threshold effect, mitotic segregation, clonal expansion and a genetic bottleneck, as well as the nuclear genome background in which it coexists or by environmental and epigenetic factors [6].

2. Laboratory Investigations and the Rationale for Muscle Biopsy

Given the complexity of mtD phenotypes and genetics, securing a diagnosis frequently requires extensive non-invasive and invasive tests including imaging, neurophysiology, metabolic and biochemical studies, muscle pathology and functional testing, followed by definitive molecular genetic confirmation. Resting and exercise induced increase in blood lactate is a useful albeit non-specific screening tool for mtD, but can be normal or minimally elevated as in mitochondrial polymerase gamma (POLG1) associated diseases, Leber Hereditary Optic Neuropathy (LHON), Leigh disease, Kearns-Sayre syndrome and Complex I deficiency [9]. The blood lactate/pyruvate ratio may increase in inborn errors of the mitochondrial respiratory chain [10]. Spurious elevation of plasma lactate and/or pyruvate may occur from poor collection or handling technique, secondary mitochondrial dysfunction in a range of systemic and metabolic diseases, and in nutritional thiamine deficiency. Blood and/or CSF pyruvate levels may increase in defects of pyruvate metabolism. Similarly, CSF lactate and/or pyruvate levels may increase without blood elevation in mtD with predominant CNS manifestations [11]. Elevated plasma/CSF amino acids, urine organic acids, and plasma acylcarnitines all suggest underlying mitochondrial dysfunction, however, normal levels do not exclude mtD [12,13]. CK values are normal or mildly elevated, unless measured in the setting of rhabdomyolysis [14]. Neurophysiology may show non-specific signs of a myopathy, or a neuropathy when present, but may be normal. Neuroimaging, encompassing several modalities beyond routine T1 and T2 magnetic resonance imaging (MRI), including volumetric analysis, diffuse tensor imaging (DTI), magnetic resonance spectroscopy (MRS), arterial spin labelling, and functional magnetic resonance imaging (fMRI) has shown its potential as an investigative tool, and in many cases, providing non-invasive and repeatable biomarker inquiry in patients with mtD [15]. MRI findings in patients with mtD can often be non-specific, including in those with clinical central nervous system involvement; however, it is the pattern of involvement that can suggest an underlying neurometabolic defect. In children, a common pattern is "global" delay in myelination early in the course of the disease [16]. The most common specific MRI findings are a symmetrical signal abnormality of deep grey matter presenting with hyperintensity on T2 and FLAIR images, and hypointensity on T1 images. Any deep structure can be involved and the character of the lesion can be either patchy or homogeneous. Cerebral and cerebellar atrophy may be present in varying degrees. These specific MRI findings are more likely to be associated with well recognised syndromic phenotypes such as Leigh disease, MERRF, MELAS, KSS, MNGIE, etc. [17]. MRS can provide valuable in vivo metabolic information to measure metabolites possessing resonating nuclei (hydrogen-1; $_1$H: phosphorous-31; $_{31}$P: carbon-13; $_{13}$C) in the mM range. mtD represent a particularly prominent set of diseases that show MRS changes due to the consequences of impaired OXPHOS. The most consistent MRS change accompanying increased lactate in mitochondrial disease is decreased *N*-acetyl-L-aspartate (NAA) normalised to creatine, suggestive of cellular compromise [12,17]. Other brain metabolites such as myo-inositol, choline, creatinine and succinate can be measured by MRS. One disease biomarker that appears highly specific to complex II disease is a large elevation in succinate in white matter [18]. MRS is being increasingly applied as a non-invasive tool to monitor the effects of therapeutic intervention in patients with mtD. Diagnostically useful patterns of selective muscle involvement are well-recognised and increasingly being used in the diagnostic algorithm for muscular dystrophies, congenital myopathies and a few other heritable neuromuscular disorders [19]. This approach appears less useful in mtD, probably reflecting the fact that clinically and biochemically affected muscles in mtD rarely ever show significant fibro-fatty infiltration when examined histologically.

There is no single "gold standard" laboratory test for diagnosing mtD. The screening tests described above broadly confirm presence of dysfunction in various organ systems and help to increase or decrease the clinical suspicion of mtD. More invasive testing is necessary to establish direct morphological, biochemical and molecular genetic evidence of mitochondrial dysfunction [5]. In principle, the relevant tissue to investigate is one that clinically expresses disease. Skeletal muscle remains the tissue of choice, and is frequently sampled in part due to the relative safety and ease with which tissue samples can be obtained. It can provide valuable diagnostic information in many cases, even without clinically overt myopathic involvement [20–22]. Skeletal muscle is a post-mitotic terminally differentiated tissue with only limited regenerative capacity via satellite cell transformation. This terminal differentiation results in a fairly stable lifelong relationship between the mutant and wild-type mtDNA ratio (heteroplasmy) in contrast to nucleated blood cells in which this ratio can decrease due to selection pressure, thereby obscuring evidence of mitochondrial dysfunction [22]. Skeletal muscle mitochondria are abundant in subsarcolemmal and intermyofibrillar locations and larger than in most other tissues. Pathological assessment of "non-muscle" components in biopsies including blood vessels and nerves can provide evidence of multi-organ dysfunction [21]. Biochemical testing of respiratory chain enzyme dysfunction typically involves determination of individual or paired respiratory chain enzyme complex activities in mitochondrial fractions or tissue homogenates prepared from fresh or frozen muscle tissue. Biochemical assays have low inter-laboratory reproducibility and a systematic program to share samples and standardise methodologies across diagnostic laboratories has not been implemented. Other confounding factors include masking of a RC defect in tissue homogenates due to low-level heteroplasmy and a physiological compensatory mitochondrial proliferative response [8]. In this context, parallel histological assessment of skeletal muscle can uniquely provide histochemical evidence of RC defect at the single cell level. Simultaneously, the biopsy can be assessed for a number of conditions in the clinical differential diagnosis that can mimic a mitochondrial myopathy or induce secondary mitochondrial dysfunction. This includes fatty acid oxidation defects, glycogen storage disorders, endocrine, congenital and inflammatory myopathies and muscular dystrophies. The reliability of detecting morphological and histochemical abnormalities in skeletal muscle in mtD has led to their inclusion as major and minor criteria in several classification schemes for diagnosing mtD in adults and children [23–26]. It is standard practice to perform a skin biopsy in parallel to a muscle biopsy primarily for establishing fibroblast cultures. While it is less invasive, it is not uncommon for patients with OXPHOS defects in skeletal muscle to have normal RC activities in fibroblasts [27,28]. This in part due to altered heteroplasmy and high tissue regeneration rate of fibroblasts compared to skeletal muscle [27]. It is equally important to recognise the limitations of muscle biopsy analysis in investigating mtD. RC deficiencies are usually tissue specific, particularly if sporadic and somatic, further influenced by the type of mutation and the peculiarities of mitochondrial genetics. Therefore even muscle samples with proven mtD mutations/phenotypes may not show pathological and/or biochemical evidence of mitochondrial dysfunction [14]. mtDNA copy number analysis in muscle tissue by real time qPCR normalised to age-matched controls gives an indication of depletion or amplification of mtDNA content. mtDNA depletion can point towards mtDNA depletion syndromes caused by a number of genes involved in mtDNA maintenance. mtDNA depletion in muscle is however not as obvious in patients with the myopathic disease form as it is in liver tissue in patients with the hepatocerebral form [29]. There are no reliable histochemical assays for demonstration of defects in complexes I, III and V [1,21]. Muscle biopsy may appear histologically normal even in the context of genetically confirmed mtD and when the biochemical defect does not involve complex IV [23,30]. The histological interpretation of paediatric muscle biopsies can be challenging in the absence of age-matched controls. Morphological and histochemical abnormalities of mitochondria are not entirely specific and are seen secondary to other myopathic processes and with ageing. Despite its invasive nature and limitations, muscle biopsy remains the gold standard for mtD, especially due to primary mtDNA mutations [31].

3. Technical Considerations

Muscle and skin biopsies must be performed and processed in a manner that optimally preserves mitochondrial morphology, enzymatic/functional activity, protein and DNA/RNA content to allow for the broadest range of tissue investigations into mitochondrial dysfunction. This requires good communication between clinicians, surgeons and pathologists and teamwork. Rigorous implementation of a standardised biopsy protocol minimises the risk of ambiguity in interpretation of results due to a myriad of artefacts resulting from improper sampling and processing. Some of these issues are discussed below. A limb muscle such as vastus lateralis, gastrocnemius, deltoid or biceps brachii is selected for sampling, depending on the institutional preference. Occasionally extraocular muscles may be sampled in patients with external ophthalmoplegia. However, these muscles may normally harbour features considered "myopathic" for limb muscles, and show a greater prevalence of ragged red fibres (RRF) and COX-negative fibres compared to limb muscles from the third decade of life [32,33]. Skeletal muscle can be harvested via an open biopsy or a needle biopsy procedure. The latter has the advantage of being performed under local anaesthesia and/or deep sedation, and producing a smaller scar. Concerns about tissue fragmentation and smaller tissue yield that may be insufficient for biochemical assays have been addressed by implementing protocols using the modified Bergström needle for sampling. Ideally, the sample should be examined under a dissecting microscope in the procedure room for adequacy and orientation. A portion of fresh unfixed muscle is immediately placed in RNase free tubes and snap frozen in liquid nitrogen for biochemistry and genetic testing. The best oriented portion can be transported to the laboratory wrapped in cling film or by placing in a closed petri dish on a piece of gauze lightly moistened in saline, and then frozen in isopentane cooled in liquid nitrogen for histology, histochemistry and immunohistochemistry. A small longitudinal piece 0.5 mm long is placed in chilled 2% glutaraldehyde for electron microscopy. Functional assays including polarographic studies require fresh, unfixed tissue. Excessive mechanical trauma to the sampled tissue, infiltration of the local aesthetic into the fascicles, drying out and excess contact with saline can render morphology and histochemistry uninterpretable. Isopentane may interfere with measurement of complexes I, II and III in biochemical assays and give falsely low activities [34]. Many laboratories routinely fix a portion of muscle in 10% formalin for paraffin embedding. Apart from providing a larger sampling field and with the exception of high-risk infectious samples, this practice has several disadvantages and its routine implementation should be discouraged. Formalin fixation causes loss of histochemical enzyme activity, yields inferior muscle morphology, enhances autofluorescence, and may require laborious antigen retrieval protocols for protein immunohistochemistry. In case of large open biopsies, an additional block can be prepared for freezing in isopentane, and any surplus tissue can be snap frozen in liquid nitrogen. A 4 × 4 mm skin punch biopsy from the thigh provides sufficient tissue for growing fibroblast cultures and should be placed in sterile conditions in a culture medium containing uridine and pyruvate to prevent the loss of cells harbouring mutant mtDNA [35]. Detailed protocols for sampling and processing of skin and muscle biopsies have been published [22,36].

4. Histochemical Assays for Detecting RC Defects

The mitochondrial RC responsible for cellular ATP generation is located on the inner mitochondrial membrane comprising five complexes: CI, NADH-coenzyme Q reductase; CII, succinate-CoQ reductase that includes the FAD-dependent succinate dehydrogenase (SDH) and iron-sulphur proteins; CIII, reduced CoQ-cytochrome C reductase; CIV, cytochrome C oxidase; and CV, ATP synthase [37]. The free energy generated via redox reactions involving electron transfer across the complexes to molecular oxygen creates a transmembrane proton gradient—protons are pumped through CI, CIII and CIV, and CV allows protons to flow back into the mitochondrial matrix, and the released energy is used to synthesise ATP. Histochemical stains are available that can demonstrate activities of CI (NADH-TR), CII (SDH) and CIV (COX). NADH-TR stain comprises reduced nicotinamide adenine dinucleotide as the substrate that is oxidized by NADH-dependent enzymes. Addition of a tetrazolium salt (NBT) results in the deposition of a reduced, insoluble blue formazan product at the reaction site. TR denotes tetrazolium reductase. However, the NADH is not only oxidized by CI of the RC, but also sarcoplasmic reticulum (SR) NADH-oxidising enzymes. In the absence of specific inhibitors of the "non-mitochondrial" NADH-oxidase activity, this stain is not specific for CI, and any CI defect is invariably masked [38]. The advantage is this stain can be used as a general marker of mitochondria and SR, and thereby is excellent in highlighting structural defects such as cores or mini-cores. SDH stain can demonstrate CII activity. Na-succinate is used as the substrate, which gets oxidized to fumarate by CII in the presence of NBT, which is reduced to insoluble blue formazan at the reaction site [38]. In the modified SDH reaction, 1-methoxyphenazine methosulphate (mPMS) or phenazine methosulphate (PMS) are added to the incubation medium as exogenous electron carriers and azide or cyanide as inhibitors of cytochrome oxidase. The mPMS and azide substitution results in substantial reduction in the non-specific reduction of NBT and a linear reaction rate, thereby allowing better histochemical quantitation [39]. Nuclear genes encode all sub-units of CII; therefore, CII is rarely affected in diseases with primary mtDNA defects. COX stain demonstrates cytochrome C oxidase activity. In this reaction, diaminobenzidine (DAB) acts as the electron donor to reduce cytochrome C. The haeme units of CIV catalyse the transfer of electrons from reduced cytochrome C to molecular oxygen to form water. In turn, the oxidized DAB forms a brown coloured indamine polymer that is deposited at the reaction site. As continuous reoxidation of cytochrome C is required for the accumulation of the visible oxidized DAB, the reaction serves to visualize CIV activity [40]. Addition of catalase prevents contamination from endogenous peroxidase activity. In keeping with their physiological properties, in skeletal muscle, type I fibres show the darkest staining, type IIA fibres intermediate staining and type IIB (IIX) fibres show the weakest staining for all three reactions, commensurate with mitochondrial enrichment in these fibres. As the SDH and COX reactions specifically demonstrate CII and CIV activities, they are regarded as specific mitochondrial markers, and the intensity and distribution of staining allows simultaneous assessment of complex activity, mitochondrial mass and distribution at the single cell level as well as within a spatial two-dimensional context of the section. A further refinement in technique is the development of the sequential COX-SDH reaction, with early studies dating back to 1968, and now widely regarded as the optimal technique for demonstrating CIV defects, particularly in instances of heteroplasmic mtDNA defects. The technique relies on the preserved activity of CII (being entirely nDNA encoded) in cells with mtDNA defects. In cells with functional CIV, the brown indamine polymer product will localize in and saturate mitochondrial cristae. Those cells with reduced or absent CIV activity will not be saturated by the brown indamine polymer product, allowing for visualization of CII activity by deposition of the blue formazan end product [41–43]. Thus, in a normal skeletal muscle cross-section, the brown COX staining overshadows the blue SDH staining in all three fibre types. However, CIV deficient fibres will stand out as varying intensities of blue, depending on the deficiency being partial or complete. A number of variations in protocols for these histochemical reactions exist, and no attempt has been made yet to standardize protocols, at least amongst larger reference laboratories. Furthermore, these tests are highly susceptible to a variety of artefacts arising from poor sampling and processing, contributing to ambiguous results, or worse, false-positives and false-negatives. Excessive trauma to fibres, excess saline contact, drying out, repeated cycles of freeze-thawing, and drawing the hydrophobic barrier pen too close to the section can cause loss of activity. Hypercontracted fibres can cause fibres to appear darker. The sequential COX-SDH reaction should not be substituted for, but rather be run in parallel with individual COX and SDH reactions. Several quality control measures to preserve the integrity of the COX-SDH reaction have been outlined [44].

5. Canonical Pathological Features

Pathological changes in skeletal muscle biopsies from individuals with mtD can be varied, depending on the underlying genotype. Although not entirely specific, ragged red fibres (RRF) and COX-negative fibres are widely regarded to be the canonical features of mtD pathology [6,7,12,21,22,26,45–48]. Another useful diagnostic feature is SDH deficiency, although rare.

5.1. Ragged Red Fibres (RRF)

The recognition of RRF as a morphological hallmark of mtD predates the molecular era. Mitochondrial myopathies were described in the 1960s when systematic histochemical and ultrastructural studies revealed excessive proliferation of normal or abnormal-looking mitochondria in skeletal muscle of patients with weakness or exercise intolerance [30,49,50]. With the development of the modified Gomori Trichrome (MGT) stain allowing visualisation of connective tissue (light green), nuclei (red/purple), mitochondria, sarcoplasmic reticulum, sarcolemma (red) and myofibrils (green) in frozen sections, the abnormal fibres in these conditions showed up as bright red accumulation of staining and "cracking" of the fibre edges, corresponding to the massive irregular proliferation of mitochondria, and were dubbed "ragged red" [51]. The reason for the red staining is the affinity of chromotrome-2R, one of the MGT constituents that is lipophilic, and binds to sphingomyelin that is in abundance in mitochondrial membranes [21]. RRF usually show ultrastructurally abnormal mitochondria that frequently contain paracrystalline inclusions [47]. RRF are difficult to identify with MGT in formalin-fixed tissue as normal myofibrils stain red post-fixation. The red proliferative zones can be identified in Haematoxylin and Eosin-stained sections as subsarcolemmal areas of amorphous, basophilic staining [21]. SDH histochemistry shows increased staining in RRF, and such fibres appear as ragged blue fibres [22]. The term ragged red fibre equivalents (RRF equivalents) has been used to describe muscle fibres with mitochondrial accumulation showing positive staining with the modified SDH reaction, and the modified SDH reaction was demonstrated to be more sensitive than MGT in highlighting myofibres with increased mitochondrial proliferation [52]. RRF accumulate a very high percentage of mutant mitochondrial genomes >80% [53]. In longitudinal sections, RRF appear as segmental abnormalities, and there is a correlation between defective OXPHOS and the segmental abnormality suggesting that the abnormal proliferation is consequent to the defective OXPHOS [45,54]. RRF are seen in syndromic presentations of defects in mitochondrial protein synthesis (mtDNA rearrangements and point mutations), being more prevalent in MELAS, MERRF, KSS, and less frequently in CPEO [7,12]. RRF are usually absent in patients with syndromic presentations of defects in mitochondrial protein coding genes (LHON, NARP), however, few RRF may be seen in myopathic forms with isolated defects of CI, CIII and CIV due to mutations in mtDNA genes encoding ND subunits, cytochrome b, and COX subunits respectively [7]. In patients with mutations in nDNA genes encoding subunits or ancillary proteins of the RC, RRF are usually absent, e.g., in autosomal recessive Leigh syndrome due to mutations in CI or CII subunits, and Mendelian defects in assembly factors of CIV causing COX-deficient Leigh syndrome [7,30]. RRF are present in myopathic and encephalomyopathic forms of primary CoQ10 deficiency [55]. RRF are also present in myopathic forms of mtDNA depletion syndromes [56]. COX-negative (ragged blue) RRF are typically seen in MERRF, KSS and CPEO, when wild type mtDNA genomes fall below the threshold required to maintain CIV activity. In contrast, in classic MELAS due to the A3243G tRNA^{Leu} gene mutation, RRF are mostly COX-positive due to an even distribution of mutant and wild-type mtDNA genomes in these fibres [12,48,57,58]. RRF are COX-positive in mtDNA encoded CI and CIII subunit gene mutations, and COX-negative in CIV subunit gene mutations [7,12]. COX-negative RRF have also been observed in mtDNA depletion syndromes [59].

5.2. COX-Negative Fibres

A majority of individuals with mitochondrial myopathies that cause isolated or combined CIV deficiency due to mtDNA mutations harbour a mix of wild-type and mutant mtDNA molecules within each myofibre giving rise to heteroplasmy, a unique aspect of mitochondrial genetics. The proportion of mutant mtDNA can vary between individual myofibres [60,61]. A myofibre segment will develop biochemical OXPHOS deficiency when the mutant mtDNA exceeds a critical threshold, i.e., the level to which the cell can tolerate defective mtDNA molecules [62,63]. Heteroplasmic mutations have a variable threshold in different tissues. Furthermore, for unknown reasons, the threshold varies among mutation types and in skeletal muscle the mutation load for any particular tRNA (~90%) is typically higher than that for large-scale partial deletions of mtDNA (~70–80%) [61,64]. The threshold for mutation load in polypeptide-coding genes can be similarly broad, with low levels of mutation causing one type of clinical presentation and higher levels causing another, e.g., m.8993T→G mutation in the ATP synthase 6 (*ATP6*) gene: at mutation loads above 90%, manifests as maternally inherited Leigh's syndrome (MILS), at mutation loads in the range of 70–90%, manifests with neuropathy, ataxia and retinitis pigmentosa (NARP), and by contrast, patients with 70% mutation in a tRNA will rarely display overt disease [48]. The consequent biochemical OXPHOS deficiency can be demonstrated histochemically in transverse sections of frozen skeletal muscle as a mosaic pattern of COX-positive and COX-negative fibres, which equally affect slow-twitch (oxidative) and fast-twitch (glycolytic) muscle fibres [63,65], and is considered a hallmark of mitochondrial disease [66]. In the stand-alone COX reaction, the COX-negative fibres appear unstained amongst the brown COX-positive fibres. In the SDH reaction, they often stain intense blue due to compensatory mitochondrial proliferation increasing the mitochondrial mass, and the fact that CII being entirely nuclear-encoded, its biochemical activity is usually intact in primary mtDNA defects. In the sequential COX-SDH reaction, COX-negative fibres appear blue amongst the brown COX-positive fibres. The broad genotypic correlations described above for RRF are also true for the presence of COX-negative fibres in mitochondrial myopathies. A mosaic pattern of COX-negative fibres is a robust marker of heteroplasmic mtDNA mutations affecting mitochondrial protein synthesis (rearrangements and point mutations in mitochondrial tRNA or ribosomal RNA genes), or rarely, affecting one of the three mtDNA encoded CIV sub-units [12,22,31]. A notable exception to this rule is a homoplasmic mutation in mitochondrial tRNA[Glu] that is associated with a severe but reversible infantile mitochondrial myopathy and profound, though not exclusive, biochemical and histochemical COX deficiency, often accompanied by RRF in skeletal muscle [67,68]. The biopsy features in reversible and irreversible, fatal COX deficiency in the neonatal period are identical, and in both conditions the histochemical defect is restricted to extrafusal myofibres, sparing intrafusal muscle fibres and vascular smooth muscle [68–71]. Fatal infantile COX deficiency also affects the heart and brain, and has been linked to autosomal recessive mutations in COX assembly factors (*SCO2*, *COX15*, *COA5*, and *COA6*) [72–75]. Muscle biopsies from patients with defects in mtDNA maintenance will also show a similar mosaic COX-negative pattern due to overlapping influence of Mendelian and mitochondrial genetics, especially in cases of PEO with multiple mtDNA deletions [31,76,77]. In contrast, Mendelian disorders such as COX-deficient Leigh syndrome, e.g., due to mutations in COX assembly factors such as *SURF1*, will show diffuse COX-deficiency [78–80]. A mosaic COX deficiency is also found in cardiac muscle [81,82], renal cells [83] and the central nervous system [84]. The percentage of COX-negative fibres often correlates with disease severity and progression [85]. The biochemical defect develops within individual muscle fibres independent of the status of adjacent myofibres, likely due to the clonal expansion of mutant mtDNA, and appears to be an intrinsic property of the intracellular mitochondrial genome, largely independent of the nuclear genome [86,87]. Heteroplasmic mtDNA mutations are unevenly distributed along longitudinal syncytial muscle fibres, such that adjacent segmental sections can have widely varying amounts of mutant mtDNA [31]. Defined regions of COX deficiency have been documented in biopsies from patients with CPEO and MERRF [88,89]. There can be a striking variation in the length of COX-negative segments in the same biopsy, and the same muscle fibre can contain multiple non-contiguous COX-negative segments [60]. The latter suggests that the COX deficit may appear at multiple sites along a diseased fibre, with the length of the COX-negative segments expanding over time to coalesce with other COX-negative segments [90]. This could be due to continuous mtDNA replication in syncytial myofibres leading to changing proportions of mutant mtDNA through random intracellular genetic drift, and its lengthwise propagation over time [87]. Long COX-negative segments may eventually cause fibre atrophy, but do not lead to acute myonecrosis. It is also apparent that in biopsies from patients with various mtDNA mutations, a spectrum of deficiency exists with presence of fibres that show staining properties between completely COX-negative (blue) and COX-positive (brown) fibres, so called COX-intermediate fibres. COX-intermediate fibres, in part, represent the transition zones between COX-positive and COX-negative segments [91]. A significant difference was observed between COX-intermediate fibres and COX-positive as well as COX-negative fibres for mutant mtDNA, even more significant for wild-type mtDNA, but not for the total mtDNA copy number, suggesting that it is the wild-type mtDNA that is the critical determinant in determining the COX activity status [92,93]. The prevalence of RRF and COX-negative fibres may vary in biopsies depending on the genotype. Frequencies of RRF and COX-negative fibres are reported to be lower in MELAS and MERRF due to mtDNA point mutations than in CPEO due to mtDNA deletion, and are usually absent in LHON due to mtDNA point mutations [94–96]. COX-negative fibres have been noted to occur more frequently than RRF in CPEO patients associated with mtDNA point mutations and single deletions, and multiple mtDNA deletions due to *POLG1* mutations [97]. Levels of mtDNA heteroplasmy appear to directly correlate with the frequencies of RRF and COX-negative fibres [98]. In patients with primary mtDNA mutations, despite high levels of mutant mtDNA genomes in mature muscle, myogenic progenitor satellite cells have low to undetectable levels of the causative mutation [85,99,100]. Resistance exercise strength training in a group of mitochondrial myopathy patients due to a single large mtDNA deletion led to improved muscle strength, exercise induced necrosis and regeneration, increased numbers of NCAM+ satellite cells, and increased oxidative capacity including decreased percentage of COX-negative fibres and increased percentage of COX-intermediate fibres. This likely reflects a satellite cell-derived genetic drift in favour of wild-type mitochondrial genotype [91]. Taken together, it appears that, in addition to the absolute number of COX-negative fibres, the length of COX-negative segments, as well as COX-intermediate fibres, are important phenotypes to assess mitochondrial disease severity, progression and the effects of therapeutic interventions on mtDNA mutation levels and biochemical activities.

5.3. SDH Deficiency

Isolated CII deficiency is a rare Mendelian mitochondrial disease due to autosomal recessive mutations in the nuclear-encoded structural sub-units and assembly factor genes of CII (*SDHA*, *SDHB*, *SDHD*, and *SDHAF1*) [101–104]. Most reported cases are of early onset, presenting with Leigh syndrome, cardiomyopathy, leukodystrophy or encephalomyopathy, with the exception of autosomal dominant mutation in *SDHA* presenting with late onset optic atrophy, ataxia and myopathy [105]. Biochemical measurement of CII in muscle is the most reliable means of diagnosis, with levels reduced to 50% or greater compared to reference mean levels. Histochemically a diffuse reduction in SDH staining with normal COX staining is demonstrable [102,106]. CII deficiency with histochemically demonstrable diffuse reduction in SDH staining in skeletal muscle, but sparing of intramuscular blood vessels is reported with autosomal recessive mutations in *ISCU* encoding for iron sulphur cluster scaffold protein, presenting with myopathy, exercise intolerance and lactic acidosis. Additional features include increased iron deposition in mitochondria and aconitase deficiency [107,108].

6. Associated Pathological Features

Muscle biopsy may appear histologically normal, e.g., in patients with CI deficiency due to recessive mutations in nuclear-encoded subunits, in patients with mild RC defects, or early on in the disease course. Even in cases with heteroplasmic mtDNA mutations, there may be little abnormality apart from the presence of canonical features. Myopathic changes such as increased fibre-size variation and internal nucleation when present are typically of mild-to-moderate severity. Inflammation is absent, and necrosis and regeneration are not seen, except in mitochondrial myopathies presenting with rhabdomyolysis. Rhabdomyolysis has been associated with mutations in *CoQ2*, mtDNA encoded CIV subunit genes (*MT-CO1*, *MT-CO2*, *MT-CO3*), and tRNA genes (*MTT1*, *MT-TL1* m.3243 A > G MELAS mutation) [109–120]. Even late in the disease course, overtly dystrophic features with necrosis, fibrosis and fatty infiltration are not seen, with the exception of *TK2*-related myopathic form of mtDNA depletion syndrome [121,122]. Variable slow/type I fibre predominance and fast/type II atrophy may be present. Increased lipid may be present in fibres with or without ragged red change and COX deficiency, e.g., in KSS and PEO due to mtDNA rearrangements [21], mtDNA depletion syndrome due to mutations in *TK2*, *RRM2B*, *SUCLA2* and *SUCLG1*, and *CoQ2* [31,56,123–125]. Secondary carnitine deficiency with lipid storage can occur in patients with primary RCE defects [126]. The lipid storage is generally less florid when compared to primary lipid storage myopathies with massive lipidosis (primary carnitine deficiency, neutral lipid storage disease and multiple acyl-coA dehydrogenase deficiency) [127], and the presence of RRF and/or COX-negative fibres is not typically seen in the latter, although rare exceptions are reported [126]. The distinction between primary mtD and primary lipid myopathy is not possible based on muscle pathology alone, particularly in the absence of canonical mitochondrial pathology, and mild or inconstant lipidosis in the biopsy.

7. Myopathology in Novel Mitochondrial Diseases

Recessive loss-of-function mutations in *CHKB* that encodes choline kinase β, an enzyme that catalyses the first de-novo biosynthetic step of phosphatidylcholine, the most abundant mitochondrial membrane phospholipid that is formed through a pathway within the mitochondria-associated endoplasmic reticulum membrane (MAM), cause a congenital muscular dystrophy with raised serum CK, severe intellectual disability with skeletal and cardiac muscle involvement, and characteristic biopsy appearances that include enlarged mitochondria at the periphery, and loss of mitochondria in the centres of myofibres, probably as a result of elimination through mitophagy and compensatory enlargement [128]. The relationship between phospholipid and mitochondrial abnormalities could be mediated via the MAM, as several proteins involved in mitochondrial dynamics are an integral part of MAM, and MAM dysfunction may mediate increase in size and intracellular displacement of mitochondria [31,129,130]. In most cases, mild dystrophic changes are consistently present in biopsies. There is variable biochemical RCE deficiency. Muscle choline kinase activity and phosphatidylcholine content are markedly reduced with aberrant remodelling of phosphatidylcholine. Loss-of-function mutations in *MICU1*, a regulator of the inner mitochondrial complex MCU, responsible for regulating mitochondrial CA^{2+} uptake and preserving normal mitochondrial CA^{2+} concentration are reported to cause a childhood-onset disease with raised CK, relatively static proximal myopathy, variable CNS involvement, and distinctive biopsy features including preserved fibre typing, mild central nucleation, mini-cores and clustered regeneration. Biochemical or histochemical RCE defects are not yet reported. There is significant loss of *MICU1* mRNA and protein in muscle, with dysfunctional mitochondrial CA^{2+} uptake in fibroblasts resulting in CA^{2+}-induced fragmentation of mitochondrial networks [131]. More recently, dominant heterozygous and recessive compound heterozygous loss-of-function mutations in *MSTO1* have been characterised by whole exome sequencing in patients presenting with a multisystem disease characterized mainly by myopathy, ataxia, endocrine dysfunction and psychiatric symptoms. Serum CK ranges from normal to moderate elevation, and biopsies show myopathic or dystrophic changes, without histochemical and biochemical RC OXPHOS deficiency. Reduced levels of MSTO1 mRNA and protein in fibroblasts is associated with abnormalities of the mitochondrial network including fragmentation, aggregation, decreased network continuity and fusion activity, pointing to a putative role for *MSTO1* in mitochondrial morphogenesis by regulating mitochondrial fusion, and loss-of-function mutations linked to a multisystem mitochondrial disease [132,133]. Defects in *CHKB*, *MICU1* and *MSTO1* are examples of novel pathomechanisms and overlapping clinicopathological features involving muscular dystrophy, lipid metabolism, congenital myopathy and mitochondrial biology, with unique and recognizable muscle pathology signatures in absence of primary OXPHOS defects involving the RCE.

8. Vascular Pathology

Mitochondrial vasculopathy can manifest in large blood vessels (macroangiopathy) or small blood vessels (microangiopathy) including small arteries, arterioles, venules and capillaries. The clinical manifestations of macroangiopathy include premature atherosclerosis, arterial ectasia, vascular malformation, spontaneous rupture and reduced flow-mediated vasodilation [134]. In a 15-year old girl with the m.3243 A > G mutation, fatal spontaneous rupture of the aorta was associated with disorganisation and reduced COX staining in the vascular smooth muscle cells (VSMCs) of the aortic vasa vasora, and 85% mutation load in the arteries compared to 40% in blood lymphocytes [135]. Microangiopathy can manifest clinically as leukoencephalopathy, migraine-like headaches, stroke-like episodes or peripheral retinopathy. Careful morphological assessment in skeletal muscle or other tissues may reveal morphological abnormalities in VSMCs, pericytes or endothelial cells suggesting a subclinical microangiopathy [136]. MELAS is a multisystem mtD with predominant involvement of the brain, skeletal muscle and endocrine organs [137]. Unique to MELAS, particularly in association with the m.3243 A > G mutation, are transient stroke-like episodes due to lesions in the temporal and occipital lobes. Histologically, these lesions resemble true infarcts in that they are pan-necrotic and demonstrate profound neuronal loss, microvacuolation, gliosis and eosinophilic change in surviving neurons, but their topographic distribution does not follow major vascular territories or their watershed [138]. However, the presence of a microangiopathy, both within the lesions and in extra-CNS tissue like skeletal muscle, has long been recognised, manifesting as strongly SDH reactive vessels (SSVs) containing increased mtDNA copy number and ultrastructurally enlarged mitochondria. Similar SSVs can be found in MERRF, but angiopathy is less prevalent. Similar to RRF, SSVs in MERRF are typically COX-negative, whereas in MELAS, they are COX-positive [57,58]. It is postulated that the absolute amount of COX in SSVs in MELAS due to compensatory proliferation is far greater than normal [58]. As COX binds to nitric oxide, a key molecular signal for vasodilation, supernormal COX levels in these vessels titrate out nitric oxide, preventing cerebral vasodilation and triggering the stroke-like episodes [139,140]. The microangiopathy is not restricted to m.3243 A > G MELAS patients, but also documented in patients with m.8344 A > G, and autosomal recessive *POLG* mutations [138]. In these patients, multiple ischaemic stroke-like lesions in the cerebellar cortex were associated with microvascular abnormalities including loss of VSMCs and endothelial cells, evidence of blood-brain-barrier breakdown with plasma protein extravasation and loss of endothelial tight junctions, with accompanying high heteroplasmy levels of mutated mtDNA in the vessel wall. Despite clear evidence of a structurally damaged or altered microvasculature in association with vascular COX-deficiency, precisely how these deficiencies lead to the cerebral vascular events is not fully understood. It is also not known if a more generalised sub-clinical microangiopathy is present in mtD with diverse genetic backgrounds. In patients with mitochondrial myopathy, muscle capillary growth was increased as a result of impaired OXPHOS by a hypoxia-independent mechanism, promoting increased blood flow to respiration-incompetent muscles and a mismatch between systemic oxygen delivery and oxygen utilization during cycle exercise. The capillary area was greatest in patients with more severe oxidative deficits, and twice higher around fibres with oxidative defects compared to fibres with preserved oxidative function [141]. Vascular proliferation is a characteristic pathological feature of Leigh's encephalopathy due to a variety of mitochondrial defects causing severe OXPHOS deficits in the developing CNS [142]. Therefore, capillary proliferation driven by impaired OXPHOS may be a common consequence of mtD in highly oxidative tissues.

9. Ultrastructure: Pathological Features and Role in Diagnostics

A range of morphological alterations has been historically documented in patients with mitochondrial myopathy with transmission electron microscopy (TEM). These include excessive numbers of mitochondria in subsarcolemmal and intermyofibrillar locations; variation in size and shape including bizarre forms, excessively large size or length exceeding 3–4 sarcomeres; abnormalities of cristae including deficient cristae, abnormal stacking or whorling; a total absence of cristae with an amorphous granular substance replacing the cristal space; electron-dense granules; and paracrystalline structures with regular geometric periodicity [21]. Of all features, paracrystalline structures are regarded as the most pathognomonic, and are frequently present in RRF. Paracrystalline structures represent mitochondrial creatine kinase crystal formation due to upregulated activity in an attempt to compensate for the energy deficit [22]. However, these morphological changes lack specificity and may be seen in a range of myopathic and dystrophic conditions. They rarely ever provide clues to the underlying biochemical and/or genetic defect [14,20–22,143]. Simplification of cristae with accumulation of homogenous material is apparently a specific change seen in mtDNA depletion syndrome [144]. In adult biopsies with normal light microscopic findings, ultrastructural examination is unlikely to provide additional evidence of disease. Based on a small series of five patients, it has been suggested that the earliest ultrastructural changes in infants are often noted in endothelial cells of intramuscular blood vessels even when the myofibres themselves do not show histological or ultrastructural abnormalities [145]. Given the overall lack of specificity and the time and expense involved, the routine application of electron microscopy in the investigation of suspected mtD is questionable, particularly in the era of advanced molecular diagnostics. A recent ultrastructural study combining TEM with serial block face scanning EM (SBF-SEM) and 3D reconstruction techniques has reported features not previously described in patients with mtD include linearisation and angular arrangement of cristae, localised membrane distension, nanotunnels, and donut-shaped mitochondria. Systematic assessment of mitochondrial morphology using quantitative EM methodologies sensitive mitochondrial size, shape, and branching complexity and particularly three-dimensional reconstruction methods such as serial block face (SBF-SEM) and focused-ion beam (FIB-SEM), could be used in the future to ascertain the role of structural remodelling in certain mitochondrial and other musculoskeletal diseases [146].

10. Secondary Mitochondrial Abnormalities

Neither presence of RRF nor focal COX deficiency is entirely specific for primary mtD. Similar changes may be seen in skeletal muscle in the context of ageing, and in a range of genetic and acquired disorders. These include infantile Pompe disease and adult-onset acid maltase deficiency [21], occasionally in muscular dystrophies such as LGMD2A [147] and FSHD [148] rarely primary lipid storage myopathies [126]. Muscle biopsies of patients with inclusion body myositis (IBM) may show increased numbers of RRF and COX-negative fibres [149]. In IBM, the on-going inflammation and cytokine environment, the associated production of reactive oxygen and nitrogen species, and the associated endoplasmic reticulum stress have a role in the initiation of mitochondrial DNA damage, leading to the accumulation of clonally-expanded mtDNA deletions and respiratory deficiency, a phenomenon that is not compensated by the malfunctioning cell repair mechanisms [150]. Increased numbers of COX-deficient and SDH-positive fibres within atrophic perifascicular zones are a common feature in dermatomyositis [151]. Histochemical and biochemical OXPHOS dysfunction can be induced by the toxic effects of a range of drugs on mitochondrial respiration, including antiretroviral agents and statins [152], antiepileptics such as valproate, immunosuppressant and cytotoxic chemotherapeutic agents [14,20]. Accumulation of multiple mtDNA deletions and tRNA point mutations has been observed in ageing human tissues [153,154] with highest levels in post-mitotic tissues such as brain and skeletal muscle. Increased numbers of RRF and COX-negative fibres are seen in skeletal muscle of older individuals and RRF comprise an average of 0.4% of all fibres by the eighth decade [155]. The age-related mtDNA mutations appear to accumulate randomly in certain myofibre segments to very high levels resulting in focal COX deficiency [156]. Overall, the amount of mutant mtDNA is very low in ageing muscle compared to patients with mitochondrial myopathy and is unlikely to cause a clinically significant OXPHOS defect [45]. Nevertheless, late-onset mitochondrial myopathy has been documented in patients over 69 years of age with multiple mtDNA deletions and increased numbers of RRF and COX-negative fibres in biopsies, possibly representing an exaggerated form of age-related mitochondrial dysfunction [157].

11. Myopathology of Paediatric mtD

In contrast to adults who more often present with well-defined syndromic mtD, paediatric presentations of mtD are harder to define. Neonatal or early infantile disease onset is often associated with severe progressive encephalomyopathy, with multi-organ involvement such as cardiomyopathy, hepatopathy, and myopathic involvement suggested by hypotonia, muscle weakness, wasting and arthrogryposis [123,158–160]. Over 90% of paediatric patients with mtD carry mutations in their nuclear genes causing defective OXPHOS [66]. This explains the long-held observation that mosaic RRF and/or COX-negative fibres are uncommon in biopsies of these patients. RRF and/or COX-negative fibres were demonstrated in 89% of biopsies with mtDNA mutations but only in 17% of biopsies without detectable mtDNA mutations in a large series of 117 children with mtD [161]. RRF and COX-negative fibres, and increased lipid are usually present in biopsies from children with mtDNA depletion syndromes secondary to defects in nuclear gene involved in mtDNA maintenance and in the myopathic form of CoQ10 deficiency [158]. COX-deficient fibres may outnumber RRF and may be the only abnormal finding in the muscle biopsy [162]. In neonates, there may be no detectable light microscopic abnormality [163]. This suggests that the compensatory proliferative response may develop over time to form RRF. The small size of fibres in biopsies from neonates and infants may make recognition of the morphological abnormality more difficult. It has been suggested that SDH-positive subsarcolemmal mitochondrial aggregates (SSMA) representing a milder form of mitochondrial proliferation is more prevalent in paediatric mtD [12]. More than 2% SSMA in patients under 16 years has been listed as a minor diagnostic criterion [23]. The sensitivity and specificity of this marker has been questioned. Such mitochondrial proliferation was absent in 35% of paediatric patients with proven mitochondrial dysfunction [164]. In 95 patients under 16 years of age, there was no difference in the frequency of SSMA between patients with and without definite mtD. Large SSMAs were observed to be more frequent in the group with definite mtD [165]. A large-scale retrospective study evaluating factors associated with SSMA in paediatric biopsies with suspected mtD found an inverse relationship between the percentage of myofibres with SSMA and RCE deficiency. Patients with low %SSMA ($\leq$4%) were significantly more likely to develop RCE deficiency than patients with higher %SSMA ($\geq$10%) [166]. However, it is important to note that the morphology of mitochondrial networks changes from birth to adolescence and SSMAs appear to develop over time, even in biopsies from patients in whom a primary neuromuscular disease has been excluded. Therefore, any diagnostic cut-off must take into account the confounding effect of age, and assessment of multi-centre large-scale cohorts will be necessary to develop age-stratified SSMA cut-offs with sufficiently high sensitivity and specificity to serve as a useful histological diagnostic indicator of RC deficiency in children.

12. Consensus Diagnostic Criteria

To facilitate the diagnosis of mitochondrial diseases, various expert groups have proposed consensus criteria and classification systems incorporating clinical, physiological, biochemical, morphological and molecular genetic criteria. Given the overlap in morphological findings associated with primary mtD and mitochondrial abnormalities secondary to ageing and various inherited and acquired non-mitochondrial neuromuscular disorders, morphological criteria essentially involve a quantitative evaluation of RRF, COX-negative fibres and SSMA in diagnostic muscle biopsies. Varying diagnostic cut-offs have been proposed including: >2% RRF and/or >2% COX-negative fibres for individuals <50 years, or >5% COX-negative fibres for individuals > 50 years [23,167–169]. Occasional COX-negative fibres are regarded normal > 40 years and the proportion increases with age. Any RRF < 30 years is regarded as being suspicious to warrant investigation of mtD [169]. A major limitation of the studies used to formalise such cut-offs is the lack of standardized methodology: differing biopsy sites, differing histological techniques and differing methods of quantitative assessment [52,170,171]. Age-matched control groups included patients with chronic myopathies and myositis confounding assessment due to prevalence of secondary mitochondrial abnormalities [157]. In a post mortem study evaluating biopsies from patients with and without mtD [172], <0.1% abnormal fibres were present in controls before the fifth decade. The proportion of abnormal fibres increased with age and there were regional differences (deltoid > quadriceps). Most patients with mtD had more than 0.5% abnormal fibres. Overall, COX-negative fibres were more numerous than RRF or SDH-positive fibres and provided a sensitive measure of mitochondrial abnormality. This study brings into question the widely used 2% cut-off given that the levels of abnormal fibres in controls were well below 1%. In the absence of other neuromuscular disease, mitochondrial abnormalities in muscle biopsies below the current 2% cut-off may be significant. Similarly, in paediatric biopsies, there are studies with findings challenging the currently used 2% SSMA cut-off as a minor diagnostic criterion; these are alluded to in the preceding paragraph. Notwithstanding various formal quantitative diagnostic cut-offs, it is important to remember that normal muscle morphology, especially in children, does not exclude mtD [173].

13. Recent Advances in Diagnostic and Research Tools: Immunoassays, Transcriptomics and Biomarkers

The absence of reliable histochemical assays to evaluate complex I, which is the largest and most commonly affected OXPHOS complex, as well as CIII and CV is a serious limitation to the histochemical analysis of RC defects in mtD. Catalytic deficiency of RC is most often associated with a decreased amount of the assembled complex. This fact underlies the application of immunohistochemistry as a tool for investigating RC defects [174–179]. Secondly, an ever-increasing array of highly specific monoclonal antibodies is available against components of the mitoproteome spanning the nDNA and mtDNA-encoded compartments. A severe and selective reduction of immunolabelled mtDNA encoded COXI and COXII subunits with normally labelled nDNA encoded COXIV and COXVa subunits in histochemically COX-negative fibres were observed in patients with mtDNA mutations. nDNA-encoded COXVIc immunostaining was however also reduced. This was thought to relate to the holoenzyme's quaternary structure with close interaction between COXII and COXVIc, while other nDNA encoded subunits with preserved immunoreactivity could form stable partial complexes in absence of mtDNA encoded subunits [180]. Different patterns of subunit expression were reported in the same study in mtDNA depletion syndrome including selective and non-selective loss of mtDNA encoded CIV subunits, suggesting differences in genetic background or the disease stage. Rapid protocols have been developed for fluorescent or peroxidase labelled immunostaining of cultured fibroblasts on coverslips or as cytospins using monoclonal antibodies [179,181]. Heteroplasmic mitochondrial tRNA mutations gave a heterogeneous immunostaining pattern for CI, CIII and CIV subunits as opposed to the uniformly reduced immunostaining seen in cell lines from patients with nuclear DNA defects [181]. Normal immunostaining despite reduced histochemical/biochemical activity of the corresponding complex may be due to the subcomplexes remaining active despite failure to assemble the holoenzyme, or formation of the holoenzyme with a kinetic defect. Activity dipstick assays, a type of lateral flow immunocapture assays that measure electron transfer activity of CI and CIV were developed as rapid, accurate and reproducible tests that combined the specificity of immunocapture monoclonal antibodies with the functionality of enzyme activity assays [182]. Immunolabelling for anti-DNA antibodies has been applied as an alternative to in situ hybridization to study mtDNA localization and distribution in cells. In mtDNA depletion, cytoplasmic labelling for mtDNA is either absent or reduced while the intensity of nDNA labelling is unchanged [183]. Subunit-specific immunohistochemistry can also provide insights into developmental regulation of tissue specific expression of respiratory chain complexes and their relevance in understanding disease mechanisms. Using a combination of isoform-specific antibodies (COX6AH, COX6AL, COX7AH, and COX7AL) for protein expression studies by immunohistochemistry on sections and immunoblotting muscle homogenates in combination with gene expression profiling, Boczonadi et al. demonstrated evidence for an isoform switch of COX6A and COX7A in skeletal muscle that occurs around three months of age, but there was no causative link between the isoform switch and clinical recovery in reversible infantile respiratory chain deficiency [184]. Rocha et al. have developed a quadruple immunofluorescent technique enabling quantification of key subunits of respiratory chain CI and CIV together with an indicator of mitochondrial mass and a cell membrane marker enabling protein quantitation in large numbers of fibres [185]. This technique is also able to demonstrate distinct biochemical signatures in association with specific genotypes providing insights into molecular mechanisms. For instance in patients with the common m.3243A > G *MT-TL1* mutation it was observed that CIV deficiency occurs only after CI deficiency is already established, and the defect is smoothly graduated from the normal to deficient levels of both complexes in contrast to the polarised pattern seen in the *MT-ND1* mutation. Several recent studies have shown that fibroblast growth factor 21 (FGF21), a growth factor with pleiotropic effects on regulating lipid and glucose metabolism is upregulated in patients with mtD, mice with RC deficiency, and mice with defective muscle autophagy/mitophagy [186–188]. mRNA and protein levels of FGF21 were robustly increased in patients with mitochondrial myopathy or MELAS. The increased FGF21 expression was shown to be a compensatory response to RC deficiency, effecting enhanced mitochondrial function through an mTOR-YY1-PGC1α-dependent pathway in skeletal muscle [189]. The accuracy of FGF21 to correctly identify muscle-manifesting mtD appeared to be higher than conventional biomarkers in one study [186]. Kalko et al. analysed the whole transcriptome of skeletal muscle in patients with *TK2* mutations and compared it to normal muscle and muscle in other mitochondrial myopathies. Bioinformatics pathway analysis identified the tumour suppressor p53 as the regulator at the centre of a network of genes responsible for a coordinated response to *TK2* mutations including induction of growth and differentiation factor 15 (GDF15), leading to its identification as a potential novel biomarker of mitochondrial dysfunction [190]. This was soon validated in two subsequent studies. One study measured the serum levels of (GDF15) against FGF21 and other conventional biomarkers in patients with mtD and healthy controls, and showed that the area under the receiver operating characteristic curve was significantly higher for GDF15 than FGF21 and other biomarkers [191]. Another study showed that elevated levels of GDF15 and FGF21 correctly identified a greater proportion of patients with mtD than GDF15 or FGF21 alone [192].

14. Conclusions

Despite rapid advances in genetic technologies and the increasing use of high-throughput next-generation-sequencing (NGS) platforms in the diagnostic pipeline for patients with suspected mtD, the laboratory investigation of mtD is still complex, and muscle biopsy remains a key tool that provides tissue for diagnostic and functional studies to direct molecular genetic testing. Demonstration of histochemical mosaic COX deficiency provides crucial evidence for a heteroplasmic mtDNA disease. The pathologist must take into account developmental, ageing-associated and secondary mitochondrial changes whilst interpreting mitochondrial pathology in muscle biopsies. Optimal handling and processing of tissue maximises the diagnostic yield in biopsies. With increasing adoption of NGS platforms in diagnostic laboratories comes the challenge of functional testing to determine pathogenicity for variants of uncertain significance found with increasing frequency. In this context, it is incumbent upon pathologists to develop novel pathology tools incorporating advances in tissue multiplexing and imaging; enabling more objective and informatics-based assessment of OXPHOS deficiency to improve the diagnostic outcome in mtD; understand pathomechanisms of mitochondrial dysfunction in primary mtD as well as in other diseases; monitor mitochondrial disease progression; and serve as biological outcome measures in clinical trials.

Acknowledgments: NHS England Highly Specialised Services Diagnostic Service for Congenital Muscular Dystrophies and Congenital Myopathies, Dubowitz Neuromuscular Centre, Great Ormond Street Hospital for Children NHS Foundation Trust, London, United Kingdom is acknowledged.

Conflicts of Interest: The author declares no conflict of interest.

References

1. Alston, C.L.; Rocha, M.C.; Lax, N.Z.; Turnbull, D.M.; Taylor, R.W. The genetics and pathology of mitochondrial disease. *J. Pathol.* **2017**, *241*, 236–250. [CrossRef] [PubMed]
2. Anderson, S.; Bankier, A.T.; Barrell, B.G.; de Bruijn, M.H.; Coulson, A.R.; Drouin, J.; Eperon, I.C.; Nierlich, D.P.; Roe, B.A.; Sanger, F.; et al. Sequence and organization of the human mitochondrial genome. *Nature* **1981**, *290*, 457–465. [CrossRef] [PubMed]
3. Calvo, S.E.; Clauser, K.R.; Mootha, V.K. MitoCarta2.0: An updated inventory of mammalian mitochondrial proteins. *Nucleic Acids Res.* **2016**, *44*, D1251–D1257. [CrossRef] [PubMed]
4. Lightowlers, R.N.; Taylor, R.W.; Turnbull, D.M. Mutations causing mitochondrial disease: What is new and what challenges remain? *Science* **2015**, *349*, 1494–1499. [CrossRef] [PubMed]
5. Pfeffer, G.; Chinnery, P.F. Diagnosis and treatment of mitochondrial myopathies. *Ann. Med.* **2013**, *45*, 4–16. [CrossRef] [PubMed]
6. Tuppen, H.A.; Blakely, E.L.; Turnbull, D.M.; Taylor, R.W. Mitochondrial DNA mutations and human disease. *Biochim. Biophys. Acta* **2010**, *1797*, 113–128. [CrossRef] [PubMed]
7. DiMauro, S.; Hirano, M. Mitochondrial encephalomyopathies: An update. *Neuromuscul. Disord.* **2005**, *15*, 276–286. [CrossRef] [PubMed]
8. Wong, L.J.; Scaglia, F.; Graham, B.H.; Craigen, W.J. Current molecular diagnostic algorithm for mitochondrial disorders. *Mol. Genet. Metab.* **2010**, *100*, 111–117. [CrossRef] [PubMed]
9. Triepels, R.H.; van den Heuvel, L.P.; Loeffen, J.L.; Buskens, C.A.; Smeets, R.J.; Rubio Gozalbo, M.E.; Budde, S.M.; Mariman, E.C.; Wijburg, F.A.; Barth, P.G.; et al. Leigh syndrome associated with a mutation in the NDUFS7 (PSST) nuclear encoded subunit of complex I. *Ann. Neurol.* **1999**, *45*, 787–790. [CrossRef]
10. Debray, F.G.; Mitchell, G.A.; Allard, P.; Robinson, B.H.; Hanley, J.A.; Lambert, M. Diagnostic accuracy of blood lactate-to-pyruvate molar ratio in the differential diagnosis of congenital lactic acidosis. *Clin. Chem.* **2007**, *53*, 916–921. [CrossRef] [PubMed]
11. Finsterer, J. Cerebrospinal-fluid lactate in adult mitochondriopathy with and without encephalopathy. *Acta Med. Austriaca* **2001**, *28*, 152–155. [CrossRef] [PubMed]
12. Haas, R.H.; Parikh, S.; Falk, M.J.; Saneto, R.P.; Wolf, N.I.; Darin, N.; Wong, L.J.; Cohen, B.H.; Naviaux, R.K.; Mitochondrial Medicine Society's Committee on Diagnosis. The in-depth evaluation of suspected mitochondrial disease. *Mol. Genet. Metab.* **2008**, *94*, 16–37. [CrossRef] [PubMed]
13. Haas, R.H.; Parikh, S.; Falk, M.J.; Saneto, R.P.; Wolf, N.I.; Darin, N.; Cohen, B.H. Mitochondrial disease: A practical approach for primary care physicians. *Pediatrics* **2007**, *120*, 1326–1333. [CrossRef] [PubMed]
14. Milone, M.; Wong, L.J. Diagnosis of mitochondrial myopathies. *Mol. Genet. Metab.* **2013**, *110*, 35–41. [CrossRef] [PubMed]
15. Gropman, A.L. Neuroimaging in mitochondrial disorders. *Neurotherapeutics* **2013**, *10*, 273–285. [CrossRef] [PubMed]
16. Dinopoulos, A.; Cecil, K.M.; Schapiro, M.B.; Papadimitriou, A.; Hadjigeorgiou, G.M.; Wong, B.; de Grauw, T.; Egelhoff, J.C. Brain MRI and proton MRS findings in infants and children with respiratory chain defects. *Neuropediatrics* **2005**, *36*, 290–301. [CrossRef] [PubMed]
17. Saneto, R.P.; Friedman, S.D.; Shaw, D.W. Neuroimaging of mitochondrial disease. *Mitochondrion* **2008**, *8*, 396–413. [CrossRef] [PubMed]

18. Brockmann, K.; Bjornstad, A.; Dechent, P.; Korenke, C.G.; Smeitink, J.; Trijbels, J.M.; Athanassopoulos, S.; Villagran, R.; Skjeldal, O.H.; Wilichowski, E.; et al. Succinate in dystrophic white matter: A proton magnetic resonance spectroscopy finding characteristic for complex II deficiency. *Ann. Neurol.* **2002**, *52*, 38–46. [CrossRef] [PubMed]

19. Simon, N.G.; Noto, Y.I.; Zaidman, C.M. Skeletal muscle imaging in neuromuscular disease. *J. Clin. Neurosci.* **2016**, *33*, 1–10. [CrossRef] [PubMed]

20. Delonlay, P.; Rötig, A.; Sarnat, H.B. Respiratory chain deficiencies. *Handb. Clin. Neurol.* **2013**, *113*, 1651–1666. [PubMed]

21. Sarnat, H.B.; Marín-García, J. Pathology of mitochondrial encephalomyopathies. *Can. J. Neurol. Sci.* **2005**, *32*, 152–166. [CrossRef] [PubMed]

22. Bourgeois, J.M.; Tarnopolsky, M.A. Pathology of skeletal muscle in mitochondrial disorders. *Mitochondrion* **2004**, *4*, 441–452. [CrossRef] [PubMed]

23. Bernier, F.P.; Boneh, A.; Dennett, X.; Chow, C.W.; Cleary, M.A.; Thorburn, D.R. Diagnostic criteria for respiratory chain disorders in adults and children. *Neurology* **2002**, *59*, 1406–1411. [CrossRef] [PubMed]

24. Wolf, N.I.; Smeitink, J.A. Mitochondrial disorders: A proposal for consensus diagnostic criteria in infants and children. *Neurology* **2002**, *59*, 1402–1405. [CrossRef] [PubMed]

25. Morava, E.; van den Heuvel, L.; Hol, F.; de Vries, M.C.; Hogeveen, M.; Rodenburg, R.J.; Smeitink, J.A. Mitochondrial disease criteria: Diagnostic applications in children. *Neurology* **2006**, *67*, 1823–1826. [CrossRef] [PubMed]

26. Wong, L.-J.C. *Biochemical and Molecular Methods for the Study of Mitochondrial Disorders*; Springer: New York, NY, USA, 2013; pp. 27–45.

27. Van den Heuvel, L.P.; Smeitink, J.A.; Rodenburg, R.J. Biochemical examination of fibroblasts in the diagnosis and research of oxidative phosphorylation (OXPHOS) defects. *Mitochondrion* **2004**, *4*, 395–401. [CrossRef] [PubMed]

28. Thorburn, D.R.; Smeitink, J. Diagnosis of mitochondrial disorders: Clinical and biochemical approach. *J. Inherit. Metab. Dis.* **2001**, *24*, 312–316. [CrossRef] [PubMed]

29. Dimmock, D.; Tang, L.Y.; Schmitt, E.S.; Wong, L.J. Quantitative evaluation of the mitochondrial DNA depletion syndrome. *Clin. Chem.* **2010**, *56*, 1119–1127. [CrossRef] [PubMed]

30. DiMauro, S. Mitochondrial diseases. *Biochim. Biophys. Acta* **2004**, *1658*, 80–88. [CrossRef] [PubMed]

31. DiMauro, S.; Schon, E.A.; Carelli, V.; Hirano, M. The clinical maze of mitochondrial neurology. *Nat. Rev. Neurol.* **2013**, *9*, 429–444. [CrossRef] [PubMed]

32. Carry, M.R.; Ringel, S.P. Structure and histochemistry of human extraocular muscle. *Bull. Soc. Belge Ophtalmol.* **1989**, *237*, 303–319. [PubMed]

33. Yu-Wai-Man, P.; Lai-Cheong, J.; Borthwick, G.M.; He, L.; Taylor, G.A.; Greaves, L.C.; Taylor, R.W.; Griffiths, P.G.; Turnbull, D.M. Somatic mitochondrial DNA deletions accumulate to high levels in aging human extraocular muscles. *Investig. Ophthalmol. Vis. Sci.* **2010**, *51*, 3347–3353. [CrossRef] [PubMed]

34. Marín-García, J.; Goldenthal, M.J.; Sarnat, H.B. Probing striated muscle mitochondrial phenotype in neuromuscular disorders. *Pediatr. Neurol.* **2003**, *29*, 26–33. [CrossRef]

35. Bourgeron, T.; Chretien, D.; Rötig, A.; Munnich, A.; Rustin, P. Fate and expression of the deleted mitochondrial DNA differ between human heteroplasmic skin fibroblast and Epstein-Barr virus-transformed lymphocyte cultures. *J. Biol. Chem.* **1993**, *268*, 19369–19376. [PubMed]

36. Dubowitz, V.; Sewry, C.A.; Oldfors, A.; Lane, R.J.M. *Muscle Biopsy: A Practical Approach*, 4th ed.; Saunders: Oxford, UK, 2013; p. 1. Available online: https://www.clinicalkey.com/dura/browse/bookChapter/3-s2.0-C2009063539X (accessed on 4 July 2017).

37. Rotig, A. Genetic bases of mitochondrial respiratory chain disorders. *Diabetes Metab.* **2010**, *36*, 97–107. [CrossRef] [PubMed]

38. Smeitink, J.A.M.; Sengers, R.C.A.; Trijbels, J.M.F. *Oxidative phosphorylation in Health and Disease*; Landes Bioscience/Eurekah.com: Georgetown, TX, USA; Great Britain, UK, 2004.

39. Blanco, C.E.; Sieck, G.C.; Edgerton, V.R. Quantitative histochemical determination of succinic dehydrogenase activity in skeletal muscle fibres. *Histochem. J.* **1988**, *20*, 230–243. [CrossRef] [PubMed]

40. Gonzalez-Lima, F. *Cytochrome Oxidase in Neuronal Metabolism and Alzheimer's Disease*; Plenum Press: New York, NY, USA, 1998.

41. Old, S.L.; Johnson, M.A. Methods of microphotometric assay of succinate dehydrogenase and cytochrome c oxidase activities for use on human skeletal muscle. *Histochem. J.* **1989**, *21*, 545–555. [CrossRef] [PubMed]
42. Seligman, A.M.; Karnovsky, M.J.; Wasserkrug, H.L.; Hanker, J.S. Nondroplet ultrastructural demonstration of cytochrome oxidase activity with a polymerizing osmiophilic reagent, diaminobenzidine (DAB). *J. Cell Biol.* **1968**, *38*, 1–14. [CrossRef] [PubMed]
43. Cottrell, D.A.; Blakely, E.L.; Johnson, M.A.; Ince, P.G.; Borthwick, G.M.; Turnbull, D.M. Cytochrome c oxidase deficient cells accumulate in the hippocampus and choroid plexus with age. *Neurobiol. Aging* **2001**, *22*, 265–272. [CrossRef]
44. Ross, J.M. Visualization of mitochondrial respiratory function using cytochrome c oxidase/succinate dehydrogenase (COX/SDH) double-labeling histochemistry. *J. Vis. Exp.* **2011**, *57*, e3266. [CrossRef] [PubMed]
45. Larsson, N.G.; Oldfors, A. Mitochondrial myopathies. *Acta Physiol. Scand.* **2001**, *171*, 385–393. [CrossRef] [PubMed]
46. Filosto, M.; Tomelleri, G.; Tonin, P.; Scarpelli, M.; Vattemi, G.; Rizzuto, N.; Padovani, A.; Simonati, A. Neuropathology of mitochondrial diseases. *Biosci. Rep.* **2007**, *27*, 23–30. [CrossRef] [PubMed]
47. Siciliano, G.; Volpi, L.; Piazza, S.; Ricci, G.; Mancuso, M.; Murri, L. Functional diagnostics in mitochondrial diseases. *Biosci. Rep.* **2007**, *27*, 53–67. [CrossRef] [PubMed]
48. Schon, E.A.; DiMauro, S.; Hirano, M. Human mitochondrial DNA: Roles of inherited and somatic mutations. *Nat. Rev. Genet.* **2012**, *13*, 878–890. [CrossRef] [PubMed]
49. Shy, G.M.; Gonatas, N.K. Human myopathy with giant abnormal mitochondria. *Science* **1964**, *145*, 493–496. [CrossRef] [PubMed]
50. Shy, G.M.; Gonatas, N.K.; Perez, M. Two childhood myopathies with abnormal mitochondria. I. Megaconial myopathy. II. Pleoconial myopathy. *Brain* **1966**, *89*, 133–158. [CrossRef] [PubMed]
51. Engel, W.K.; Cunningham, G.G. Rapid examination of muscle tissue. An improved trichrome method for fresh-frozen biopsy sections. *Neurology* **1963**, *13*, 919–923. [CrossRef] [PubMed]
52. Rifai, Z.; Welle, S.; Kamp, C.; Thornton, C.A. Ragged red fibers in normal aging and inflammatory myopathy. *Ann. Neurol.* **1995**, *37*, 24–29. [CrossRef] [PubMed]
53. Nishigaki, Y.; Tadesse, S.; Bonilla, E.; Shungu, D.; Hersh, S.; Keats, B.J.; Berlin, C.I.; Goldberg, M.F.; Vockley, J.; DiMauro, S.; et al. A novel mitochondrial tRNA(Leu(UUR)) mutation in a patient with features of MERRF and Kearns-Sayre syndrome. *Neuromuscul. Disord.* **2003**, *13*, 334–340. [CrossRef]
54. Larsson, N.G.; Wang, J.; Wilhelmsson, H.; Oldfors, A.; Rustin, P.; Lewandoski, M.; Barsh, G.S.; Clayton, D.A. Mitochondrial transcription factor A is necessary for mtDNA maintenance and embryogenesis in mice. *Nat. Genet.* **1998**, *18*, 231–236. [CrossRef] [PubMed]
55. Quinzii, C.M.; DiMauro, S.; Hirano, M. Human coenzyme Q10 deficiency. *Neurochem. Res.* **2007**, *32*, 723–727. [CrossRef] [PubMed]
56. El-Hattab, A.W.; Scaglia, F. Mitochondrial DNA depletion syndromes: Review and updates of genetic basis, manifestations, and therapeutic options. *Neurotherapeutics* **2013**, *10*, 186–198. [CrossRef] [PubMed]
57. Hasegawa, H.; Matsuoka, T.; Goto, Y.; Nonaka, I. Cytochrome c oxidase activity is deficient in blood vessels of patients with myoclonus epilepsy with ragged-red fibers. *Acta Neuropathol.* **1993**, *85*, 280–284. [CrossRef] [PubMed]
58. Naini, A.; Kaufmann, P.; Shanske, S.; Engelstad, K.; De Vivo, D.C.; Schon, E.A. Hypocitrullinemia in patients with MELAS: An insight into the "MELAS paradox". *J. Neurol. Sci.* **2005**, *229–230*, 187–193. [CrossRef] [PubMed]
59. Mancuso, M.; Filosto, M.; Bonilla, E.; Hirano, M.; Shanske, S.; Vu, T.H.; DiMauro, S. Mitochondrial myopathy of childhood associated with mitochondrial DNA depletion and a homozygous mutation (T77M) in the TK2 gene. *Arch. Neurol.* **2003**, *60*, 1007–1009. [CrossRef] [PubMed]
60. Elson, J.L.; Samuels, D.C.; Johnson, M.A.; Turnbull, D.M.; Chinnery, P.F. The length of cytochrome c oxidase-negative segments in muscle fibres in patients with mtDNA myopathy. *Neuromuscul. Disord.* **2002**, *12*, 858–864. [CrossRef]
61. Sciacco, M.; Bonilla, E.; Schon, E.A.; DiMauro, S.; Moraes, C.T. Distribution of wild-type and common deletion forms of mtDNA in normal and respiration-deficient muscle fibers from patients with mitochondrial myopathy. *Hum. Mol. Genet.* **1994**, *3*, 13–19. [CrossRef] [PubMed]

62. Schon, E.A.; Bonilla, E.; DiMauro, S. Mitochondrial DNA mutations and pathogenesis. *J. Bioenerg. Biomembr.* **1997**, *29*, 131–149. [CrossRef] [PubMed]

63. Johnson, M.A.; Bindoff, L.A.; Turnbull, D.M. Cytochrome c oxidase activity in single muscle fibers: Assay techniques and diagnostic applications. *Ann. Neurol.* **1993**, *33*, 28–35. [CrossRef] [PubMed]

64. Yoneda, M.; Miyatake, T.; Attardi, G. Heteroplasmic mitochondrial tRNA(Lys) mutation and its complementation in MERRF patient-derived mitochondrial transformants. *Muscle Nerve* **1995**, *3*, S95–S101. [CrossRef] [PubMed]

65. Müller-Höcker, J.; Seibel, P.; Schneiderbanger, K.; Kadenbach, B. Different in situ hybridization patterns of mitochondrial DNA in cytochrome c oxidase-deficient extraocular muscle fibres in the elderly. *Virchows Arch. A* **1993**, *422*, 7–15. [CrossRef]

66. Taylor, R.W.; Schaefer, A.M.; Barron, M.J.; McFarland, R.; Turnbull, D.M. The diagnosis of mitochondrial muscle disease. *Neuromuscul. Disord.* **2004**, *14*, 237–245. [CrossRef] [PubMed]

67. DiMauro, S.; Nicholson, J.F.; Hays, A.P.; Eastwood, A.B.; Papadimitriou, A.; Koenigsberger, R.; DeVivo, D.C. Benign infantile mitochondrial myopathy due to reversible cytochrome c oxidase deficiency. *Ann. Neurol.* **1983**, *14*, 226–234. [CrossRef] [PubMed]

68. Horvath, R.; Kemp, J.P.; Tuppen, H.A.; Hudson, G.; Oldfors, A.; Marie, S.K.; Moslemi, A.R.; Servidei, S.; Holme, E.; Shanske, S.; et al. Molecular basis of infantile reversible cytochrome c oxidase deficiency myopathy. *Brain* **2009**, *132*, 3165–3174. [CrossRef] [PubMed]

69. Boczonadi, V.; Bansagi, B.; Horvath, R. Reversible infantile mitochondrial diseases. *J. Inherit. Metab. Dis.* **2015**, *38*, 427–435. [CrossRef] [PubMed]

70. Bresolin, N.; Zeviani, M.; Bonilla, E.; Miller, R.H.; Leech, R.W.; Shanske, S.; Nakagawa, M.; DiMauro, S. Fatal infantile cytochrome C oxidase deficiency: Decrease of immunologically detectable enzyme in muscle. *Neurology* **1985**, *35*, 802–812. [CrossRef] [PubMed]

71. DiMauro, S.; Lombes, A.; Nakase, H.; Mita, S.; Fabrizi, G.M.; Tritschler, H.J.; Bonilla, E.; Miranda, A.F.; De Vivo, D.C.; Schon, E.A. Cytochrome c oxidase deficiency. *Pediatr. Res.* **1990**, *28*, 536–541. [CrossRef] [PubMed]

72. Papadopoulou, L.C.; Sue, C.M.; Davidson, M.M.; Tanji, K.; Nishino, I.; Sadlock, J.E.; Krishna, S.; Walker, W.; Selby, J.; Glerum, D.M.; et al. Fatal infantile cardioencephalomyopathy with COX deficiency and mutations in *SCO2*, a COX assembly gene. *Nat. Genet.* **1999**, *23*, 333–337. [PubMed]

73. Alfadhel, M.; Lillquist, Y.P.; Waters, P.J.; Sinclair, G.; Struys, E.; McFadden, D.; Hendson, G.; Hyams, L.; Shoffner, J.; Vallance, H.D. Infantile cardioencephalopathy due to a COX15 gene defect: Report and review. *Am. J. Med. Genet. A* **2011**, *155*, 840–844. [CrossRef] [PubMed]

74. Huigsloot, M.; Nijtmans, L.G.; Szklarczyk, R.; Baars, M.J.; van den Brand, M.A.; Hendriksfranssen, M.G.; van den Heuvel, L.P.; Smeitink, J.A.; Huynen, M.A.; Rodenburg, R.J. A mutation in C2orf64 causes impaired cytochrome c oxidase assembly and mitochondrial cardiomyopathy. *Am. J. Hum. Genet.* **2011**, *88*, 488–493. [CrossRef] [PubMed]

75. Ghosh, A.; Trivedi, P.P.; Timbalia, S.A.; Griffin, A.T.; Rahn, J.J.; Chan, S.S.; Gohil, V.M. Copper supplementation restores cytochrome c oxidase assembly defect in a mitochondrial disease model of COA6 deficiency. *Hum. Mol. Genet.* **2014**, *23*, 3596–3606. [CrossRef] [PubMed]

76. Kornblum, C.; Nicholls, T.J.; Haack, T.B.; Schöler, S.; Peeva, V.; Danhauser, K.; Hallmann, K.; Zsurka, G.; Rorbach, J.; Iuso, A.; et al. Loss-of-function mutations in MGME1 impair mtDNA replication and cause multisystemic mitochondrial disease. *Nat. Genet.* **2013**, *45*, 214–219. [CrossRef] [PubMed]

77. Ronchi, D.; Garone, C.; Bordoni, A.; Gutierrez Rios, P.; Calvo, S.E.; Ripolone, M.; Ranieri, M.; Rizzuti, M.; Villa, L.; Magri, F.; et al. Next-generation sequencing reveals DGUOK mutations in adult patients with mitochondrial DNA multiple deletions. *Brain* **2012**, *135*, 3404–3415. [CrossRef] [PubMed]

78. Sue, C.M.; Karadimas, C.; Checcarelli, N.; Tanji, K.; Papadopoulou, L.C.; Pallotti, F.; Guo, F.L.; Shanske, S.; Hirano, M.; De Vivo, D.C.; et al. Differential features of patients with mutations in two COX assembly genes, SURF-1 and SCO2. *Ann. Neurol.* **2000**, *47*, 589–595. [CrossRef]

79. Zhu, Z.; Yao, J.; Johns, T.; Fu, K.; De Bie, I.; Macmillan, C.; Cuthbert, A.P.; Newbold, R.F.; Wang, J.; Chevrette, M.; et al. SURF1, encoding a factor involved in the biogenesis of cytochrome c oxidase, is mutated in Leigh syndrome. *Nat. Genet.* **1998**, *20*, 337–343. [PubMed]

80. Teraoka, M.; Yokoyama, Y.; Ninomiya, S.; Inoue, C.; Yamashita, S.; Seino, Y. Two novel mutations of SURF1 in Leigh syndrome with cytochrome c oxidase deficiency. *Hum. Genet.* **1999**, *105*, 560–563. [CrossRef] [PubMed]

81. Müller-Höcker, J.; Johannes, A.; Droste, M.; Kadenbach, B.; Pongratz, D.; Hübner, G. Fatal mitochondrial cardiomyopathy in Kearns-Sayre syndrome with deficiency of cytochrome-c-oxidase in cardiac and skeletal muscle. An enzymehistochemical—Ultra-immunocytochemical—Fine structural study in longterm frozen autopsy tissue. *Virchows Arch. B* **1986**, *52*, 353–367. [CrossRef] [PubMed]

82. Moslemi, A.R.; Selimovic, N.; Bergh, C.H.; Oldfors, A. Fatal dilated cardiomyopathy associated with a mitochondrial DNA deletion. *Cardiology* **2000**, *94*, 68–71. [CrossRef] [PubMed]

83. Tulinius, M.H.; Oldfors, A.; Holme, E.; Larsson, N.G.; Houshmand, M.; Fahleson, P.; Sigström, L.; Kristiansson, B. Atypical presentation of multisystem disorders in two girls with mitochondrial DNA deletions. *Eur. J. Pediatr.* **1995**, *154*, 35–42. [CrossRef] [PubMed]

84. Sparaco, M.; Schon, E.A.; DiMauro, S.; Bonilla, E. Myoclonic epilepsy with ragged-red fibers (MERRF): An immunohistochemical study of the brain. *Brain Pathol.* **1995**, *5*, 125–133. [CrossRef] [PubMed]

85. Weber, K.; Wilson, J.N.; Taylor, L.; Brierley, E.; Johnson, M.A.; Turnbull, D.M.; Bindoff, L.A. A new mtDNA mutation showing accumulation with time and restriction to skeletal muscle. *Am. J. Hum. Genet.* **1997**, *60*, 373–380. [PubMed]

86. Chinnery, P.F.; Howel, D.; Turnbull, D.M.; Johnson, M.A. Clinical progression of mitochondrial myopathy is associated with the random accumulation of cytochrome c oxidase negative skeletal muscle fibres. *J. Neurol. Sci.* **2003**, *211*, 63–66. [CrossRef]

87. Chinnery, P.F.; Samuels, D.C. Relaxed replication of mtDNA: A model with implications for the expression of disease. *Am. J. Hum. Genet.* **1999**, *64*, 1158–1165. [CrossRef] [PubMed]

88. Yamamoto, M.; Nonaka, I. Skeletal muscle pathology in chronic progressive external ophthalmoplegia with ragged-red fibers. *Acta Neuropathol.* **1988**, *76*, 558–563. [CrossRef] [PubMed]

89. Matsuoka, T.; Goto, Y.; Yoneda, M.; Nonaka, I. Muscle histopathology in myoclonus epilepsy with ragged-red fibers (MERRF). *J. Neurol. Sci.* **1991**, *106*, 193–198. [CrossRef]

90. Bogenhagen, D.; Clayton, D.A. Mouse L cell mitochondrial DNA molecules are selected randomly for replication throughout the cell cycle. *Cell* **1977**, *11*, 719–727. [CrossRef]

91. Murphy, J.L.; Blakely, E.L.; Schaefer, A.M.; He, L.; Wyrick, P.; Haller, R.G.; Taylor, R.W.; Turnbull, D.M.; Taivassalo, T. Resistance training in patients with single, large-scale deletions of mitochondrial DNA. *Brain* **2008**, *131*, 2832–2840. [CrossRef] [PubMed]

92. Durham, S.E.; Bonilla, E.; Samuels, D.C.; DiMauro, S.; Chinnery, P.F. Mitochondrial DNA copy number threshold in mtDNA depletion myopathy. *Neurology* **2005**, *65*, 453–455. [CrossRef] [PubMed]

93. Murphy, J.L.; Ratnaike, T.E.; Shang, E.; Falkous, G.; Blakely, E.L.; Alston, C.L.; Taivassalo, T.; Haller, R.G.; Taylor, R.W.; Turnbull, D.M. Cytochrome c oxidase-intermediate fibres: Importance in understanding the pathogenesis and treatment of mitochondrial myopathy. *Neuromuscul. Disord.* **2012**, *22*, 690–698. [CrossRef] [PubMed]

94. Choi, B.O.; Hwang, J.H.; Cho, E.M.; Jeong, E.H.; Hyun, Y.S.; Jeon, H.J.; Seong, K.M.; Cho, N.S.; Chung, K.W. Mutational analysis of whole mitochondrial DNA in patients with MELAS and MERRF diseases. *Exp. Mol. Med.* **2010**, *42*, 446–455. [CrossRef] [PubMed]

95. De Vivo, D.C. The expanding clinical spectrum of mitochondrial diseases. *Brain Dev.* **1993**, *15*, 1–22. [CrossRef]

96. Tarnopolsky, M.A.; Baker, S.K.; Myint, T.; Maxner, C.E.; Robitaille, J.; Robinson, B.H. Clinical variability in maternally inherited leber hereditary optic neuropathy with the G14459A mutation. *Am. J. Med. Genet. A* **2004**, *124*, 372–376. [CrossRef] [PubMed]

97. Zierz, C.M.; Joshi, P.R.; Zierz, S. Frequencies of myohistological mitochondrial changes in patients with mitochondrial DNA deletions and the common m.3243A > G point mutation. *Neuropathology* **2015**, *35*, 130–136. [CrossRef] [PubMed]

98. Goto, Y.; Koga, Y.; Horai, S.; Nonaka, I. Chronic progressive external ophthalmoplegia: A correlative study of mitochondrial DNA deletions and their phenotypic expression in muscle biopsies. *J. Neurol. Sci.* **1990**, *100*, 63–69. [CrossRef]

99. Fu, K.; Hartlen, R.; Johns, T.; Genge, A.; Karpati, G.; Shoubridge, E.A. A novel heteroplasmic tRNAleu(CUN) mtDNA point mutation in a sporadic patient with mitochondrial encephalomyopathy segregates rapidly in skeletal muscle and suggests an approach to therapy. *Hum. Mol. Genet.* **1996**, *5*, 1835–1840. [CrossRef] [PubMed]

100. Moraes, C.T.; DiMauro, S.; Zeviani, M.; Lombes, A.; Shanske, S.; Miranda, A.F.; Nakase, H.; Bonilla, E.; Werneck, L.C.; Servidei, S. Mitochondrial DNA deletions in progressive external ophthalmoplegia and Kearns-Sayre syndrome. *N. Engl. J. Med.* **1989**, *320*, 1293–1299. [CrossRef] [PubMed]

101. Hoekstra, A.S.; Bayley, J.P. The role of complex II in disease. *Biochim. Biophys. Acta* **2013**, *1827*, 543–551. [CrossRef] [PubMed]

102. Alston, C.L.; Davison, J.E.; Meloni, F.; van der Westhuizen, F.H.; He, L.; Hornig-Do, H.T.; Peet, A.C.; Gissen, P.; Goffrini, P.; Ferrero, I.; et al. Recessive germline *SDHA* and *SDHB* mutations causing leukodystrophy and isolated mitochondrial complex II deficiency. *J. Med. Genet.* **2012**, *49*, 569–577. [CrossRef] [PubMed]

103. Jackson, C.B.; Nuoffer, J.M.; Hahn, D.; Prokisch, H.; Haberberger, B.; Gautschi, M.; Häberli, A.; Gallati, S.; Schaller, A. Mutations in *SDHD* lead to autosomal recessive encephalomyopathy and isolated mitochondrial complex II deficiency. *J. Med. Genet.* **2014**, *51*, 170–175. [CrossRef] [PubMed]

104. Ghezzi, D.; Goffrini, P.; Uziel, G.; Horvath, R.; Klopstock, T.; Lochmüller, H.; D'Adamo, P.; Gasparini, P.; Strom, T.M.; Prokisch, H.; et al. *SDHAF1*, encoding a LYR complex-II specific assembly factor, is mutated in SDH-defective infantile leukoencephalopathy. *Nat. Genet.* **2009**, *41*, 654–656. [CrossRef] [PubMed]

105. Birch-Machin, M.A.; Taylor, R.W.; Cochran, B.; Ackrell, B.A.; Turnbull, D.M. Late-onset optic atrophy, ataxia, and myopathy associated with a mutation of a complex II gene. *Ann. Neurol.* **2000**, *48*, 330–335. [CrossRef]

106. Vladutiu, G.D.; Heffner, R.R. Succinate dehydrogenase deficiency. *Arch. Pathol. Lab. Med.* **2000**, *124*, 1755–1758. [PubMed]

107. Haller, R.G.; Henriksson, K.G.; Jorfeldt, L.; Hultman, E.; Wibom, R.; Sahlin, K.; Areskog, N.H.; Gunder, M.; Ayyad, K.; Blomqvist, C.G. Deficiency of skeletal muscle succinate dehydrogenase and aconitase. Pathophysiology of exercise in a novel human muscle oxidative defect. *J. Clin. Investig.* **1991**, *88*, 1197–1206. [CrossRef] [PubMed]

108. Sanaker, P.S.; Toompuu, M.; Hogan, V.E.; He, L.; Tzoulis, C.; Chrzanowska-Lightowlers, Z.M.; Taylor, R.W.; Bindoff, L.A. Differences in RNA processing underlie the tissue specific phenotype of ISCU myopathy. *Biochim. Biophys. Acta* **2010**, *1802*, 539–544. [CrossRef] [PubMed]

109. Nance, J.R.; Mammen, A.L. Diagnostic evaluation of rhabdomyolysis. *Muscle Nerve* **2015**, *51*, 793–810. [CrossRef] [PubMed]

110. Ogasahara, S.; Engel, A.G.; Frens, D.; Mack, D. Muscle coenzyme Q deficiency in familial mitochondrial encephalomyopathy. *Proc. Natl. Acad. Sci. USA* **1989**, *86*, 2379–2382. [CrossRef] [PubMed]

111. Sobreira, C.; Hirano, M.; Shanske, S.; Keller, R.K.; Haller, R.G.; Davidson, E.; Santorelli, F.M.; Miranda, A.F.; Bonilla, E.; Mojon, D.S.; et al. Mitochondrial encephalomyopathy with coenzyme Q10 deficiency. *Neurology* **1997**, *48*, 1238–1243. [CrossRef] [PubMed]

112. Di Giovanni, S.; Mirabella, M.; Spinazzola, A.; Crociani, P.; Silvestri, G.; Broccolini, A.; Tonali, P.; Di Mauro, S.; Servidei, S. Coenzyme Q10 reverses pathological phenotype and reduces apoptosis in familial CoQ10 deficiency. *Neurology* **2001**, *57*, 515–518. [CrossRef] [PubMed]

113. McFarland, R.; Taylor, R.W.; Chinnery, P.F.; Howell, N.; Turnbull, D.M. A novel sporadic mutation in cytochrome c oxidase subunit II as a cause of rhabdomyolysis. *Neuromuscul. Disord.* **2004**, *14*, 162–166. [CrossRef] [PubMed]

114. Marotta, R.; Chin, J.; Kirby, D.M.; Chiotis, M.; Cook, M.; Collins, S.J. Novel single base pair COX III subunit deletion of mitochondrial DNA associated with rhabdomyolysis. *J. Clin. Neurosci.* **2011**, *18*, 290–292. [CrossRef] [PubMed]

115. Keightley, J.A.; Hoffbuhr, K.C.; Burton, M.D.; Salas, V.M.; Johnston, W.S.; Penn, A.M.; Buist, N.R.; Kennaway, N.G. A microdeletion in cytochrome c oxidase (COX) subunit III associated with COX deficiency and recurrent myoglobinuria. *Nat. Genet.* **1996**, *12*, 410–416. [CrossRef] [PubMed]

116. Andreu, A.L.; Hanna, M.G.; Reichmann, H.; Bruno, C.; Penn, A.S.; Tanji, K.; Pallotti, F.; Iwata, S.; Bonilla, E.; Lach, B.; et al. Exercise intolerance due to mutations in the cytochrome b gene of mitochondrial DNA. *N. Engl. J. Med.* **1999**, *341*, 1037–1044. [CrossRef] [PubMed]

117. Karadimas, C.L.; Greenstein, P.; Sue, C.M.; Joseph, J.T.; Tanji, K.; Haller, R.G.; Taivassalo, T.; Davidson, M.M.; Shanske, S.; Bonilla, E.; et al. Recurrent myoglobinuria due to a nonsense mutation in the COX I gene of mitochondrial DNA. *Neurology* **2000**, *55*, 644–649. [CrossRef] [PubMed]

118. Emmanuele, V.; Sotiriou, E.; Shirazi, M.; Tanji, K.; Haller, R.G.; Heinicke, K.; Bosch, P.E.; Hirano, M.; DiMauro, S. Recurrent myoglobinuria in a sporadic patient with a novel mitochondrial DNA tRNA(Ile) mutation. *J. Neurol. Sci.* **2011**, *303*, 39–42. [CrossRef] [PubMed]

119. Blum, S.; Robertson, T.; Klingberg, S.; Henderson, R.D.; McCombe, P. Atypical clinical presentations of the A3243G mutation, usually associated with MELAS. *Intern. Med. J.* **2011**, *41*, 199–202. [CrossRef] [PubMed]

120. Vissing, C.R.; Duno, M.; Olesen, J.H.; Rafiq, J.; Risom, L.; Christensen, E.; Wibrand, F.; Vissing, J. Recurrent myoglobinuria and deranged acylcarnitines due to a mutation in the mtDNA *MT-CO2* gene. *Neurology* **2013**, *80*, 1908–1910. [CrossRef] [PubMed]

121. Béhin, A.; Jardel, C.; Claeys, K.G.; Fagart, J.; Louha, M.; Romero, N.B.; Laforêt, P.; Eymard, B.; Lombès, A. Adult cases of mitochondrial DNA depletion due to TK2 defect: An expanding spectrum. *Neurology* **2012**, *78*, 644–648. [CrossRef] [PubMed]

122. Paradas, C.; Gutiérrez Ríos, P.; Rivas, E.; Carbonell, P.; Hirano, M.; DiMauro, S. TK2 mutation presenting as indolent myopathy. *Neurology* **2013**, *80*, 504–506. [CrossRef] [PubMed]

123. Tulinius, M.; Oldfors, A. Neonatal muscular manifestations in mitochondrial disorders. *Semin. Fetal Neonatal Med.* **2011**, *16*, 229–235. [CrossRef] [PubMed]

124. Kollberg, G.; Darin, N.; Benan, K.; Moslemi, A.R.; Lindal, S.; Tulinius, M.; Oldfors, A.; Holme, E. A novel homozygous *RRM2B* missense mutation in association with severe mtDNA depletion. *Neuromuscul. Disord.* **2009**, *19*, 147–150. [CrossRef] [PubMed]

125. Morava, E.; Steuerwald, U.; Carrozzo, R.; Kluijtmans, L.A.; Joensen, F.; Santer, R.; Dionisi-Vici, C.; Wevers, R.A. Dystonia and deafness due to *SUCLA2* defect; Clinical course and biochemical markers in 16 children. *Mitochondrion* **2009**, *9*, 438–442. [CrossRef] [PubMed]

126. Köller, H.; Stoll, G.; Neuen-Jacob, E. Postpartum manifestation of a necrotising lipid storage myopathy associated with muscle carnitine deficiency. *J. Neurol. Neurosurg. Psychiatry* **1998**, *64*, 407–408. [CrossRef] [PubMed]

127. Laforêt, P.; Vianey-Saban, C. Disorders of muscle lipid metabolism: Diagnostic and therapeutic challenges. *Neuromuscul. Disord.* **2010**, *20*, 693–700. [CrossRef] [PubMed]

128. Mitsuhashi, S.; Nishino, I. Megaconial congenital muscular dystrophy due to loss-of-function mutations in choline kinase β. *Curr. Opin. Neurol.* **2013**, *26*, 536–543. [CrossRef] [PubMed]

129. Gutiérrez Ríos, P.; Kalra, A.A.; Wilson, J.D.; Tanji, K.; Akman, H.O.; Area Gómez, E.; Schon, E.A.; DiMauro, S. Congenital megaconial myopathy due to a novel defect in the choline kinase Beta gene. *Arch. Neurol.* **2012**, *69*, 657–661. [PubMed]

130. Schon, E.A.; Area-Gomez, E. Mitochondria-associated ER membranes in Alzheimer disease. *Mol. Cell Neurosci.* **2013**, *55*, 26–36. [CrossRef] [PubMed]

131. Logan, C.V.; Szabadkai, G.; Sharpe, J.A.; Parry, D.A.; Torelli, S.; Childs, A.M.; Kriek, M.; Phadke, R.; Johnson, C.A.; Roberts, N.Y.; et al. Loss-of-function mutations in MICU1 cause a brain and muscle disorder linked to primary alterations in mitochondrial calcium signaling. *Nat. Genet.* **2014**, *46*, 188–193. [CrossRef] [PubMed]

132. Gal, A.; Balicza, P.; Weaver, D.; Naghdi, S.; Joseph, S.K.; Várnai, P.; Gyuris, T.; Horváth, A.; Nagy, L.; Seifert, E.L.; et al. MSTO1 is a cytoplasmic pro-mitochondrial fusion protein, whose mutation induces myopathy and ataxia in humans. *EMBO Mol. Med.* **2017**. [CrossRef] [PubMed]

133. Nasca, A.; Scotton, C.; Zaharieva, I.; Neri, M.; Selvatici, R.; Magnusson, O.T.; Gal, A.; Weaver, D.; Rossi, R.; Armaroli, A.; et al. Recessive mutations in MSTO1 cause mitochondrial dynamics impairment, leading to myopathy and ataxia. *Hum. Mutat.* **2017**. [CrossRef] [PubMed]

134. Finsterer, J.; Zarrouk-Mahjoub, S. Mitochondrial vasculopathy. *World J. Cardiol.* **2016**, *8*, 333–339. [CrossRef] [PubMed]

135. Tay, S.H.; Nordli, D.R.; Bonilla, E.; Null, E.; Monaco, S.; Hirano, M.; DiMauro, S. Aortic rupture in mitochondrial encephalopathy, lactic acidosis, and stroke-like episodes. *Arch. Neurol.* **2006**, *63*, 281–283. [CrossRef] [PubMed]

136. Coquet, M.; Fontan, D.; Vital, C.; Tudesq, N.; Baronnet, R. Muscle and brain biopsy in a case of mitochondrial encephalomyopathy. Demonstration of a mitochondrial vasculopathy. *Ann. Pathol.* **1990**, *10*, 181–186. [PubMed]

137. Kaufmann, P.; Engelstad, K.; Wei, Y.; Kulikova, R.; Oskoui, M.; Sproule, D.M.; Battista, V.; Koenigsberger, D.Y.; Pascual, J.M.; Shanske, S.; et al. Natural history of MELAS associated with mitochondrial DNA m.3243A > G genotype. *Neurology* **2011**, *77*, 1965–1971. [CrossRef] [PubMed]

138. Lax, N.Z.; Pienaar, I.S.; Reeve, A.K.; Hepplewhite, P.D.; Jaros, E.; Taylor, R.W.; Kalaria, R.N.; Turnbull, D.M. Microangiopathy in the cerebellum of patients with mitochondrial DNA disease. *Brain* **2012**, *135*, 1736–1750. [CrossRef] [PubMed]

139. Shiva, S.; Brookes, P.S.; Patel, R.P.; Anderson, P.G.; Darley-Usmar, V.M. Nitric oxide partitioning into mitochondrial membranes and the control of respiration at cytochrome c oxidase. *Proc. Natl. Acad. Sci. USA* **2001**, *98*, 7212–7217. [CrossRef] [PubMed]

140. Torres, J.; Darley-Usmar, V.; Wilson, M.T. Inhibition of cytochrome c oxidase in turnover by nitric oxide: Mechanism and implications for control of respiration. *Biochem. J.* **1995**, *312*, 169–173. [CrossRef] [PubMed]

141. Taivassalo, T.; Ayyad, K.; Haller, R.G. Increased capillaries in mitochondrial myopathy: Implications for the regulation of oxygen delivery. *Brain* **2012**, *135*, 53–61. [CrossRef] [PubMed]

142. DiMauro, S.; Schon, E.A. Mitochondrial disorders in the nervous system. *Ann. Rev. Neurosci.* **2008**, *31*, 91–123. [CrossRef] [PubMed]

143. Vogel, H. Mitochondrial myopathies and the role of the pathologist in the molecular era. *J. Neuropathol. Exp. Neurol.* **2001**, *60*, 217–227. [CrossRef] [PubMed]

144. Gilkerson, R.W.; Margineantu, D.H.; Capaldi, R.A.; Selker, J.M. Mitochondrial DNA depletion causes morphological changes in the mitochondrial reticulum of cultured human cells. *FEBS Lett.* **2000**, *474*, 1–4. [CrossRef]

145. Sarnat, H.B.; Flores-Sarnat, L.; Casey, R.; Scott, P.; Khan, A. Endothelial ultrastructural alterations of intramuscular capillaries in infantile mitochondrial cytopathies: "Mitochondrial angiopathy". *Neuropathology* **2012**, *32*, 617–627. [CrossRef] [PubMed]

146. Vincent, A.E.; Ng, Y.S.; White, K.; Davey, T.; Mannella, C.; Falkous, G.; Feeney, C.; Schaefer, A.M.; McFarland, R.; Gorman, G.S.; et al. The Spectrum of mitochondrial ultrastructural defects in mitochondrial myopathy. *Sci. Rep.* **2016**, *6*, 30610. [CrossRef] [PubMed]

147. Cotta, A.; Carvalho, E.; da-Cunha-Júnior, A.L.; Paim, J.F.; Navarro, M.M.; Valicek, J.; Menezes, M.M.; Nunes, S.V.; Xavier Neto, R.; Takata, R.I.; et al. Common recessive limb girdle muscular dystrophies differential diagnosis: Why and how? *Arq. Neuropsiquiatr.* **2014**, *72*, 721–734. [CrossRef] [PubMed]

148. Sacconi, S.; Salviati, L.; Bourget, I.; Figarella, D.; Péréon, Y.; Lemmers, R.; van der Maarel, S.; Desnuelle, C. Diagnostic challenges in facioscapulohumeral muscular dystrophy. *Neurology* **2006**, *67*, 1464–1466. [CrossRef] [PubMed]

149. Machado, P.M.; Dimachkie, M.M.; Barohn, R.J. Sporadic inclusion body myositis: New insights and potential therapy. *Curr. Opin. Neurol.* **2014**, *27*, 591–598. [CrossRef] [PubMed]

150. Rygiel, K.A.; Miller, J.; Grady, J.P.; Rocha, M.C.; Taylor, R.W.; Turnbull, D.M. Mitochondrial and inflammatory changes in sporadic inclusion body myositis. *Neuropathol. Appl. Neurobiol.* **2015**, *41*, 288–303. [CrossRef] [PubMed]

151. Alhatou, M.I.; Sladky, J.T.; Bagasra, O.; Glass, J.D. Mitochondrial abnormalities in dermatomyositis: Characteristic pattern of neuropathology. *J. Mol. Histol.* **2004**, *35*, 615–619. [CrossRef] [PubMed]

152. Mastaglia, F.L.; Needham, M. Update on toxic myopathies. *Curr. Neurol. NeuroSci. Rep.* **2012**, *12*, 54–61. [CrossRef] [PubMed]

153. Zhang, C.; Baumer, A.; Maxwell, R.J.; Linnane, A.W.; Nagley, P. Multiple mitochondrial DNA deletions in an elderly human individual. *FEBS Lett.* **1992**, *297*, 34–38. [CrossRef]

154. Münscher, C.; Rieger, T.; Müller-Höcker, J.; Kadenbach, B. The point mutation of mitochondrial DNA characteristic for MERRF disease is found also in healthy people of different ages. *FEBS Lett.* **1993**, *317*, 27–30. [CrossRef]

155. Müller-Höcker, J.; Schneiderbanger, K.; Stefani, F.H.; Kadenbach, B. Progressive loss of cytochrome c oxidase in the human extraocular muscles in ageing—A cytochemical-immunohistochemical study. *Mutat. Res.* **1992**, *275*, 115–124. [CrossRef]

156. Cao, Z.; Wanagat, J.; McKiernan, S.H.; Aiken, J.M. Mitochondrial DNA deletion mutations are concomitant with ragged red regions of individual, aged muscle fibers: Analysis by laser-capture microdissection. *Nucleic Acids Res.* **2001**, *29*, 4502–4508. [CrossRef] [PubMed]

157. Johnston, W.; Karpati, G.; Carpenter, S.; Arnold, D.; Shoubridge, E.A. Late-onset mitochondrial myopathy. *Ann. Neurol.* **1995**, *37*, 16–23. [CrossRef] [PubMed]

158. Nascimento, A.; Ortez, C.; Jou, C.; O'Callaghan, M.; Ramos, F.; Garcia-Cazorla, A. Neuromuscular manifestations in mitochondrial diseases in children. *Semin. Pediatr. Neurol.* **2016**, *23*, 290–305. [CrossRef] [PubMed]

159. Koenig, M.K. Presentation and diagnosis of mitochondrial disorders in children. *Pediatr. Neurol.* **2008**, *38*, 305–313. [CrossRef] [PubMed]

160. Goldstein, A.C.; Bhatia, P.; Vento, J.M. Mitochondrial disease in childhood: Nuclear encoded. *Neurotherapeutics* **2013**, *10*, 212–226. [CrossRef] [PubMed]

161. Lamont, P.J.; Surtees, R.; Woodward, C.E.; Leonard, J.V.; Wood, N.W.; Harding, A.E. Clinical and laboratory findings in referrals for mitochondrial DNA analysis. *Arch. Dis. Child.* **1998**, *79*, 22–27. [CrossRef] [PubMed]

162. Yamamoto, M.; Koga, Y.; Ohtaki, E.; Nonaka, I. Focal cytochrome c oxidase deficiency in various neuromuscular diseases. *J. Neurol. Sci.* **1989**, *91*, 207–213. [CrossRef]

163. Gire, C.; Girard, N.; Nicaise, C.; Einaudi, M.A.; Montfort, M.F.; Dejode, J.M. Clinical features and neuroradiological findings of mitochondrial pathology in six neonates. *Child's Nerv Syst.* **2002**, *18*, 621–628.

164. Scaglia, F.; Towbin, J.A.; Craigen, W.J.; Belmont, J.W.; Smith, E.O.; Neish, S.R.; Ware, S.M.; Hunter, J.V.; Fernbach, S.D.; Vladutiu, G.D.; et al. Clinical spectrum, morbidity, and mortality in 113 pediatric patients with mitochondrial disease. *Pediatrics* **2004**, *114*, 925–931. [CrossRef] [PubMed]

165. Miles, L.; Wong, B.L.; Dinopoulos, A.; Morehart, P.J.; Hofmann, I.A.; Bove, K.E. Investigation of children for mitochondriopathy confirms need for strict patient selection, improved morphological criteria, and better laboratory methods. *Hum. Pathol.* **2006**, *37*, 173–184. [CrossRef] [PubMed]

166. Miles, L.; Miles, M.V.; Horn, P.S.; Degrauw, T.J.; Wong, B.L.; Bove, K.E. Importance of muscle light microscopic mitochondrial subsarcolemmal aggregates in the diagnosis of respiratory chain deficiency. *Hum. Pathol.* **2012**, *43*, 1249–1257. [CrossRef] [PubMed]

167. Chaturvedi, S.; Bala, K.; Thakur, R.; Suri, V. Mitochondrial encephalomyopathies: Advances in understanding. *Med. Sci. Monit.* **2005**, *11*, RA238–RA246. [PubMed]

168. McFarland, R.; Turnbull, D.M. Batteries not included: Diagnosis and management of mitochondrial disease. *J. Intern. Med.* **2009**, *265*, 210–228. [CrossRef] [PubMed]

169. Walker, U.A.; Collins, S.; Byrne, E. Respiratory chain encephalomyopathies: A diagnostic classification. *Eur. Neurol.* **1996**, *36*, 260–267. [CrossRef] [PubMed]

170. Byrne, E.; Dennett, X. Respiratory chain failure in adult muscle fibres: Relationship with ageing and possible implications for the neuronal pool. *Mutat. Res.* **1992**, *275*, 125–131. [CrossRef]

171. Müller-Höcker, J. Cytochrome c oxidase deficient fibres in the limb muscle and diaphragm of man without muscular disease: An age-related alteration. *J. Neurol. Sci.* **1990**, *100*, 14–21. [CrossRef]

172. Sleigh, K.; Ball, S.; Hilton, D.A. Quantification of changes in muscle from individuals with and without mitochondrial disease. *Muscle Nerve* **2011**, *43*, 795–800. [CrossRef] [PubMed]

173. Patterson, K. Mitochondrial muscle pathology. *Pediatr. Dev. Pathol.* **2004**, *7*, 629–632. [CrossRef] [PubMed]

174. De Meirleir, L.; Seneca, S.; Lissens, W.; De Clercq, I.; Eyskens, F.; Gerlo, E.; Smet, J.; Van Coster, R. Respiratory chain complex V deficiency due to a mutation in the assembly gene ATP12. *J. Med. Genet.* **2004**, *41*, 120–124. [CrossRef] [PubMed]

175. Van Coster, R.; Seneca, S.; Smet, J.; Van Hecke, R.; Gerlo, E.; Devreese, B.; Van Beeumen, J.; Leroy, J.G.; De Meirleir, L.; Lissens, W. Homozygous Gly555Glu mutation in the nuclear-encoded 70 kDa flavoprotein gene causes instability of the respiratory chain complex II. *Am. J. Med. Genet. A* **2003**, *120*, 13–18. [CrossRef] [PubMed]

176. De Meirleir, L.; Seneca, S.; Damis, E.; Sepulchre, B.; Hoorens, A.; Gerlo, E.; García Silva, M.T.; Hernandez, E.M.; Lissens, W.; Van Coster, R. Clinical and diagnostic characteristics of complex III deficiency due to mutations in the BCS1L gene. *Am. J. Med. Genet. A* **2003**, *121*, 126–131. [CrossRef] [PubMed]

177. Antonicka, H.; Ogilvie, I.; Taivassalo, T.; Anitori, R.P.; Haller, R.G.; Vissing, J.; Kennaway, N.G.; Shoubridge, E.A. Identification and characterization of a common set of complex I assembly intermediates in mitochondria from patients with complex I deficiency. *J. Biol. Chem.* **2003**, *278*, 43081–43088. [CrossRef] [PubMed]

178. Tiranti, V.; Galimberti, C.; Nijtmans, L.; Bovolenta, S.; Perini, M.P.; Zeviani, M. Characterization of SURF-1 expression and Surf-1p function in normal and disease conditions. *Hum. Mol. Genet.* **1999**, *8*, 2533–2540. [CrossRef] [PubMed]

179. De Paepe, B.; Smet, J.; Leroy, J.G.; Seneca, S.; George, E.; Matthys, D.; van Maldergem, L.; Scalais, E.; Lissens, W.; de Meirleir, L.; et al. Diagnostic value of immunostaining in cultured skin fibroblasts from patients with oxidative phosphorylation defects. *Pediatr. Res.* **2006**, *59*, 2–6. [CrossRef] [PubMed]
180. Rahman, S.; Lake, B.D.; Taanman, J.W.; Hanna, M.G.; Cooper, J.M.; Schapira, A.H.; Leonard, J.V. Cytochrome oxidase immunohistochemistry: Clues for genetic mechanisms. *Brain* **2000**, *123*, 591–600. [CrossRef] [PubMed]
181. Hanson, B.J.; Capaldi, R.A.; Marusich, M.F.; Sherwood, S.W. An immunocytochemical approach to detection of mitochondrial disorders. *J. Histochem. Cytochem.* **2002**, *50*, 1281–1288. [CrossRef] [PubMed]
182. Willis, J.H.; Capaldi, R.A.; Huigsloot, M.; Rodenburg, R.J.; Smeitink, J.; Marusich, M.F. Isolated deficiencies of OXPHOS complexes I and IV are identified accurately and quickly by simple enzyme activity immunocapture assays. *Biochim. Biophys. Acta* **2009**, *1787*, 533–538. [CrossRef] [PubMed]
183. Tanji, K.; Bonilla, E. Light microscopic methods to visualize mitochondria on tissue sections. *Methods* **2008**, *46*, 274–280. [CrossRef] [PubMed]
184. Boczonadi, V.; Giunta, M.; Lane, M.; Tulinius, M.; Schara, U.; Horvath, R. Investigating the role of the physiological isoform switch of cytochrome c oxidase subunits in reversible mitochondrial disease. *Int. J. Biochem. Cell Biol.* **2015**, *63*, 32–40. [CrossRef] [PubMed]
185. Rocha, M.C.; Grady, J.P.; Grünewald, A.; Vincent, A.; Dobson, P.F.; Taylor, R.W.; Turnbull, D.M.; Rygiel, K.A. A novel immunofluorescent assay to investigate oxidative phosphorylation deficiency in mitochondrial myopathy: Understanding mechanisms and improving diagnosis. *Sci. Rep.* **2015**, *5*, 15037. [CrossRef] [PubMed]
186. Suomalainen, A.; Elo, J.M.; Pietiläinen, K.H.; Hakonen, A.H.; Sevastianova, K.; Korpela, M.; Isohanni, P.; Marjavaara, S.K.; Tyni, T.; Kiuru-Enari, S.; et al. FGF-21 as a biomarker for muscle-manifesting mitochondrial respiratory chain deficiencies: A diagnostic study. *Lancet Neurol.* **2011**, *10*, 806–818. [CrossRef]
187. Tyynismaa, H.; Carroll, C.J.; Raimundo, N.; Ahola-Erkkilä, S.; Wenz, T.; Ruhanen, H.; Guse, K.; Hemminki, A.; Peltola-Mjøsund, K.E.; Tulkki, V.; et al. Mitochondrial myopathy induces a starvation-like response. *Hum. Mol. Genet.* **2010**, *19*, 3948–3958. [CrossRef] [PubMed]
188. Kim, K.H.; Jeong, Y.T.; Oh, H.; Kim, S.H.; Cho, J.M.; Kim, Y.N.; Kim, S.S.; Kim, D.H.; Hur, K.Y.; Kim, H.K.; et al. Autophagy deficiency leads to protection from obesity and insulin resistance by inducing Fgf21 as a mitokine. *Nat. Med.* **2013**, *19*, 83–92. [CrossRef] [PubMed]
189. Ji, K.; Zheng, J.; Lv, J.; Xu, J.; Ji, X.; Luo, Y.B.; Li, W.; Zhao, Y.; Yan, C. Skeletal muscle increases FGF21 expression in mitochondrial disorders to compensate for energy metabolic insufficiency by activating the mTOR-YY1-PGC1alpha pathway. *Free Radic. Biol. Med.* **2015**, *84*, 161–170. [CrossRef] [PubMed]
190. Kalko, S.G.; Paco, S.; Jou, C.; Rodríguez, M.A.; Meznaric, M.; Rogac, M.; Jekovec-Vrhovsek, M.; Sciacco, M.; Moggio, M.; Fagiolari, G.; et al. Transcriptomic profiling of TK2 deficient human skeletal muscle suggests a role for the p53 signalling pathway and identifies growth and differentiation factor-15 as a potential novel biomarker for mitochondrial myopathies. *BMC Genom.* **2014**, *15*, 91. [CrossRef] [PubMed]
191. Yatsuga, S.; Fujita, Y.; Ishii, A.; Fukumoto, Y.; Arahata, H.; Kakuma, T.; Kojima, T.; Ito, M.; Tanaka, M.; Saiki, R.; et al. Growth differentiation factor 15 as a useful biomarker for mitochondrial disorders. *Ann. Neurol.* **2015**, *78*, 814–823. [CrossRef] [PubMed]
192. Montero, R.; Yubero, D.; Villarroya, J.; Henares, D.; Jou, C.; Rodriguez, M.A.; Ramos, F.; Nascimento, A.; Ortez, C.I.; Campistol, J.; et al. GDF-15 Is elevated in children with mitochondrial diseases and Is induced by mitochondrial dysfunction. *PLoS ONE* **2016**, *11*, e0148709. [CrossRef] [PubMed]

Journal of
Clinical Medicine

MDPI

Review

Use of the Ketogenic Diet to Treat Intractable Epilepsy in Mitochondrial Disorders

Eleni Paleologou, Naila Ismayilova and Maria Kinali *

Chelsea and Westmister Hospital, 369 Fulham Road, Chelsea, London SW10 9NH, UK;
elenipal13@yahoo.com (E.P.); ismayilova@doctors.org.uk (N.I.)
* Correspondence: m.kinali@imperial.ac.uk; Tel.: +44-020-33158645

Academic Editor: Iain P. Hargreaves
Received: 31 March 2017; Accepted: 22 May 2017; Published: 26 May 2017

Abstract: Mitochondrial disorders are a clinically heterogeneous group of disorders that are caused by defects in the respiratory chain, the metabolic pathway of the adenosine tri-phosphate (ATP) production system. Epilepsy is a common and important feature of these disorders and its management can be challenging. Epileptic seizures in the context of mitochondrial disease are usually treated with conventional anti-epileptic medication, apart from valproic acid. However, in accordance with the treatment of intractable epilepsy where there are limited treatment options, the ketogenic diet (KD) has been considered as an alternative therapy. The use of the KD and its more palatable formulations has shown promising results. It is especially indicated and effective in the treatment of mitochondrial disorders due to complex **I** deficiency. Further research into the mechanism of action and the neuroprotective properties of the KD will allow more targeted therapeutic strategies and thus optimize the treatment of both epilepsy in the context of mitochondrial disorders but also in other neurodegenerative disorders.

Keywords: ketogenic diet; mitochondrial disorders; intractable epilepsy; treatment

1. Introduction in Mitochondrial Disorders and Mitochondrial Epilepsy

Mitochondrial disorders are a clinically heterogeneous group of disorders arising from defects in the respiratory chain, the metabolic pathway of the mitochondrial adenosine tri-phosphate (ATP) production system via oxidative phosphorylation (OXPHOS). ATP is commonly referred to as the "molecular energy unit" of cells providing the energy source for metabolic functions. Mitochondrial disorders can affect any tissue or organ, but those most affected are those with the highest energy demands, such as the central nervous system, skeletal and cardiac muscles, kidney, liver and endocrine system.

Although mitochondrial diseases have traditionally been considered rare diseases, epidemiological studies suggest otherwise. A United Kingdom (UK) study in 2008 showed that 9.2 people in 100,000 of working age (between 16 and 65 years of age) had a clinically manifested mitochondrial DNA (mtDNA) disease. A further 16.5 children and adults younger than retirement age per 100,000 were at risk of developing one [1]. Other studies quote a minimum prevalence of at least one in 10,000 children and adults, with one in 200 children with a congenital mitochondrial DNA mutation [2]. Bearing in mind that several mtDNA mutations are being discovered each year, these numbers likely underestimate the true prevalence.

Diagnosing mitochondrial disorders can be challenging given their extremely broad clinical spectrum and the lack of consistent phenotype-genotype correlations [3]. Tests that would help with diagnosis range from simple laboratory blood tests, electrocardiograms, genetic testing, hearing and ophthalmology assessments, brain imaging, lumbar puncture, and muscle biopsy, while new tests are being evaluated to facilitate the diagnosis.

A complete laboratory diagnostic work-up, and evaluation of the results in the context of the clinical phenotype and family history are preferred. Elevated blood or cerebrospinal fluid (CSF) lactate can be suggestive of mitochondrial disease, but their respective normality cannot exclude this diagnosis. Likewise, electrocardiograms (ECG) can show heart block in Kearns-Sayre syndrome (KSS) or mitochondrial encephalopathy, lactic acidosis and stroke-like episodes (MELAS), or pre-excitation in MELAS or myoclonus epilepsy with ragged red fibres (MERFF), but a normal ECG does not exclude the diagnosis. In addition, brain magnetic resonance imaging (MRI) can reveal basal ganglia calcification or atrophy or involvement of the occipital lobes, the thalamus and inferior olivary nuclei. Magnetic resonance spectroscopy can demonstrate elevated lactate in the brain, whereas positron emission tomography (PET) scanning can suggest lowered ATP production. While these tests can be suggestive of mitochondrial disorders they are neither sensitive nor specific [3].

Ultimately, the biochemical examination of a skeletal muscle biopsy to evaluate the functional state of mitochondria and to look for the characteristic ultrastructural abnormalities seen in mitochondrial disease remains the optimal method of diagnosing mitochondrial disorders, even in the absence of myopathy. Biopsies provide material for respiratory chain enzyme assays to look for characteristic abnormalities like the presence of ragged red fibres (RRF) with abnormal mitochondria in MERFF, KSS or MELAS using the Gomori trichrome stain or succinate dehydrogenase (SDH)-reactive vessels, and the preservation of cytochrome c oxidase (COX) staining in RRFs often seen in MELAS [3,4]. Skeletal muscle biopsies can also be analyzed for mtDNA. Histochemical stains for mitochondrial enzymes can reveal specific enzyme abnormalities and blue native polyacrylamide gel electrophoresis has been shown to be very useful in analyzing the function of individual enzyme complexes and detecting assembly defects [5]. It is important to bear in mind that on rare occasions the histological changes described may be absent.

Novel biomarkers, like fibroblast growth factor 21 (FGF-21) which has been shown to be a sensitive and specific biomarker for mitochondrial disease affecting skeletal muscles, may play a more crucial role in the future. FGF-21 has also been shown to be useful for monitoring of disease progression and to assess the effect of therapeutic interventions, but is not yet commercially available [6]. There is a urine test currently only available in research laboratories for the diagnosis of progressive external ophthalmoplegia (PEO) and KSS, which would revolutionize the way we diagnose these conditions if made commercially available [7].

The metabolic pathway that generates ATP via oxidative phosphorylation is regulated both by mtDNA, which is maternally inherited, as well as nuclear DNA (nDNA), which is inherited in an autosomal or X-linked manner. Mutations in either of these genomes can result in mitochondrial disorders. mtDNA diseases, therefore, assume a complex pattern of inheritance, which includes maternal, X-linked, recessive and dominant transmission modes [2,3]. Over 270 mutations in mtDNA have been described with mutations in the mitochondrial transfer ribonucleic acid (tRNA) genes being particularly common. Epilepsy remains a common feature in conditions that result from these mutations and notably status epilepticus, which can be the presenting feature. Polymerase Gamma (POLG) is a gene that codes for the catalytic subunit of the mitochondrial DNA polymerase gamma and is crucial for replicating mitochondrial DNA and for DNA repair. Although many mitochondrial nuclear gene defects can cause epilepsy, POLG mutations have been studied the most [2,3].

With the rapid advances in the field of genetics and the emergence of next generation sequencing, new genes are continuously discovered as linked to mitochondrial disease. This technique allows sequencing of multiple candidate genes at the same time, increasing the diagnostic yield, but is still mainly a research tool [8]. Moreover, whole mtDNA sequencing of blood samples is now possible in some diagnostic laboratories and could be a useful screen for patients whose family history suggests maternal inheritance. Patients with clinical suspicion of Alpers-Huttenlocher syndrome, sensory ataxic neuropathy dysarthria, ophathalmoplegia (SANDO), or PEO should be screened for the POLG mutations [7].

J. Clin. Med. **2017**, *6*, 56

To date, there is no cure for mitochondrial disorders, but treatment aims at improving function of the respiratory chain. A "cocktail" of supplements is used by clinicians which includes Co-enzyme Q10 (CoQ10), alpha-lipoic acid, riboflavin, folic acid, creatine monohydrate, thiamin, niacin, pantothenic acid, vitamins B12, C, E, biotin and nitric oxide precursors. Although these supplements are safe and well tolerated, their efficacy is limited [3]. A Cochrane review from 2012 did not find enough evidence to support the use of any treatment in mitochondrial disease [9]. However, there are examples where specific treatment can prove useful, like Co-enzyme Q10 (CoQ10) supplementation in primary CoQ10 deficiency or idebenone, a CoQ analogue and cofactor for NADPH dehydrogenase, in patients with Leber's hereditary optic neuropathy (LHON) [3]. In addition, aerobic exercise is known to improve mitochondrial biogenesis and lead to mitochondrial proliferation [3]. Lastly, treatments specifically aimed at the various symptoms of mitochondrial disease, like physiotherapy for hypotonia or cochlear implants for hearing loss, also form part of the management of MD.

The central nervous system (CNS) is the second organ most frequently affected by mitochondrial diseases. Epilepsy is a common feature [3,10], although preceded by other clinical presentations in most cases [11]. In fact, mitochondrial dysfunction has been shown to both generate seizures as well as result in neuronal cell death [12]. Seizures can manifest at any age and be the presenting feature of an underlying biochemical defect. Despite a genetic aetiology, they may even occur in the absence of a clear family history. The clinical spectrum of epilepsy from mitochondrial respiratory chain defects is variable, ranging from early-onset Ohtahara syndrome to late-onset Landau-Kleffner syndrome, as well as focal and generalized types of epilepsy [13]. Although there are no data on the exact prevalence of epilepsy in mitochondrial disorders, seizures are more common in some mitochondrial disorders than in others [10]. For example, in children with mitochondrial disorders the reported prevalence of seizures ranged between 35% and 61% [12]. Whittaker et al., found an overall prevalence of epilepsy of 23.1% in a prospective cohort of 182 adults with mitochondrial disease followed up for 7 years [14].

Mitochondrial disorders that occur due to defects in the respiratory chain usually have epilepsy as part of their clinical phenotype. Moreover, the mitochondrial syndromes with epilepsy as a leading clinical feature include MERFF, Alpers-Huttenlocher syndrome, Leigh syndrome, myoclonic epilepsy myopathy sensory ataxia (MEMSA), MELAS and others [12]. The seizures most commonly encountered in these disorders are myoclonic and focal, with a predilection to occipital lobes (at least initially), but can be generalized tonic–clonic seizures and status epilepticus [2,12]. In MELAS specifically, the seizures encountered are often associated with migraine-like headache and affected individuals frequently present in status epilepticus. They also commonly present with symptomatic focal seizures secondary to strokes. When myoclonus is present, it is less severe and more infrequent than in MERFF [2]. Myoclonus, which can be epileptic or non-epileptic, can be seen in all types of mitochondrial disease. It can be practically constant or intermittent, photosensitive or intensified by actions like writing.

There is no single anti-epileptic drug (AED) specifically indicated in mitochondrial disease. In addition, a lot of the commonly used antiepileptic drugs are mitochondrial toxic and could worsen the condition or even prove fatal [15]. While most of the seizure types (including generalized tonic-clonic seizures and status epilepticus) in mitochondrial disease are generally amenable to traditional anti-epileptic treatment, myoclonus often remains refractory to medical treatment and is likely to be progressive. This could suggest a non-epileptic origin of myoclonus.

Usually, a combination of AEDs (including benzodiazepines) is necessary to achieve the best control of the seizures. Aggressive treatment is usually advised to try and prevent secondary damage. In the context of mitochondrial disease, it is crucial to establish whether a disorder is due to a nuclear-encoded POLG mutation, such as Alpers-Huttenlocher syndrome, as sodium valproate is absolutely contraindicated in patients with this disease [15]. Sodium valproate could accelerate a tendency to liver failure and expedite death. It is well known that numerous AEDs disrupt mitochondrial function by interfering with the respiratory chain and can be toxic in mitochondrial disorders. However, the precise mechanism of toxicity is not fully understood and there is considerable individual variability on how well these drugs are tolerated. AEDs that are known to be toxic in mitochondrial disorders, but less than sodium valproate, include phenobarbital, carbamazepine, phenytoin, oxcarbazepine and ethosuximide [15]. Mitochondrial dysfunction leads to disruption in calcium homeostasis, which can increase neuronal excitability causing seizures. Levetiracetam is one of the most effective AEDs in mitochondrial epilepsy as it modulates intracellular calcium influx [16].

It is also important to note that drug interventions that target a single biological pathway will only help the specific individuals where that drug's mechanism of action is relevant to their disorder. Since we know that epilepsy in mitochondrial disorders is complex and includes numerous types of seizures, it is no surprise that a single AED almost never results in seizure control. Challenges in effectively controlling the epilepsy in mitochondrial disorders have prompted the use of the ketogenic diet (KD) either in first-line or in refractory epilepsy.

2. Pathogenesis of Mitochondrial Epilepsy

Mitochondria are organelles commonly denoted as the "power house of the cell" where ATP is generated via oxidative phosphorylation, mainly by using pyruvate derived from glycolysis. Ketone bodies generated by fatty acid oxidation can serve as alternative metabolites for aerobic energy production [3,17]. Figure 1 summarizes the various proposed mechanisms through which mitochondrial dysfunction is linked to epileptogenesis.

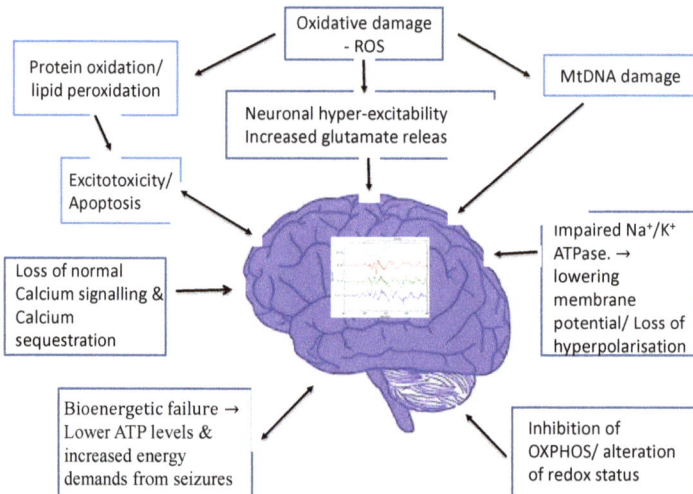

Figure 1. Potential mechanisms of epileptogenesis in mitochondrial disease. (ROS: Reactive Oxygen Species; ATP: Adenosine Tri-Phosphate; Na+: Sodium; K+: Potassium; MtDNA: mitochondrial DNA; OXPHOS: oxidative phosphorylation).

Increasingly, mitochondrial dysfunction has been recognized as a common mechanism underlying many neurological disorders [18]. Since neurons have high-energy demands and no significant capacity to regenerate, they are particularly vulnerable to mitochondrial dysfunction. Mitochondria, in turn, are the primary site of reactive oxygen species (ROS) making them particularly vulnerable to oxidative damage [19]. The latter could contribute to neuronal hyper-excitability and seizures. Mitochondrial dysfunction may also be an important cause of therapy-resistant types of severe epilepsy. Studies have confirmed the disruption of the anti-oxidant protection system and increased production of ROS in patients with epilepsy. Animal seizure models have also shown neuroprotective effects following both endogenous and exogenous anti-oxidants [20].

It is known that mitochondrial impairment occurs acutely, due to precipitating injuries such as status epilepticus. However, mitochondrial dysfunction is also implicated in epileptogenesis and acquired, chronic, epilepsy, like that seen in temporal lobe epilepsy (TLE) [19,21]. Experimental models of TLE have suggested mitochondrial dysfunction and oxidative stress may play a crucial role in epileptogenesis, through various mechanisms which include mitochondrial DNA damage, protein oxidation, lipid peroxidation and changes in redox status [21].

Defective cellular energy production seems to play an important role in the development of epilepsy, where mitochondrial dysfunction can induce cortical excitation of neurons. There is a complex interplay between oxidative stress, mitochondrial dysfunction and epileptogenesis. Inhibition of oxidative phosphorylation enzyme complexes in mitochondria result in decreased intracellular ATP levels. This in turn causes hyper-excitability of neurons by impairing sodium-potassium ATPase activity and decreasing membrane potential. Calcium sequestration, normally occurring in mitochondria, can no longer occur, rendering the neurons vulnerable to further excitotoxicity. This theory is backed by the fact that mutations in the mitochondrial respiratory chain complexes are known to result in neuronal hyperexcitability and epileptogenesis [22]. Experiments have also shown that the direct inhibition of enzymes in the mitochondrial respiratory chain using toxins can induce seizures in a dose-dependent manner [23]. Although dysfunction in any of the respiratory complexes can result in epileptogenesis, defects in complex I seems to be the most likely to result in seizures [2].

Hyper-excitability of neurons has also been proposed to be the result of excess glutamate release, which results from mitochondrial bioenergetics failure, dysfunction of the mitochondrial glutamate-aspartate transporter and loss of normal calcium signalling [2]. Excess release of glutamate has also been shown to have a direct effect on epileptogenesis and spread of epileptic activity across the cortex [24].

Furthermore, there is evidence to support that mitochondrial dysfunction can also play a role in seizure-related cell death. Kovac et al. [25] showed a direct association between blocking mitochondrial complex I, complex V or mitochondrial oxidative phosphorylation and rates of apoptosis, while conversely supplementing with the complex I substrate pyruvate led to decreased rates of apoptosis [25]. If the mitochondrial dysfunction can be prevented or treated, a level of neuroprotection would be expected, suggesting possible novel therapeutic avenues.

3. The Ketogenic Diet

Use of dietary manipulation and forms of fasting as a means of controlling epileptic seizures can be traced back to the time of Hippocrates, who was first to propose that seizures were not supernatural in origin [26]. In the 1920s, both in France and in the United States, significant discoveries were made regarding physiological changes linked with anti-seizure properties of fasting. The first to publish results showing the effectiveness of the ketogenic diet in epilepsy was Dr. Peterman, a paediatrician from the Mayo clinic in 1924. At the time the use of the KD was hyped due to limited alternatives, falling out of favour in the late 1930s due to its unpalatability and the emergence of AEDs like phenytoin [26–28].

The interest in the diet was re-ignited in the sixties due to growing evidence of its broad neuroprotective effects. The use of the KD is being studied in numerous other neurological diseases, which are felt to be arising partly from cellular energy failure and are characterized by neuronal death [18,27–29].

The KD is currently being used therapeutically for intractable epilepsy and for rare diseases of glucose metabolism, where it is the preferred first-line treatment. For instance, it is considered as the first-line treatment for glucose transporter type 1, pyruvate dehydrogenase deficiency and phosphofructokinase deficiency [17,30]. There is growing evidence to suggest that the KD should also be considered early in the treatment of other epilepsy syndromes like Dravet, West syndrome and myoclonic-astatic epilepsy (Doose syndrome) [27,30]. One variant of the KD, the low glycaemic index diet, has recently shown promise in the seizure control of patients with Angelman syndrome [31].

While there is now ample evidence of the benefits of the KD, one also needs to be aware of its limitations. The list of absolute contraindications includes fatty acid oxidation defects, pyruvate carboxylase deficiency and other gluconeogenesis defects, glycogen storage diseases, ketolysis defects, ketogenesis defects, porphyria, prolonged QT or other cardiac diseases, liver/kidney/or pancreatic insufficiency and hyperinsulinism. In addition, there are relative contraindications limiting its widespread use, like poor compliance, growth retardation, severe gastro-oesophageal reflux disease, inability to maintain adequate nutrition and a surgical focus identified by imaging or electroencephalogram [26,32].

The implementation of the KD can be arduous for families and requires close supervision by a specialized dietician, who will monitor the progress of the treatment and tailor it to each patient's individual needs. This is especially important in the younger population where growth prevents protein intake to be below a certain minimum. Children under 6 months traditionally required an inpatient stay for the initiation of the treatment, although currently in the UK the aim is to start the KD on an outpatient basis.

Prior to initiation, baseline monitoring is advised by blood and urine tests along with measurement of growth parameters. During diet initiation monitoring occurs daily, so that changes can be made if side effects are intolerable. Side effects include gastrointestinal disturbance, impaired growth, hypoglycaemia, acidosis and dehydration at initiation, kidney stones, nutritional deficiencies, and hyperlipidaemia. Most of them can be alleviated by adjustments in the diet and vitamin, mineral and trace element supplementation [26,32]. Importantly, in the largest randomized trial to date the most frequent side-effects reported at 3-month review were constipation, vomiting, lack of energy, and hunger [33].

Finally, the KD therapy is not available in all epilepsy centres in the UK, despite evidence of its efficacy, due to funding restrictions and lack of resources, further limiting its use [27].

4. Mechanism of Action

The KD is high in fat and low in carbohydrates, which mimics the metabolic state of starvation, forcing the body to utilize fat as its primary source of energy instead of carbohydrates. In the setting of elevated fatty acids and low dietary carbohydrate content while on a KD, the liver produces ketone bodies by shunting excess acetyl-CoA to ketogenesis (Figure 2). Thus, the two primary ketone bodies produced are acetoacetate and β-hydroxybutyrate (BHB). Acetoacetate metabolizes to acetone, the other major ketone elevated in patients on a KD.

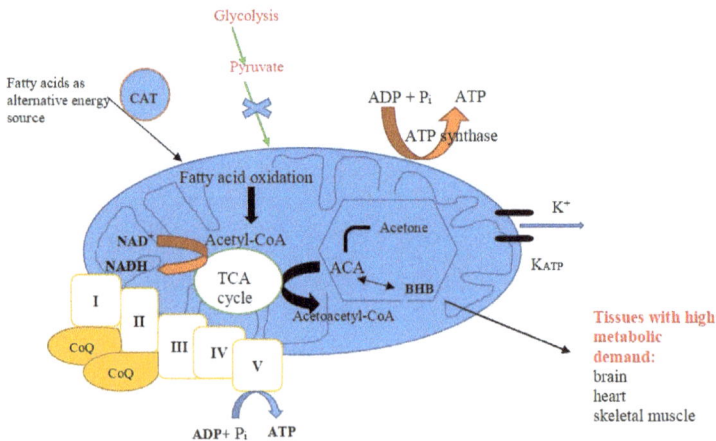

Figure 2. Mitochondri in hepatocyte. Ketone bodies alter mitochondrial metabolism at the cellular level; ketones reduce mitochondrial NAD coupling, oxidize Co-enzyme Q, increase the rate of ATP hydrolysis and increase metabolic efficiency. Oxidation of Co-enzyme Q decreases the major source reactive oxygen species (ROS). Increased rate of ATP hydrolysis widens the extra/intracellular ionic gradients, leading to hyperpolarization of cells. Potassium (K^+): Hyperpolarization of cell membrane reducing excitotoxicity; CAT: carnitine-acylcarnitine translocase; BHB: β-hydroxybutyrate; ACA: acetoacetate; I–V: electron transport chain complexes; ADP: Adenosine Di-Phosphate; ATP: Adenosine Tri-Phosphate; CoQ10: Co-enzyme Q10; K_{ATP}: ATP-sensitive potassium channels.

However, for each of the proposed mechanisms in Table 1 below, there is evidence suggesting that it is unlikely to be the sole explanation for the efficacy of the ketogenic diet. It is more likely that the efficacy of the KD depends on many different mechanisms acting synergistically [28,34]. We will look at some of these theories in more detail.

Table 1. Proposed mechanisms of action of the ketogenic diet (KD).

1.	Anti-inflammatory action and neuroprotection against excitotoxicity and oxidative stress (enhanced bioenergetics reserves, antioxidant mechanisms).
2.	Direct anticonvulsant action of ketones (e.g., reduces hyperexcitability, supports synaptic vesicle recycling, inhibits glutamate release).
3.	Direct anticonvulsant action of polyunsaturated fatty acids and protein and glycolytic restriction.
4.	Modulation of sodium and potassium and calcium balance (e.g., opening of voltage-gated potassium channels via poly-unsaturated fatty acids, ketone modulation of ATP-sensitive potassium channels, blockage of voltage-gated calcium channels).
5.	Direct anticonvulsant action of medium-chain fatty acids (MCTs) (e.g., decanoic acid (CA10), direct inhibition of a-amino-3-hydroxy-5-methyl-4-isoxazolepropionic acid receptor (AMPA) receptors, MCT modulation of amino acid metabolism).
6.	Disease-modifying anticonvulsant effects/epigenetic mechanisms (e.g., upregulation of transcription of genes involved in fatty acid metabolism while down regulating those involved in glucose metabolism, improved mitochondrial biogenesis and increase in mitochondrial numbers via peroxisome proliferator-activated receptor gamma (PPAR-γ), MCT modulation of astrocyte metabolism).

Initially there was the theory that the antiepileptic effect of the KD was directly related to the ketosis, which we know now to be incorrect. BHB was thought to be the main ketone responsible for the anticonvulsant effect of the KD, with proposed mechanisms including reducing glycolysis, promoting endocytosis of synaptic vesicles more than exocytosis and conferring neuroprotection via hydroxylcarboxylic acid receptor 2 (HCA2) activation in macrophages [34]. Acetone was also shown to suppress acutely provoked seizures in animals, but the same has not been proven in vivo [28]. There is now plenty of evidence that shows poor correlation between plasma ketone levels and degree of seizure control [35,36].

Recent studies by Chang et al. [37] on hippocampal rat slices demonstrated that decanoic acid (CA10—a major component of the medium-chain fatty acid (MCT) diet), but not ketone bodies, has direct anti-epileptic properties at doses that produce similar plasma levels as those seen in patients on the MCT diet. Decanoic acid modifies the excitatory post-synaptic currents via direct inhibition of excitatory AMPA receptors. This effect is achieved through non-competitive binding to the M3 helix sites of the AMPA-GluA2 transmembrane domain. The antagonism of these AMPA receptors is essential in epilepsy treatment, as they are widespread throughout the brain, play a crucial role in generating and spreading epileptic activity and are implicated in seizure-induced cortical damage [38]. The co-administration of octanoic acid (CA8) enhances these effects and is independent of ketones bodies [39]. CA10 has also been associated with increased mitochondrial numbers, via the PPAR-γ receptor, and improved biogenesis, also suggesting that ketones bodies per se are not essential to bring about the anticonvulsant effects of the MCT diet [40]. Decanoic acid has also been shown to upregulate transcription of genes involved in fatty acid metabolism while down regulating those involved in glucose metabolism. Furthermore, medium-chain fatty acids have also been shown to modulate astrocyte metabolism (providing fuel in the form of lactate and ketones) and amino acid metabolism (increasing tryptophan), resulting in reduced neuronal excitability [34]. Various other medium chain triglycerides, like heptanoic acid and tridecanoin, are also being investigated for their anticonvulsant properties [34].

There is growing evidence that the KD alters the fundamental biochemistry of neurons in a manner that not only inhibits neuronal hyperexcitability, but also protects against a range of neurological disorders where cellular energy failure is felt to play a key role in pathogenesis [18,29]. In a rat kindling model, the KD was shown to reduce oxidative stress, reduce glycolysis, diminish spontaneous firing of ATP-sensitive potassium channels and delay epileptogenesis [41]. The KD has been shown to be neuroprotective in animal models of several CNS disorders, including Alzheimer's disease, Parkinson's disease, hypoxia, glutamate toxicity, ischemia, and traumatic brain injury. This neuroprotective effect may be possible through antioxidant and anti-inflammatory actions [29]. This is supported by studies in animal models and isolated cells that showed neuroprotection against many types of cellular injury is afforded by ketone bodies, especially BHB [29]. Thus, the KD may ultimately prove useful in the treatment of a variety of neurological disorders [18,41].

While initially there were concerns about KD initiation in the younger population due to its possible side-effects on growth, the KD has been proven to be very effective in this age group. Its efficacy can be partly explained by the higher levels of ketone metabolizing enzymes and monocarboxylic acid transporters produced during infancy, which transfer ketone bodies across the blood-brain barrier [18]. Thus, the brain is much more efficient at extracting and utilizing ketone bodies from the blood in this age group.

Current evidence suggests that a fundamental shift from glycolysis to intermediary metabolism induced by the KD is both necessary and sufficient for clinical efficacy. This notion is supported by a growing number of studies indicating that metabolic changes likely related to the KD's anticonvulsant properties include ketosis, reduced glycolysis, protein restriction, elevated fatty acid levels, and enhanced bioenergetic reserves [34,41,42]. Direct neuronal effects induced by the diet may involve ATP-sensitive potassium channel modulation, enhanced purinergic (i.e., adenosine) and GABAergic (gabba-aminobutyric acid) neurotransmission, increased brain-derived neurotrophic factor (BDNF) expression, attenuation of neuroinflammation, and expansion in energy reserves and stabilization of the neuronal membrane potential through improved mitochondrial function [43].

The observation that the full anticonvulsant effect of the KD can take weeks to develop, suggested that altered gene expression could be implicated [44]. It has been proposed that the KD may increase mitochondrial biogenesis, by inducing transcription of some electron transport chain subunit messenger ribonucleic acids (mRNAs) [44]. This results in increased ATP levels, leading to increased neuronal "energy reserves", allowing neurons to withstand metabolic challenges and stabilize neuronal membrane potential [28,29,44]. There is a theory that suggests that the disease-modifying effects of the KD on epilepsy are exerted via an adenosine-dependent epigenetic mechanism [34].

Ketone bodies may alter the behaviour of vesicular glutamate transporters (VGLUTs) responsible for filling pre-synaptic vesicles with glutamate. In Juge et al. ketone bodies were shown to inhibit glutamate release by competing with Cl^- at the site of VGLUT regulation [45].

Potassium-ATP channels are of particular interest given their close relationship with cellular metabolism. Their hyperpolarization was shown to have seizure suppressing effects and they were seen to increase in numbers when ATP was low [46]. Ketones have been shown to exert their neuroprotective and anticonvulsant effects by genetically modulating potassium-ATP channels [47].

5. Different Variants of the Ketogenic Diet

There are five main categories of ketogenic diets, initially proposed in 1921 [18].

The first is the classic KD, which uses a 4:1 or 3:1 fat-to-carbohydrate and protein ratio. This means that about 90% of a daily caloric intake comes from fat and the rest includes a small amount of protein, which ensures adequate growth, especially important in the paediatric population. The ratio can be altered to 2.5:1, 2:1 or 1:1 based on the patient's needs. A dietician will implement and closely monitor the diet and tailor it accordingly. This type of diet uses saturated long-chain fatty acids as the main source of energy [18,27].

Seizure control on the classic KD slowly increases within days to weeks of starting the diet. However, the serum levels of the ketone bodies do not correlate tightly with seizure control. The mechanism of their anti-seizure properties remains unclear. Interestingly, the ingestion of carbohydrates more than what is allowed rapidly reverses the therapeutic effect. Seizures can start within minutes of ingesting carbohydrates.

The second is the MCT diet, which was introduced in 1971 to surpass the severe restrictions of the classic KD and to make it more palatable. The main fatty acids used are caprylic acid (CA8), capric acid (CA10) and, to a lesser extent, caproic acid and lauric acid. This diet is not based on diet ratios, but uses a percentage of calories from MCT oils to create ketones. The main advantage of this diet over long-chain triglycerides (LCTs) is that MCTs are quickly transported to the liver by albumin and are more efficiently absorbed across mitochondrial membranes, without binding on carnitine [28]. Thus, MCT metabolism is faster, requires less energy expenditure, produces a higher level of ketosis than LCTs and is more palatable, as less total fat is required and more protein and carbohydrates can be consumed.

The trial designed to compare the efficacy of this diet with the classic KD in children with intractable epilepsy at 3, 6 and 12 months, found no differences between the dietary groups in the percent reduction of seizures. Importantly, there was no difference in the percentages of children achieving reduction in seizures greater than 50% or 90% [48]. The authors concluded that the diets were comparable in efficacy and tolerability, which offers a degree of choice to the patient and their family as to which diet they will use [27,47]. Chang et al. compared the effects of several MCTs with valproic acid both in vivo and in vitro and found that decanoic acid specifically was superior in controlling seizures and had fewer adverse side effects [49].

The third is the modified Atkins diet, which was originally designed and investigated at Johns Hopkins Hospital [50] and proposed as a less restrictive and more palatable dietary treatment. Contrary to other KDs, it does not overly restrict protein intake or daily calories to make it more palatable and increases compliance, especially in adults [51]. This diet restricts carbohydrates to 10 g/day for children (15 g/day in adults) while encouraging high-fat foods [52,53]. There is scope for modifying the carbohydrate content depending on seizure control [18].

There have been about 400 children and adults on the modified Atkins diet, as treatment for intractable epilepsy, taking part in over thirty prospective and retrospective studies published worldwide in the past 10 years. The studies have consistently shown similar efficacy of the modified Atkins diet to the classical KD and improved tolerability. Further studies would clarify its role in the management of intractable epilepsy and in other neurological disorders [52,53].

The fourth is the low glycaemic index diet, which allows a more moderate intake of carbohydrates, as long as they have a glycaemic index lower than 50, without increasing ketone levels [54]. The diet was implemented based on clinical observation that preventing large postprandial increases in blood glucose and maintaining as stable as possible blood glucose levels resulted in improved seizure control in a broad range of patients [51].

In a study of 20 paediatric and adult patients treated with the low glycaemic index diet, 50% experienced a greater than 90% reduction in seizure frequency, despite achieving lower blood levels of ketone bodies [51]. In another study where 89% of patients had not responded to 3 AEDs, 76 children were treated with this diet and followed up for up to 1 year. There was a greater than 50% seizure reduction in 66% of patients at 12 months. The study also showed an inverse relationship between efficacy of the diet and serum glucose, but did not find a correlation between seizure reduction and circulating levels of BHB. The commonest reason for discontinuing the diet was diet restrictiveness [55].

Lastly, the fifth is the calorie restriction or intermittent fasting diet, which involves the reduction in total caloric intake without a risk of malnutrition. No clinical studies exist yet which assess its effect on seizure suppression. Some authors believe that it is the restriction of calories, not the composition of the diet per se, that has the anti-convulsive effects, but this remains to be proven [18].

6. Evidence: Use of the Ketogenic Diet in Intractable Epilepsy in Mitochondrial Disorders

The use of the KD has been proven to be a safe and effective treatment in intractable epilepsy in numerous studies [42]. However, its specific use in mitochondrial disorders has not been extensively studies yet. The evidence of KD benefits and in the management of mitochondrial diseases, and specifically in the treatment of seizure disorders due to either nuclear or mitochondrial DNA defects is growing [11,13,56,57]. The efficacy of the KD has also been investigated in a variety of other CNS disorders, including Alzheimer's disease (AD), Parkinson's disease, headache, hypoxia, glutamate toxicity, autism, ischemia, and traumatic brain injury [18].

Jarrett et al. [58] demonstrated that the KD confers protection to the mitochondrial genome against oxidative insults, increasing the levels and stimulating de novo biosynthesis of mitochondrial glutathione and improving mitochondrial redox status. This results in improved cellular metabolism and may explain the KD's efficacy in mitochondrial diseases.

Lee et al. [13] looked at 48 patients with confirmed mitochondrial respiratory chain defects and epilepsy. Out of all the patients with medically intractable epilepsy (four or more seizures per month, resistant to three or more antiepileptic drugs), they treated 24 children with the classic ketogenic diet (4:1) and assessed its efficacy. They found that 75% of individuals had seizure reduction of more than 50%, and half of people became seizure free [13]. However, of the 31 children (64.5%) with medically intractable epilepsy only 24 received the ketogenic diet, without explaining what happened to the other 7. Furthermore, three (12.5%) patients had to stop the ketogenic diet due to serious infections or persistent metabolic acidosis.

In the retrospective study, Kang H.C. et al. [57] evaluated the clinical efficacy and safety of the classic KD for 14 patients (1.7–11.8 years of age) with intractable epilepsy and mitochondrial respiratory chain complex defects. Seven patients (50%) became seizure-free after KD initiation, three of whom successfully completed the diet without relapse. One patient had a greater than 90% seizure reduction, and two patients with a seizure reduction between 50% and 90%, remained on the diet. They concluded that the KD was a safe and effective therapy for seizures in children with intractable epilepsy and respiratory chain complex defects.

Furthermore, the KD was also reported to be a useful adjunct to traditional pharmaceutical agents. In Martikainen et al., a low glycaemic index diet was used in conjunction with three anti-epileptic drugs in a 26-year-old woman and was found to be effective and well tolerated as treatment for severe episodes of POLG-related mitochondrial epilepsy [56]. Furthermore, in a case of Ohtahara syndrome with mitochondrial respiratory chain complex I deficiency the seizures stopped and the burst suppressions pattern disappeared following 3 months on the KD therapy with vitamins, coenzyme therapy and antioxidant treatment [59]. Also, a five-year-old female with Landau-Kleffner and respiratory chain complex I deficiency was successfully treated with the KD in combination with coenzyme Q10, riboflavin, L-carnitine, and high-dose multivitamins [60]. However, these are single cases and lack statistical power.

In Barnerias et al., they studied 22 patients (early infancy to 30 years old), with confirmed pyruvate dehydrogenase complex deficiency [61]. Nine out of the 22 (41%) had epilepsy as one of the clinical manifestations. Of those treated with the KD, five showed clear benefit on childhood-onset epilepsy (two patients) or paroxysmal dystonia (three patients). This study showed that the KD was more efficacious against paroxysmal energy dysfunction seen in late-onset epilepsy and paroxysmal dystonia or ataxia than early onset epilepsy associated with infantile spasms [61].

In a randomized controlled trial over 5 years, Neal tested the efficacy of the KD given for 3 months on children aged 2–16 years with refractory epilepsy of different aetiologies [33]. Children were included if they had either daily seizures or at least seven seizures per week and had failed therapy with at least two AEDs. Twenty-eight children (38%) on the diet had greater than 50% seizure reduction compared with four (6%) controls ($p < 0.0001$), and five children (7%) on the diet had greater than 90% seizure reduction compared with none in the control group. The results favoured the use of the KD in intractable epilepsy with efficacy rates comparable to that of new AEDs [33]. Although this trial provided definitive evidence of the efficacy of the KD in intractable epilepsy, it did not look specifically at epilepsy in the context of mitochondrial disorders.

It has also been demonstrated that children benefitting on the KD show an improved neurodevelopmental outcome as early as three months after diet initiation [62]. However, Dressler et al. did not utilize a formal neuropsychological assessment in their study [62]. Instead, the improved psychomotor development was based on clinical neurological examination and EEG. In Zhu et al. [63], assessment using Gesell developmental scales assessment revealed improved neurodevelopmental progress in 42 children (6 months to 6 years of age) treated with the classic KD for intractable epilepsy. They were assessed at 3, 6, 12 and 18 months after starting the KD. They compared their data with those before the KD treatment and found statistically significant differences. It was concluded that the improvement in neurobehavioral outcome was dependent on clinical seizure control and the benefit was more pronounced with prolonged length of treatment with the KD [63]. Of note, both these studies included children with intractable epilepsy due to a variety of aetiologies and not specifically mitochondrial disorders. More research is needed to determine the effect of the KD on neurodevelopment in children with mitochondrial disorders and epilepsy.

7. Conclusions

There is growing evidence of the benefits of the KD and its variants in the treatment of a variety of neurological and neurodegenerative disorders. This suggests a possible common central mechanism that improves cellular metabolism and allows cells to resist metabolic challenges that lead to apoptosis. Identification of a correlation between caloric restriction with ketosis and protection from oxidative stress could reveal new therapeutic paths in age-related neurological disorders, like Alzheimer's. Although the full mechanism by which the KD and its variants improve oncological and neurological conditions remains to be elucidated, their clinical efficacy has attracted many new followers, and has re-ignited research interest in this field [27–29,54]. The epigenetic mechanism of action of the KD means it can alter the course of the epilepsy [1], and may explain why its anticonvulsant effects are long-lasting, even after the KD is stopped.

Further understanding of how this diet exerts its effects would allow its use to become broader and perhaps the development of a more palatable and less strict diet regimen. Research interest is also centred around the possibility of replacing the strict diet with its challenging adherence rules with supplements, like a "ketogenic pill". Ketone esters given orally were effective anticonvulsants in mouse models of Angelmans and in pentylenetetrazole-model of seizures [64,65]. In addition, in Sada et al., lactate dehydrogensae (LDH) inhibition was seen to be effective in suppressing seizures. The inhibition of this metabolic pathway was seen to mimic the KD effects, providing further hope that one day a pill may be devised that will be able to provide the benefits of the KD without the strict adherence rules [66]. In the future, novel antiepileptic drugs may contain antioxidant properties that target neurodegeneration and mimic the effects of the KD [19].

The safety of the diet also needs to be investigated further in the younger population as the randomized controlled study by Neal et al. only looked at children older than 2 years [33]. More work needs to be done to establish which patients would benefit the most, especially when the diet is being considered for a wider range of neurological disorders.

Although there is a significant body of evidence that supports the use of the KD in intractable epilepsy, more studies need to be done looking specifically at its effect in intractable epilepsy in the context of mitochondrial disorders.

Author Contributions: All three authors contributed to the writing and design of this article, as well as the review of the literature. Kinali and Ismayilova provided the critical revision of the final document.

Conflicts of Interest: The authors declare no conflict of interest.

References

1. Schaefer, A.M.; McFarland, R.; Blakely, E.L.; He, L.; Whittaker, R.G.; Taylor, R.W.; Chinnery, P.F.; Turnbull, D.M. Prevalence of mitochondrial DNA disease in adults. *Ann. Neurol.* **2008**, *63*, 35–39. [CrossRef] [PubMed]
2. Bindoff, L.A.; Engelsen, B.A. Mitochondrial diseases and epilepsy. *Epilepsia* **2012**, *53*, 92–97. [CrossRef] [PubMed]
3. El-Hattab, A.W.; Scaglia, F. Mitochondrial cytopathies. *Cell Calcium* **2016**, *60*, 199–206. [CrossRef] [PubMed]
4. Engel, W.K.; Cunningham, C.G. Rapid examination of muscle tissue: An improved trichrome stain method for fresh frozen biopsy sections. *Neurology* **1963**, *13*, 919–923. [CrossRef] [PubMed]
5. Leary, S.C. Blue native polyacrylamide gel electrophoresis: A powerful diagnostic tool for the detection of assembly defects in the enzyme complexes of oxidative phosphorylation. *Methods Mol. Biol.* **2012**, *837*, 195–206. [PubMed]
6. Suomalainen, A. Fibroblast Growth Factor 21: A novel biomarker for human muscle-manifesting mitochondrial disorders. *Expert Opin. Med. Diagn.* **2013**, *7*, 313–317. [CrossRef] [PubMed]
7. Hirano, M. (Columbia University Medical Centre, New York, NY, USA). Personal communication. 2017.
8. Ng, Y.S.; Turnbull, D.M. Mitochondrial disease: Genetics and management. *J. Neurol.* **2016**, *263*, 179–191. [CrossRef] [PubMed]
9. Pfeffer, G.; Majamaa, K.; Turnbull, D.; Thorburn, D.R.; Chinnery, P.F. Treatment for mitochondrial disorders. *Cochrane Database Syst. Rev.* **2012**. [CrossRef]
10. Finsterer, J.; Zarrouk Mahjoub, S. Epilepsy in mitochondrial disorders. *Seizure* **2012**, *21*, 316–321. [CrossRef] [PubMed]
11. El Sabbagh, S.; Lebre, A.S.; Bahi-Buisson, N.; Delonlay, P.; Soufflet, C.; Boddaert, N.; Rio, M.; Rotig, A.; Dulac, O.; Munnich, A.; et al. Epileptic phenotypes in children with respiratory chain disorders. *Epilepsia* **2010**, *51*, 1225–1235. [CrossRef] [PubMed]
12. Chevallier, J.A.; Von Allmen, G.; Koenig, M.K. Seizure semiology and EEG findings in mitochondrial diseases. *Epilepsia* **2014**, *55*, 707–712. [CrossRef] [PubMed]
13. Lee, Y.M.; Kang, H.C.; Lee, J.S.; Kim, S.H.; Kim, E.Y.; Lee, S.K.; Slama, A.; Kim, H.D. Mitochondrial respiratory chain defects: Underlying etiology in various epileptic conditions. *Epilepsia* **2008**, *49*, 685–690. [CrossRef] [PubMed]
14. Whittaker, R.G.; Devine, H.E.; Gorman, G.S.; Schaefer, A.M.; Horvath, R.; Ng, Y.; Nesbitt, V.; Lax, N.Z.; McFarland, R.; Cunningham, M.O.; et al. Epilepsy in adults with mitochondrial disease: A cohort study. *Ann. Neurol.* **2015**, *78*, 949–957. [CrossRef] [PubMed]
15. Finsterer, J.; Zarrouk Mahjoub, S. Mitochondrial toxicity of antiepileptic drugs and their tolerability in mitochondrial disorders. *Expert Opin. Drug Metab. Toxicol.* **2012**, *8*, 71–79. [CrossRef] [PubMed]
16. Niespodziany, I.; Klitgaard, H.; Margineanu, D.G. Levetiracetam inhibits the high voltage activated Ca^{2+} current in pyramidal neurones of rat hippocampal slices. *Neurosci. Lett.* **2001**, *306*, 5–8. [CrossRef]
17. Vidali, S.; Aminzadeh, S.; Lambert, B.; Rutherford, T.; Sperl, W.; Kofler, B.; Feichtinger, R.G. Mitochondria: The ketogenic diet—A metabolism-based therapy. *Int. J. Biochem. Cell Biol.* **2015**, *63*, 55–59. [CrossRef] [PubMed]
18. Gano, L.B.; Patel, M.; Rho, J.M. Ketogenic diets, mitochondria, and neurological diseases. *J. Lipid Res.* **2014**, *55*, 2211–2228. [CrossRef] [PubMed]
19. Waldbaum, S.; Patel, M. Mitochondrial dysfunction and oxidative stress: A contributing link to acquired epilepsy? *J. Bioenerg. Biomembr.* **2010**, *42*, 449–455. [CrossRef] [PubMed]
20. Martinc, B.; Grabnar, I.; Vovk, T. Antioxidants as a Preventive Treatment for Epileptic Process: A Review of the Current Status. *Curr. Neuropharmacol.* **2014**, *12*, 527–550. [CrossRef] [PubMed]
21. Rowley, S.; Patel, M. Mitochondrial involvement and oxidative stress in temporal lobe epilepsy. *Free Radic. Biol. Med.* **2013**, *62*, 121–131. [CrossRef] [PubMed]
22. Martinc, B.; Grabnar, I.; Vovk, T. The Role of Reactive Species in Epileptogenesis and Influence of Antiepileptic Drug Therapy on Oxidative Stress. *Curr. Neuropharmacol.* **2012**, *10*, 328–343. [CrossRef] [PubMed]
23. Urbanska, E.M.; Blaszczak, P.; Saran, T.; Kleinrok, Z.; Turski, W.A. Mitochondrial toxin 3-nitropropionic acid evokes seizures in mice. *Eur. J. Pharmacol.* **1998**, *359*, 55–58. [CrossRef]
24. Chapman, A.G. Glutamate and epilepsy. *J. Nutr.* **2000**, *130*, 1043S–1045S. [PubMed]

25. Kovac, S.; Domijan, A.M.; Walker, M.C.; Abramov, A.Y. Prolonged seizure activity impairs mitochondrial bioenergetics and induces cell death. *J. Cell Sci.* **2012**, *125*, 1796–1806. [CrossRef] [PubMed]
26. Dhamija, R.; Eckert, S.; Wirrell, E. Ketogenic diet. *Can. J. Neurol. Sci.* **2013**, *40*, 158–167. [CrossRef] [PubMed]
27. Cross, J.H.; Mclellan, A.M.; Neal, E.G.; Philip, S.; Williams, E.; Williams, R.E. The ketogenic diet in childhood epilepsy: Where are we now? *Arch. Dis. Child* **2010**, *95*, 550–553. [CrossRef] [PubMed]
28. Cross, J.H. Dietary therapies—An old idea with a new lease of life. *Seizure* **2010**, *19*, 671–674. [CrossRef] [PubMed]
29. Gasior, M.; Rogawski, M.A.; Hartman, A.L. Neuroprotective and disease-modifying effects of the ketogenic diet. *Behav. Pharmacol.* **2006**, *17*, 431–439. [CrossRef] [PubMed]
30. Baranano, K.W.; Hartman, A.L. The ketogenic diet: Uses in epilepsy and other neurologic illnesses. *Curr. Treat. Options Neurol.* **2008**, *10*, 410–419. [CrossRef] [PubMed]
31. Grocott, O.R.; Herrington, K.S.; Pfeifer, H.H.; Thiele, E.A.; Thibert, R.L. Low glycemic index treatment for seizure control in Angelman syndrome: A case series from the Center for Dietary Therapy of Epilepsy at the Massachusetts General Hospital. *Epilepsy Behav.* **2017**, *68*, 45–50. [CrossRef] [PubMed]
32. Van der Louw, E.; van den Hurk, D.; Neal, E.; Leiendecker, B.; Fitzsimmon, G.; Dority, L.; Thompson, L.; Marchio, M.; Dudzinska, M.; Dressler, A.; et al. Ketogenic diet guidelines for infants with refractory epilepsy. *Eur. J. Paediatr. Neurol.* **2016**, *20*, 798–809. [CrossRef] [PubMed]
33. Neal, E.G.; Chaffe, H.; Schwatrz, R.H.; Lawson, M.S.; Edwards, N.; Fitzsimmons, G.; Whitney, A.; Cross, J.H. The ketogenic diet for the treatment of childhood epilepsy: A randomised controlled trial. *Lancet Neurol.* **2008**, *7*, 500–506. [CrossRef]
34. Boison, D. New insights into the mechanisms of the ketogenic diet. *Curr. Opin. Neurol.* **2017**, *30*, 187–192. [CrossRef] [PubMed]
35. Likhodii, S.S.; Musa, K.; Mendonca, A.; Dell, C.; Burnham, W.M.; Cunnane, S.C. Dietary fat, ketosis, and seizure resistance in rats on the ketogenic diet. *Epilepsia* **2000**, *41*, 1400–1410. [CrossRef] [PubMed]
36. Thavendiranathan, P.; Mendonca, A.; Dell, C.; Likhodii, S.S.; Musa, K.; Iracleous, C.; Cunnane, S.C.; Burnham, W.M. The MCT ketogenic diet: Effects on animal seizure models. *Exp. Neurol.* **2000**, *161*, 696–703. [CrossRef] [PubMed]
37. Chang, P.; Augustin, A.; Boddum, K.; Williams, S.; Sun, M.; Terschak, J.A.; Hardege, J.D.; Chen, P.E.; Walker, M.C.; Williams, R.S.B. Seizure control by decanoic acid through direct AMPA receptor inhibition. *Brain* **2016**, *139*, 431–443. [CrossRef] [PubMed]
38. Rogawski, M.A.; Donevan, S.D. AMPA receptors in epilepsy and as targets for antiepileptic drugs. *Adv. Neurol.* **1999**, *79*, 947–963. [PubMed]
39. Wlaz, P.; Socala, K.; Nieoczym, D.; Zarnowski, T.; Zarnowska, I.; Czuczwar, S.J.; Gasior, M. Acute anticonvulsant effects of capric acid in seizure tests in mice. *Prog. Neuropsychopharmacol. Biol. Psychiatry* **2005**, *57*, 110–116. [CrossRef] [PubMed]
40. Hughes, S.D.; Kanabus, M.; Anderson, G.; Hargreaves, I.P.; Rutherford, T.; O'Donnell, M.; Cross, J.H.; Eaton, S.; Rahman, S.; Heales, S.J. The ketogenic diet component decanoic acid increases mitochondrial citrate synthase and complex I activity in neuronal cells. *J. Neurochem.* **2014**, *129*, 426–433. [CrossRef] [PubMed]
41. Kim, D.Y.; Rho, J.M. The ketogenic diet and epilepsy. *Curr. Opin. Clin. Nutr. Metab. Care* **2008**, *11*, 113–120. [CrossRef] [PubMed]
42. Clanton, R.M.; Wu, G.; Akabani, G.; Aramayo, R. Control of seizures by ketogenic diet-induced modulation of metabolic pathways. *Amino Acids* **2017**, *49*, 1–20. [CrossRef] [PubMed]
43. Masino, S.A.; Rho, J.M. Mechanisms of Ketogenic Diet Action. In *Jasper's Basic Mechanisms of the Epilepsies*, 4th ed.; Noebels, J.L., Avoli, M., Rogawski, M.A., Olsen, R., Delgado-Escueta, A., Eds.; National Center for Biotechnology Information: Bethesda, MD, USA, 2012.
44. Bough, K.J.; Wetherington, J.; Hassel, B.; Pare, J.F.; Gawryluk, J.W.; Greene, J.G.; Shaw, R.; Smith, Y.; Geiger, J.D.; Dingledine, R.J. Mitochondrial biogenesis in the anticonvulsant mechanism of the ketogenic diet. *Ann. Neurol.* **2006**, *60*, 223–235. [CrossRef] [PubMed]
45. Juge, N.; Gray, J.A.; Omote, H.; Miyaji, T.; Inoue, T.; Hara, C.; Uneyama, H.; Edwards, R.H.; Nicoll, R.A.; Moriyama, Y. Metabolic Control of Vesicular Glutamate Transport and Release. *Neuron* **2010**, *68*, 99–112. [CrossRef] [PubMed]
46. McNally, M.A.; Hartman, A.L. Ketone Bodies in Epilepsy. *J. Neurochem.* **2012**, *121*, 28–35. [CrossRef] [PubMed]

47. Kim, D.Y.; Abdelwahab, M.G.; Lee, S.H.; O'Neil, D.; Thompson, R.J.; Duff, H.J.; Sullivan, P.G.; Rho, J.M. Ketones prevent oxidative impairment of hippocampal synaptic integrity through KATP channels. *PLoS ONE* **2015**, *10*, e0119316. [CrossRef] [PubMed]

48. Neal, E.G.; Chaffe, H.; Schwatrz, R.H.; Lawson, M.S.; Edwards, N.; Fitzsimmons, G.; Whitney, A.; Cross, J.H. A randomised trial of classical and medium-chained Triglyceride ketogenic diets in the treatment of childhood epilepsy. *Epilepsia* **2009**, *50*, 1109–1117.

49. Chang, P.; Terbach, N.; Plant, N.; Chen, P.E.; Walker, M.C.; Williams, R.S.B. Seizure control by ketogenic diet-associated medium chain fatty acids. *Neuropharnacology* **2013**, *69*, 105–114. [CrossRef]

50. Kossoff, E.H.; Hartman, A.L. Ketogenic diets: New advances for metabolism-based therapies. *Curr. Opin. Neurol.* **2012**, *25*, 173–178. [CrossRef] [PubMed]

51. Pfeiffer, H.H.; Thiele, E.A. Low glycaemic index treatment: A liberalized ketogenic diet for treatment if intractable epilepsy. *Neurology* **2005**, *65*, 1810–1812. [CrossRef] [PubMed]

52. Kossoff, E.H.; Dorward, J.L. The modified Atkins diet. *Epilepsia* **2008**, *9*, 37–41. [CrossRef] [PubMed]

53. Kossoff, E.H.; Cervenca, M.C.; Henry, B.J.; Haney, C.A.; Turner, Z. A decade of the modified Atkins diet (2003–2013): Results, insights and future directions. *Epilepsy Behav.* **2013**, *29*, 437–442. [CrossRef] [PubMed]

54. Branco, A.F.; Ferreira, A.; Simões, R.F.; Magalhães-Novais, S.; Zehowski, C.; Cope, E.; Silva, A.M.; Pereira, D.; Sardão, V.A.; Cunha-Oliveira, T. Ketogenic diets: From cancer to mitochondrial diseases and beyond. *Eur. J. Clin. Investig.* **2016**, *46*, 285–298. [CrossRef] [PubMed]

55. Muzykewicz, D.A.; Luczkowski, D.A.; Memon, N.; Conant, K.D.; Pfeiffer, H.H.; Thiele, E.A. Efficacy, safety and tolerability of the low glycaemic index treatment in paediatric epilepsy. *Epilepsia* **2009**, *50*, 1118–1126. [CrossRef] [PubMed]

56. Martikainen, M.H.; Päivärinta, M.; Jääskeläinen, S.; Majamaa, K. Successful treatment of POLG-related mitochondrial epilepsy with antiepileptic drugs and low glycaemic index diet. *Epileptic Disord.* **2012**, *14*, 438–441. [PubMed]

57. Kang, H.C.; Lee, Y.M.; Kim, H.D.; Lee, J.S.; Slama, A. Safe and effective use of the ketogenic diet in children with epilepsy and mitochondrial respiratory chain complex defects. *Epilepsia* **2007**, *48*, 82–88. [CrossRef] [PubMed]

58. Jarrett, S.G.; Milder, J.B.; Liang, L.P.; Patel, M. The ketogenic diet increases mitochondrial glutathione levels. *J. Neurochem.* **2008**, *106*, 1044–1051. [CrossRef] [PubMed]

59. Seo, J.H.; Lee, Y.M.; Lee, J.S.; Kim, S.H.; Kim, H.D. A case of Ohtahara syndrome with mitochondrial respiratory chain complex I deficiency. *Brain Dev.* **2010**, *32*, 253–257. [CrossRef] [PubMed]

60. Kang, H.C.; Kim, H.D.; Lee, Y.M.; Han, S.H. Landau Kleffner syndrome with mitochondrial respiratory chain-complex I deficiency. *Pediatr. Neurol.* **2006**, *35*, 158–161. [CrossRef] [PubMed]

61. Barnerias, C.; Saudubray, J.M.; Touati, G.; De Lonlay, P.; Dulac, O.; Ponsot, G.; Marsac, C.; Brivet, M.; Desguerre, I. Pyruvate dehydrogenase complex deficiency: Four neurological phenotypes with differing pathogenesis. *Dev. Med. Child Neurol.* **2010**, *52*, e1–e9. [CrossRef] [PubMed]

62. Dressler, A.; Stocklin, B.; Reithofer, E.; Benninger, F.; Freilinger, M.; Hauser, E.; Reiter-Fink, E.; Seidl, R.; Trimmel-Schwahofer, P.; Feucht, M. Long-term outcome and tolerability of the ketogenic diet in drug resistant childhood epilepsy—The Austrian experience. *Seizure* **2010**, *19*, 404–408. [CrossRef] [PubMed]

63. Zhu, D.; Wang, M.; Wang, J.; Yuan, J.; Niu, G.; Zhang, G.; Sun, L.; Xiong, H.; Xie, M.; Zhao, Y. Ketogenic diet effects on neurobehavioral development of children with intractable epilepsy: A prospective study. *Epilepsy Behav.* **2016**, *55*, 87–91. [CrossRef] [PubMed]

64. Ciarlone, S.L.; Grieco, J.C.; D'Agostino, D.P.; Weeber, E.J. Ketone ester supplementation attenuates seizure activity, and improves behavior and hippocampal synaptic plasticity in an Angelman syndrome mouse model. *Neurobiol. Dis.* **2016**, *96*, 38–46. [CrossRef] [PubMed]

65. Viggiano, A.; Pilla, R.; Arnold, P.; Monda, M.; D'Agostino, D.; Coppola, G. Anticonvulsant properties of an oral ketone ester in a pentylenetetrazole-model of seizure. *Brain Res.* **2015**, *1618*, 50–54. [CrossRef] [PubMed]

66. Sada, N.; Lee, S.; Katsu, T.; Inoue, T.; Otsuki, T. Epilepsy treatment. Targeting LDH enzymes with a stiripentol analog to treat epilepsy. *Science* **2015**, *347*, 1362–1367. [CrossRef] [PubMed]

Journal of
Clinical Medicine

MDPI

Review

Riboflavin Responsive Mitochondrial Dysfunction in Neurodegenerative Diseases

Tamilarasan Udhayabanu [1], Andreea Manole [2], Mohan Rajeshwari [1], Perumal Varalakshmi [3], Henry Houlden [2] and Balasubramaniem Ashokkumar [1,*]

[1] Department of Genetic Engineering, School of Biotechnology, Madurai Kamaraj University,
    Madurai 625021, India; udhaya13banu@gmail.com (T.U.); ponderofbiotech@gmail.com (M.R.)
[2] Department of Molecular Neuroscience and Neurogenetics Laboratory, UCL Institute of Neurology,
    Queen Square, London WC1N 3BG, UK; andreea.manole.13@ucl.ac.uk (A.M.); h.houlden@ucl.ac.uk (H.H.)
[3] Department of Molecular Microbiology, School of Biotechnology, Madurai Kamaraj University,
    Madurai 625021, India; vara5277@gmail.com
*   Correspondence: rbashokkumar@yahoo.com; Tel.: +91-452-2459115; Fax: +91-452-2459105

Academic Editor: Mark S. Sands
Received: 16 March 2017; Accepted: 2 May 2017; Published: 5 May 2017

Abstract: Mitochondria are the repository for various metabolites involved in diverse energy-generating processes, like the TCA cycle, oxidative phosphorylation, and metabolism of amino acids, fatty acids, and nucleotides, which rely significantly on flavoenzymes, such as oxidases, reductases, and dehydrogenases. Flavoenzymes are functionally dependent on biologically active flavin adenine dinucleotide (FAD) or flavin mononucleotide (FMN), which are derived from the dietary component riboflavin, a water soluble vitamin. Riboflavin regulates the structure and function of flavoenzymes through its cofactors FMN and FAD and, thus, protects the cells from oxidative stress and apoptosis. Hence, it is not surprising that any disturbance in riboflavin metabolism and absorption of this vitamin may have consequences on cellular FAD and FMN levels, resulting in mitochondrial dysfunction by reduced energy levels, leading to riboflavin associated disorders, like cataracts, neurodegenerative and cardiovascular diseases, etc. Furthermore, mutations in either nuclear or mitochondrial DNA encoding for flavoenzymes and flavin transporters significantly contribute to the development of various neurological disorders. Moreover, recent studies have evidenced that riboflavin supplementation remarkably improved the clinical symptoms, as well as the biochemical abnormalities, in patients with neuronopathies, like Brown-Vialetto-Van-Laere syndrome (BVVLS) and Fazio-Londe disease. This review presents an updated outlook on the cellular and molecular mechanisms of neurodegenerative disorders in which riboflavin deficiency leads to dysfunction in mitochondrial energy metabolism, and also highlights the significance of riboflavin supplementation in aforementioned disease conditions. Thus, the outcome of this critical assessment may exemplify a new avenue to enhance the understanding of possible mechanisms in the progression of neurodegenerative diseases and may provide new rational approaches of disease surveillance and treatment.

Keywords: riboflavin; FAD; FMN; BVVLS; motor neuronopathy; mitochondrial dysfunction

1. Introduction

Energy metabolism generally takes place across the plasma membrane in prokaryotes, whereas eukaryotes have a well-defined specialized organelle called the mitochondrion. Mitochondria are the energy-transducing mobile organelles in eukaryotic cells that produce ATP through the process of oxidative phosphorylation, which drives cellular metabolism [1]. In addition, it acts as a site of various metabolic processes, like the breakdown of sugars and long-chain fatty acids, the synthesis of amino acids, lipids, and steroids, along with numerous other reactions that are essential for the survival of the organism. Mitochondria consist of four components: (i) an outer membrane that has porins, which allows small molecules to enter; (ii) an inner membrane which is impermeable to ions while the transport is mediated by a specific transport system; (iii) cristae; and (iv) the mitochondrial matrix, which contains enzymes involved in the Krebs cycle and electron transport chain (ETC). ETC has evolved to contain the molecular machinery for energy production in the inner mitochondrial membranes, which consists of five protein complexes, among them three of the complexes (I, III, and IV) pump protons (H^+) to generate a H^+ gradient for ATP production at complex V.

Mitochondria have their own genome, the mitochondrial DNA (mtDNA), which is located in the mitochondrial matrix. In humans, the mitochondrial genome is a small circular DNA with a size of 16.5 kb [2] that contains 13 polypeptides encoding seven subunits of complex I, one subunit of complex III, three subunits of complex IV, and two subunits of complex V in respiratory chain while genes involved in complex II are encoded by the nuclear DNA. It also contains 22 tRNA and 2 rRNA for its translational mechanism. The biochemistry of mitochondria has been well studied, however, its implications in the development of inborn errors of metabolism have only recently been established and the understanding of mitochondrial diseases caused by the inheritance of genetic variations have gained much significance in recent years. In this review, we discuss the new insights of mitochondrial biology in neurodegenerative diseases.

2. Mitochondria—The Power House

The mitochondrial matrix serves as a host for a wide variety of metabolites involved in three major processes, such as citric acid cycle, urea cycle, and electron transport chain. Additionally, mitochondrion contains several electrochemically-active species (flavin adenine dinucleotide (FAD), flavin mononucleotide (FMN), ubiquinone, and cytochrome c4), so it is involved in the production of energy by means of electrochemical mechanisms. The predominant role of mitochondria is the production of ATP, which is known as the energy currency for the proper functioning of the cells. ATP is produced in the cytosol by the oxidation of glucose and pyruvate from dietary food sources by means of cellular respiration (otherwise named as aerobic respiration) [3].

During aerobic respiration, 1%–2% of the consumed oxygen is involved in the production of reactive oxygen species (ROS) that, in excessive amount cause, oxidative damage to DNA and proteins, leads to mitochondrial damage [4]. The continuous production of ROS in mitochondria leads to age-related oxidative stress that result in cellular aging. Furthermore, mtDNA is susceptible to oxidative damage and it is interesting to note that oxidative damage of mtDNA is inversely related to the life span of humans, whereas oxidative damage of nuclear DNA is not related to the life span [5]. Oxidative damage of mtDNA was found to be reduced by reduced glutathione (GSH), however, the oxidation of glutathione increases with ageing in mitochondria of various organs than in whole cells. Thus, it is clear that mitochondria participates in oxidative damage associated with aging.

3. Riboflavin in Mitochondrial Pathways

Riboflavin, a water soluble vitamin, acts as a precursor of FMN and FAD, which are involved in key regulatory pathways of mitochondria, such as metabolism of amino acids, fatty acids, and purines, and the oxidation-reduction reaction essential for normal cellular growth and development [5]. Riboflavin consist of an isoalloxazine ring and a ribityl side chain, it is converted to FMN by the addition of phosphate group to the ribityl side chain, and further converted to FAD by the addition of ADP. Enzymes that utilize FMN and FAD are collectively known as flavor coenzymes or flavo proteins.

Riboflavin is considered a vital component of mitochondrial energy production mediated by ETC [6]. It is particularly important for the normal production of ATP, which leads to membrane stability and sustaining adequate energy-related cellular functions. In addition, flavo coenzymes are also involved in drug and toxin metabolism, along with cytochrome P450 [7].

Riboflavin obtained from exogenous dietary sources enters mitochondria from the cytosol by means of specific transporters and is converted to FAD by riboflavin kinase and FAD synthetase which can, be converted to riboflavin by FAD pyrophosphatase. This process is collectively known as the Rf-FAD cycle (Figure 1). In contrast, yeast mitochondria are devoid of FAD synthetase activity, hence, it has to obtain FAD from the cytosol through a specific transporter (FLX1) [8]. Experiments conducted in rat liver mitochondria proved that mitochondria have its own FAD transporter that carries FAD from the cytosol across the mitochondrial membrane for the flavinylation process [9].

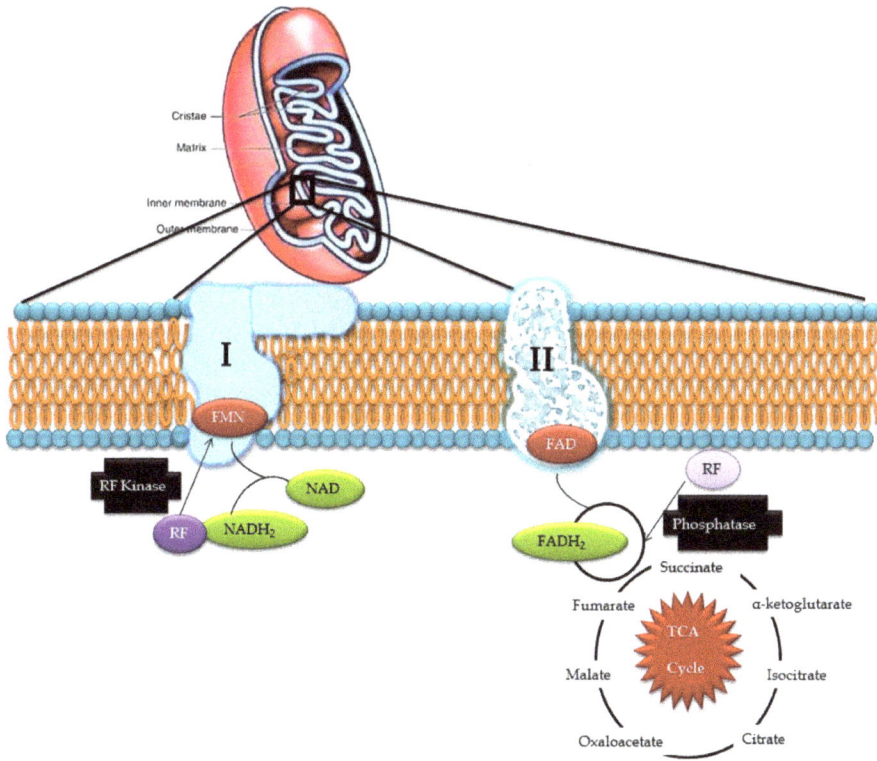

Figure 1. Riboflavin in mitochondrial pathways (RF- riboflavin; FMN- flavin mononucleotide; FAD/FADH$_2$- flavin adenine dinucleotide; and NAD/NADH$_2$- nicotinamide adenine dinucleotide).

In the electron transport chain, FMN acts as a co-factor for NADH-Coenzyme Q reductase which catalyzes the conversion of NADH to CoQ in complex I while FAD is involved in the activity of complex II where it acts as an electron carrier and cofactor for succinate dehydrogenase, which catalyzes the conversion of succinate to fumarate in the Kreb's cycle and oxidative phosphorylation. Defects in ETC produce free radical superoxide O_2^-, hydrogen peroxide, nitric oxide, and highly-reactive hydroxyl radicals, which induce membrane lipid peroxidation and DNA damage. Exposure to excessive reactive oxygen species (ROS) potentially damages nuclear DNA and mtDNA, which is highly deleterious in post-mitotic cells, such as neurons, where cells stop differentiation and thus, cell division is not possible to replace the damaged DNA. Such impairments in the mtDNA lead to bioenergetic dysfunctions that could ultimately cause neuronal cell death [10] and may be involved in the development of mitochondria-associated neurodegenerative disorders [11].

In mitochondrial fatty acid beta oxidation, fatty acids are activated in the cytosol and transported into the inner mitochondrial membrane as carnitine derivatives mediated by carnitine palmitoyl transferase I (CPT I), acylcarnitine translocase (CAT), and carnitine palmitoyl transferase II (CPT II). Within the mitochondrial matrix, the acyl-CoA fatty acids undergo dehydrogenation by acyl-CoA dehydrogenases, which are flavin-dependent and thus, flavins regulate the fatty acid beta oxidation pathway [12].

4. Riboflavin Pathogenesis in Mitochondrial Dysfunction

FMN and FAD are chief prosthetic groups that activate various flavoproteins, such as nitric oxide synthase, nitric oxide reductase, and NADPH oxidase, to protect the cell from oxidative stress and apoptosis [13]. Although riboflavin is stored in the liver, spleen, kidney, and cardiac muscle in the form of FAD, and protects these organs against riboflavin deficiency, its half-life is one hour; hence, riboflavin deficiency would impair the proper functioning of these organelles. Furthermore, energy depletion and a decrease in riboflavin kinase activity leads to the insufficient conversion of riboflavin into flavocoenzyme and results in various anomalies, like cataracts, preeclampsia, various types of cancers, and neurological disorders. In addition, deficiency of ETC enzymes, like NADH-CoQ reductase, cytochrome c oxidase, and creatine kinase, leads to infantile mitochondrial myopathy [14].

Recent studies have proved riboflavin supplementation as a therapy to alleviate or reduce the worsening of disease conditions, especially in Brown-Vialetto-Van Laere syndrome (BVVLS) and multiple acyl-CoA dehydrogenase deficiency (MADD) [15]. Although most of the flavo enzymes are encoded by the nuclear genome, surprisingly, they are synthesized and stored in different components of the mitochondria for their active participation in vital pathways, like glycolysis, Kreb's cycle, beta oxidation, urea cycle, and ETC of mitochondria. When riboflavin is accumulated in the cytoplasm instead of entering into mitochondria, there will be a shortage of riboflavin availability for the flavo coenzymes present in the mitochondria. Hence, riboflavin transporters are essential for the maintenance of riboflavin homeostasis. Till date, only mutations in riboflavin transporters were correlated with neurologically defective phenotypes, while most of the flavoproteins also take part in key regulatory pathways that determine the fate of the cell to undergo either normal or abnormal physiological functions in neurological prospects.

5. Riboflavin Related Mitochondrial Dysfunction in Neurological Disorders

Mitochondria play a key role in the interconnected network to transmit and receive signals where the central nervous system is highly dependent on energy. Furthermore, during neurogenesis for the differentiation and development of axons and dendrites, high amounts of mitochondrial mass is necessary to produce ATP in large quantities [16]. Hence, mitochondrial dysfunction due to any defect in the reduction or oxidative phosphorylation reaction results in impaired oxidative metabolism and diminished energy production which, consequently, leads to neurological disorders. Particularly in amyotrophic lateral sclerosis (ALS), mitochondrial dysfunctions, like abnormal mitochondrial morphology [17], mitochondria-mediated apoptosis [18], and disruption of the axonal transport of mitochondria [19], are the primary reasons for the disease etiology. Riboflavin-related mitochondrial dysfunction in neurological disorders are summarized in Figure 2. In particular, riboflavin deficiency in rats resulted in reduced levels of myelin lipids, cerebrosides, sphingomyelin, and phosphatidylethanolamine in the cerebrum and cerebellum and, consequently, led to the impairment of brain development and maturation [20]. These observations suggested that riboflavin plays a crucial role in the metabolism of essential fatty acids in the brain. Some of the flavo coenzymes involved in mitochondrial dysfunction of neurological diseases are listed in Table 1.

In addition, experiments conducted in chickens showed that riboflavin deficiency resulted in demyelination of peripheral nerve cells. Severity of demyelination was particularly high in Schwann cells, whereas the severity was reduced in spinal nerve roots and distal nerve branches due to nutrient accessibility [21]. Likewise, riboflavin deficiency resulted in the swelling of peripheral nerve trunks and led to peripheral neuropathy in racing pigeons [22]. Henceforth, a detailed knowledge about the role of riboflavin in the mitochondrial dysfunction of neurological disorders is a prerequisite for the discovery of drug targets and treatment.

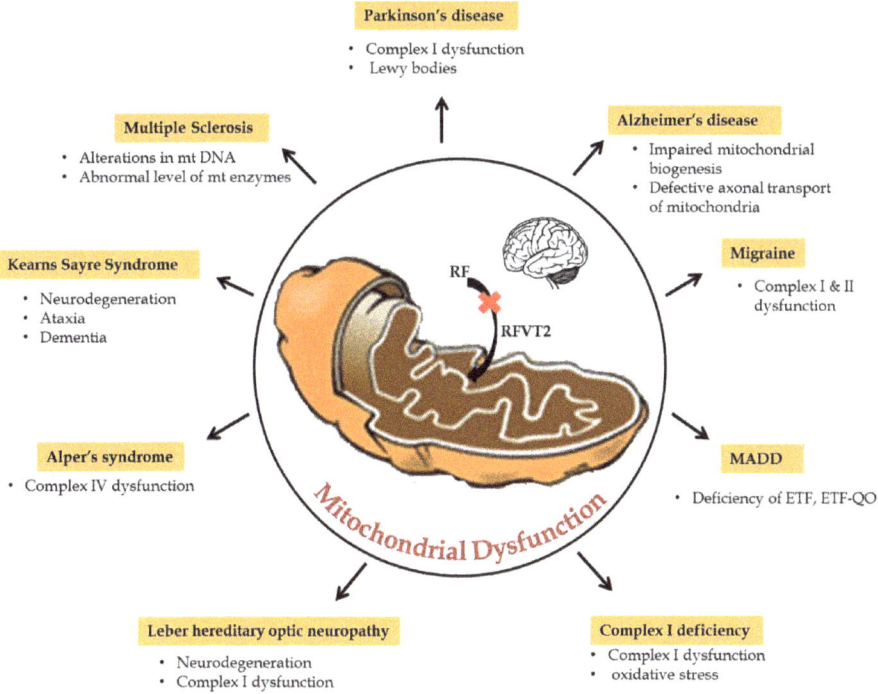

Figure 2. Riboflavin related mitochondrial dysfunction in neurological disorders.

Table 1. Enzymes involved in mitochondrial dysfunction of neurological diseases.

Enzyme	Neurological Disease	Metabolic Function	Location
Succinate dehydrogenase	Complex II deficiency	Krebs cycle	Mitochondrial inner membrane
Acyl Co-A dehydrogenase	Acyl Co-A dehydrogenase deficiency	Beta oxidation	Mitochondrial matrix
Electron transferring flavo protein—Ubiquinone oxidoreductase	Glutamic academia II C	Electron transport chain	Mitochondrial inner membrane
Electron transferring flavo protein	Glutamic academia II A and II B	Electron transport chain	Mitochondrial matrix
NADH - Ubiquinone oxidoreductase	Complex I deficiency	Electron transport chain	Mitochondrial inner membrane
Dihydrolipoyl dehydrogenase, Succinate dehydrogenase and NADH-Ubiquinone oxidoreductase	Leigh Syndrome	Energy metabolism	Mitochondrial matrix
Riboflavin transporter	BVVLS	Riboflavin uptake	Plasma/Mitochondrial membrane

6. Neurological Disorders of Mitochondrial Dysfunction

Disruption of the mitochondrial electron transport chain and other mitochondrial damage leads to several neurological disorders. Some of the riboflavin responsive neurological disorders due to mitochondrial dysfunction are discussed below.

6.1. Multiple Acyl-CoA Dehydrogenase Deficiency (OMIM 231680)

Multiple acyl-CoA dehydrogenase deficiency (MADD), also called as glutaric aciduria type II, ethylmalonic-adipic aciduria, and riboflavin-responsive C6-C10 dicarboxylic aciduria, is caused due to the deficiency of the electron transfer flavoprotein (ETF) or its dehydrogenase and ubiquinone oxidoreductase (ETF-QO) [23]. MADD affects various metabolic pathways involving fatty acids and branched amino acids, lysine and tryptophan, and results in a discharge of a variety of distinctive organic acids comprising glutaric, ethylmalonic, 3-hydroxyisovaleric, 2-hydroxyglutaric, 5-hydroxyhexanoic, adipic, suberic, sebacic, and dodecanedioic acids and glycine conjugates due to the impairment of ATP biosynthesis and the accumulation of excessive fatty acids [24]. The blood plasma acylcarnitine pattern also shows distinctive elevation of short-, medium-, and long-chain acylcarnitines ranging from C8 to C16. Features of MADD include muscle weakness, non-epileptic seizures, and atypical migraine with abnormal creatinine level. MADD is diagnosed with mutations in the alpha and beta subunits of electron transfer flavoprotein (*ETFA* and *ETFB*), ETF dehydrogenase (*ETFDH*), FAD synthase (*FADS1*), riboflavin transporters (*SLC52A1-3*), and mitochondrial FAD transporter (*SLC25A32*) [15,25–28]. Studies have documented that riboflavin supplementation reduced the abnormal behavior and normalized the biochemical profile by regulating the mitochondrial flavo proteome [26,27,29] and are termed as riboflavin-responsive forms of MADD [30].

6.2. Brown-Vialetto-Van Laere Syndrome (OMIM 211530)

BVVLS is a rare, progressive, childhood neurodegenerative disorder characterized by progressive pontobulbar palsy associated with sensorineural deafness. BVVLS has a prominent familial component, consistent with an autosomal-recessive mode of inheritance in most of the patients, while autosomal dominant [31] and X-linked inheritance [32] have also been suggested in a few cases. In BVVLS, bilateral nerve deafness is accompanied by involvement of various motor cranial nerve palsies with VII, IX, and XII, and rarely III, V, and VI, which develop over a relatively short period of time in a previously-healthy individual [32,33]. Generally, BVVLS is clinically heterogeneous, presenting as early as infancy and as late as the third decade of life [33]. Recently, defects in riboflavin transporters *SLC52A3* (formerly C20orf54) [34,35] and *SLC52A2* [35,36] have been identified as the etiology in a large proportion of BVVLS cases. Blood plasma levels of riboflavin and its active coenzyme forms, FAD and FMN, were significantly reduced in BVVLS patients [15]. Moreover, metabolic studies of BVVLS patients revealed the accumulation of acyl-CoA and carnitine esters in the plasma, as well as a urine organic acid profile which both mimic the fatty acid β-oxidation defect seen in patients with MADD. Meanwhile, oral supplementation of riboflavin showed improvement in the clinical symptoms, as well as the biochemical abnormalities in BVVLS patients, signifying that a high dose of riboflavin is a potential treatment for BVVLS [26]. Thus, riboflavin is found to have a critical role in the production of substrates used for the ETC, so it is obvious that any defect in riboflavin transport would impair ETC and consequently lead to neurodegeneration. The overall summary of riboflavin deficiency leading to mitochondrial oxidative stress-mediated neurodegeneration is given in Figure 3.

6.3. Complex I Deficiency (OMIM 252010)

The mitochondrial respiratory chain tends to decline with age by affecting complex I and IV of ETC, which leads to mitochondrial myopathies, like cardiomyopathies, encephalomyopathies, and neurological myopathies [37]. Human complex I (NADH-ubiquinone reductase) consists of at least 36 nuclear-encoded and seven mitochondrial-encoded subunits and clinical mutations in any of these subunits are diagnosed to cause this disorder [38]. Functional characterization studies carried out in *Caenorhabditis elegans* with mutation in the active site subunit of complex I revealed that supplementation of riboflavin assembled complex I and reduced oxidative stress, lactic acidosis, and increased metabolic functions [39]. Additionally, riboflavin supplementation normalized the biochemical abnormalities and muscle weakness in an infant with a complex I defect by increasing the cellular availability of FAD [40,41]. Furthermore, mutations in mitochondrial and nuclear genes encoding proteins that are required for proper assembly and stability of the mitochondrial respiratory complex also lead to complex I deficiency. ACAD9 (acyl-CoA dehydrogenase-9), a flavin-dependent acyl carrier, is involved in the proper assembly of complex I through binding with assembly factors NDUFAF-1 and Ecsit [42]. Recently, a missense mutation (Arg532Trp) was diagnosed in the active site of ACAD9 in a Dutch consanguineous family with complex I deficiency (OMIM 611126—complex I deficiency due to ACAD-9), where riboflavin supplementation improved the complex I activity from 17% to 47% in the proband [43].

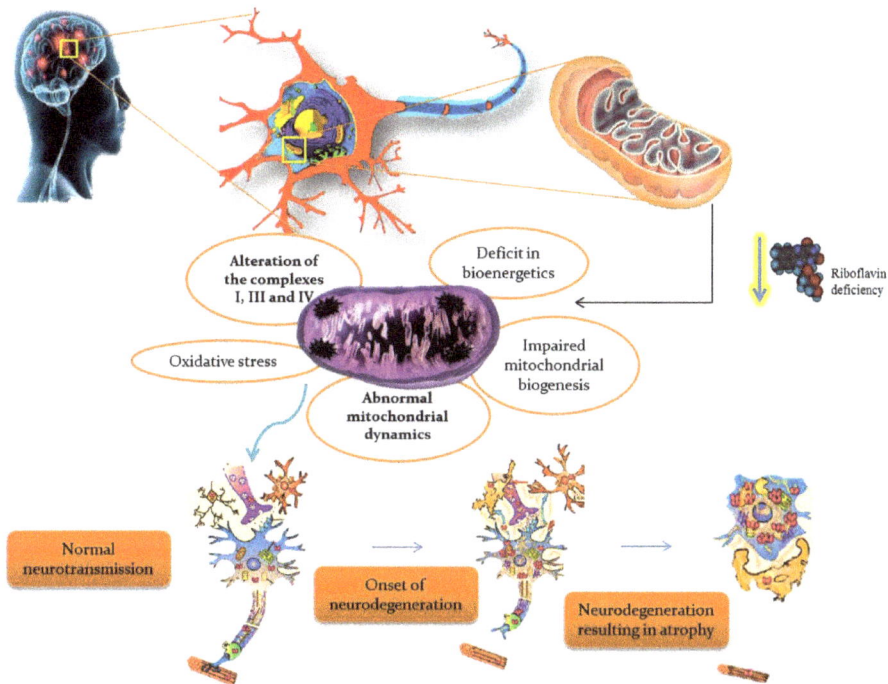

Figure 3. Riboflavin deficiency leading to mitochondrial oxidative stress-mediated neurodegeneration.

6.4. Leber Hereditary Optic Neuropathy (LHON; OMIM 535000)

LHON is a neurodegenerative disease characterized by acute or subacute loss of central vision and optic atrophy. It arises due to the neurodegeneration of retino-ganglion cells and dysfunction of respiratory chain complex I. Furthermore, it is the first human mtDNA disease identified to be caused by deletion of mtDNA. LHON cases are primarily identified with mutations in any of mitochondrial genes, including MT-ND1, MT-ND4, MT-ND4L, and MT-ND6, and over 95% of cases harbored one of three mtDNA point mutations, G3460A (ND1), G11778A (ND4), and T14484C (ND6), which encodes complex I subunits of the respiratory chain [44]. Studies have documented that supplementation of riboflavin, along with vitamin C and idebenone, in 28 LHON patients reduced the recovery period of dysfunction [45].

6.5. KearnsSayre Syndrome (OMIM 530000)

Kearns-Sayre Syndrome (KSS) is a rare neuromuscular disorder characterized by ophthalmoplegia, retinitis pigmentosa, chronic inflammation, cortico spinal dysfunction, bulbar palsies, limb girdle muscle weakness, sensory neural hearing loss, progressive neurodegeneration with ataxia, and dementia. Large deletions of mtDNA ranged in size from 2.0 to 7.0 kb [46] are known to cause KSS by the defective oxidative phosphorylation, and the deletions are heteroplasmic. Patients also showed deficiency of cytochome-c oxidase (COX) due to large deletions in the specific region of mtDNA corresponding to the *COX* gene, which was clinically observed as ragged-red fibers in muscle biopsies [47]. Most of the cases are sporadic since mtDNA deletions are inherited very rarely. Deficiency of complex II of the mitochondrial respiratory chain, especially a deficiency of succinic dehydrogenase has been revealed by enzymatic analysis [48]. Since complex II dysfunction is noticed, a combined therapy containing cytochrome c, flavin mononucleotide, and thiamine diphosphate was attempted, which alleviated fatigability, motor disability, corneal edema, and chilblains in the patients, while no improvements were recorded with opthalmoplegia, blepharoptosis, or hearing loss [49]. Recent follow-up study carried out with three complex II-deficient patients showed an improvement in neurological conditions and delayed the early onset. In addition, supplementation of riboflavin to the fibroblast culture showed a two-fold increase in the activities of complex II and succinate dehydrogenase (SDH) [50].

6.6. Alper's Syndrome (OMIM 203700)

Alper's syndrome is an autosomal recessive disorder characterized by a clinical trial of symptoms, including psychomotor retardation, refractory seizures, and liver failure. It is a mitochondrial DNA depletion disease of the brain that arises due to the degeneration of cerebral gray matter in infancy, characterized by neurodegeneration of basal ganglia. It is caused due to the dysfunction in complex IV of ETC, nuclear-encoded mitochondrial polymerase γ (*PolG1*) deficiency [51], and Twinkle helicase [51,52]. Alper's syndrome patients with mutations in *POLG* may also undergo complex I deficiency [53]. Since it involves complex I and IV dysfunction, riboflavin could play a possible role in its regulation, which corroborates with the study carried out in *C. elegans* (having mutations in NADH-ubiquinone oxidoreductase) where riboflavin supplementation enhanced the assembly of complex I and IV that further resulted in reduced oxidative stress and increased metabolic functions [39].

6.7. Multiple Sclerosis

Multiple sclerosis is an autoimmune disorder that affects the central nervous system through immune cells, potentially also due to alterations in mitochondrial DNA, defective mitochondrial DNA repair mechanisms, abnormal mitochondrial dynamics (fragmentation), impaired trafficking, defective Ca^+-mediated axonal degeneration, and abnormal levels of mitochondrial enzymes (phosphofructokinase-2 and complex I enzymes) [54]. Earlier, administration of interleukin 6 was found to act against ROS and protect against neuronal cell death, while recent studies carried out in encephalomyelitis C57BL/6 mice showed that riboflavin supplementation reduced the neuronal disability by 26.4% while the use of placebo reduced the risk by 15.4% [55]. Furthermore, riboflavin supplementation leads to a reduction in the expression of BDNF and IL-6 in the brain of an experimental autoimmune encephalomyelitis model of multiple sclerosis, which was correlated with the observed beneficial effects of riboflavin on neurological motor disability and also suggested possible targets of new rational therapeutic strategies for MS [56].

6.8. Parkinson's Disease (OMIM 168600)

Parkinson's disease (PD) is a progressive movement disorder that is associated with the death of vital nerve cells in the brain. Primary symptoms include tremor, bradykinesia, stiffness of the limbs, and postural instability. It is primarily due to the accumulation of alpha-synuclein protein in the brain as Lewy bodies. In some instances, it is characterized by complex I ETC deficiency where, due to endogenous oxidative damage, the respiratory chain protein complex is affected, which subsequently leads to decreased ATP production, increased free radical production, and results in apoptosis. Deficiency of riboflavin was shown to have impaired oxidative metabolism through reduced glutathione reductase, pyridoxine phosphate oxidase, NADH-ubiquinone reductase, and

NADH cytochrome c reductase [57]. Further, supplementation of riboflavin for six months with PD patients showed improvement in the motor capacity from 41% to 71%, and demonstrated to overcome the complex I deficiency. Thus, it is evident that riboflavin could play a role in the conversion of oxidized glutathione to reduced glutathione by catalyzing glutathione reductase, and in the assembly of mitochondrial protein complexes [58]. Recently, a bacterial metabolite produced by *Streptomyces venezuelae* caused dopaminergic neurodegeneration in a PD model of *C. elegans* expressing human α-synuclein due to the impairment of mitochondrial complex I activity. Meanwhile, mitochondrial complex I activators, such as riboflavin and D-β-hydroxybutyrate (DβHB), rescued dopamine neurodegeneration in *C. elegans* by improving both complex I and complex IV activities [59].

6.9. Alzheimer's Disease (OMIM 104300)

Alzheimer's disease (AD) is a progressive neurodegenerative disorder characterized by short-term memory loss and dementia due to the accumulation of Tau proteins in neurofibrillary tangles, the loss of connection between nerve cells, and extracellular amyloid plaque which leads to mitochondrial fragmentation [60]. Defects in electron transport chain enzymes, such as cytochrome c oxidase (COX) and F(1)F(0)-ATPase, have also been implicated in the progression of AD [61]. Neurodegeneration observed in AD has been suggested to be due to impaired mitochondrial biogenesis, defective axonal transport of mitochondria, and increased DRP1-mediated mitochondrial fission [62]. Hyperhomocysteinemia has been explained as one of the possible mechanisms for neurotoxicity in AD, which is responsible for induced cellular oxidative stress leading to the formation of ROS to cause neuronal cell death. Elevated plasma levels of homocysteine have been documented due to the impaired activity of methylenetetrahydrofolate reductase (MTHFR) in one carbon metabolism of the homocysteine remethylation pathway, which is a FAD-dependent flavoenzyme [63]. Deficiency of cellular FAD has been shown to contribute to the functional impairment of the MTHFR 677T variant genotype and an increase the homocysteine levels, particularly in individuals with low-folate status [64]. Moreover, accumulation of homocysteine for longer periods is thought to be involved in the failure of β-amyloid clearance and damage to the blood brain barrier, which develops into cerebrovascular dysfunction, leading to AD development [65].

7. Conclusions

There is a gathering body of evidence which links the interaction between riboflavin and flavoproteins to the protection of neuronal cells from death by oxidative stress and apoptosis. Any anomalous expression and regulation of mtDNA and nDNA encoding of functional proteins in the mitochondria can affect the intracellular levels of FAD and FMN, which are functionally implicated in various pathological conditions leading to neurological disorders. Mitochondrial defects may lead to axonal dysfunction and degeneration through a lack of ATP, increased ROS production, and by modulating the function of various dehydrogenases of the respiratory chain. The understanding of this relationship between mitochondria, rate of neuronal degeneration by oxidative stress, and the protection by riboflavin is at an early stage. Thus, a comprehensive knowledge and new experimental strategies are essential to elucidate the interplay between mitochondrial metabolism, mitochondrial stress, riboflavin transport and metabolism, mtDNA mutation and deletion mechanisms, and the complex neurodegeneration pathways. Such knowledge may provide new targets for combating neurodegenerative diseases.

Acknowledgments: B.A. acknowledges ICMR, INDIA for the International Fellowship for Young Biomedical Scientist for the year 2016–2017 (No. INDO/FRC/452/(Y-83)/2016–2017/IHD). T.U. acknowledges the DST-PURSE Program (India), Madurai Kamaraj University for the fellowship to carry out this research.

Author Contributions: All of the authors have significantly contributed to the design of the study and preparation of the manuscript. T.U. and B.A. were responsible for reviewing the technical aspects of each method, the collection of data, and drafting the manuscript. B.A. and H.H. were responsible for the study conception and design. A.M., P.V. and M.R. were involved in the data analysis and critical revision of the manuscript. All of the authors read and approved the final manuscript.

Conflicts of Interest: The authors declare no conflict of interest.

References

1. Wallace, D.C. A Mitochondrial Paradigm of Metabolic and Degenerative Diseases, Aging, and Cancer: A Dawn for Evolutionary Medicine. *Annu. Rev. Genet.* **2005**, *39*, 359–410. [CrossRef] [PubMed]
2. Nass, M.M.K.; Nass, S. Intramitochondrial fibers with DNA characteristics. I. Fixation and electron staining reactions. *J. Cell Biol.* **1962**, *19*, 593–612. [CrossRef]
3. Donald, V.; Voet, J.G.; Pratt, C.W. *Fundamentals of Biochemistry*, 2nd ed.; John Wiley and Sons, Inc.: Somerset, NJ, USA, 2006; Volume 547, p. 556.
4. Boveris, A.; Chance, B. The mitochondrial generation of hydrogen peroxide. *Biochem. J.* **1973**, *134*, 707–716. [CrossRef] [PubMed]
5. Barja, G.; Herrero, A. Oxidative damage to mitochondrial DNA is inversely related to maximum life span in the heart and brain of mammals. *FASEB J.* **2000**, *14*, 312–318. [PubMed]
6. Powers, H.J. Riboflavin (Vitamin B-2) and health[1,2]. *Am. J. Clin. Nutr.* **2003**, *77*, 1352–1360. [PubMed]
7. Powers, H.J. Current knowledge concerning optimum nutritional status of riboflavin, niacin and pyridoxine. *Proc. Nutr. Soc.* **1999**, *58*, 435–440. [CrossRef] [PubMed]
8. Massey, V. The chemical and biological versatility of Riboflavin. *Biochem. Soc. Trans.* **2000**, *28*, 283–296. [CrossRef] [PubMed]
9. Bafunno, V.; Giancaspero, T.A.; Brizio, C.; Bufano, D.; Passarella, S.; Boles, E.; Barile, M. Riboflavin uptake and FAD synthesis in saccharomyces cerevisiae mitochondria. Involvement of the FLX1p carrier in FAD export. *J. Biol. Chem.* **2004**, *279*, 95–102. [CrossRef] [PubMed]
10. Barile, M.; Brizio, C.; Valenti, D.; De Virgilio, C.; Passarella, S. The riboflavin/FAD cycle in rat liver mitochondria. *Eur. J. Biochem.* **2000**, *267*, 4888–4900. [CrossRef] [PubMed]
11. Mancuso, M.; Coppede, F.; Migliore, L.; Siciliano, G.; Murri, L. Mitochondrial dysfunction, oxidative stress and neurodegeneration. *J. Alzheimers Dis.* **2006**, *10*, 59–73. [CrossRef] [PubMed]
12. Stewart, V.C.; Heales, S.J.R. Nitric oxide-induced mitochondrial dysfunction: Implications for neurodegeneration. *Free Radic. Biol. Med.* **2003**, *34*, 287–303. [CrossRef]
13. Wanders, R.J.A.; Ruiter, J.P.N.; IJLst, L.; Waterham, H.R.; Houten, S.M. The enzymology of mitochondrial fatty acid beta-oxidation and its application to follow-up analysis of positive neonatal screening results. *J. Inherit. Metab. Dis.* **2010**, *33*, 479–494. [CrossRef] [PubMed]
14. Lamattina, L.; García-Mata, C.; Graziano, M.; Pagnussat, G. Nitric oxide: The versatility of an extensive signal molecule. *Annu. Rev. Plant. Biol.* **2003**, *54*, 109–136. [CrossRef] [PubMed]
15. Bosch, A.M.; Abeling, N.G.; IJLst, L.; Knoester, H.; van der Pol, W.L.; Stroomer, A.E.; Wanders, R.J.; Visser, G.; Wijburg, F.A.; Duran, M.; et al. Brown-Vialetto-Van Laere and Fazio Londe syndrome is associated with a riboflavin transporter defect mimicking mild MADD: A new inborn error of metabolism with potential treatment. *J. Inherit. Metab. Dis.* **2011**, *34*, 159–164. [CrossRef] [PubMed]
16. Roodhooft, A.M.; Van Acker, K.J.; Martin, J.J.; Ceuterick, C.; Scholte, H.R.; Luyt Houwen, I.E.M. Benign mitochondrial myopathy with deficiency of NADH-CoQ reductase and cytochrome c oxidase. *Neuropediatrics* **1986**, *17*, 221–226. [CrossRef] [PubMed]
17. Sasaki, S.; Iwata, M. Ultrastructural study of synapses in the anterior horn neurons of patients with amyotrophic lateral sclerosis. *Neurosci. Lett.* **1996**, *204*, 53–56. [CrossRef]
18. Pasinelli, P.; Belford, M.E.; Lennon, N.; Bacskai, B.J.; Hyman, B.T.; Trotti, D.; Brown, R.H. Amyotrophic lateral sclerosis-associated SOD1 mutant proteins bind and aggregate with Bcl-2 in spinal cord mitochondria. *Neuron* **2004**, *43*, 19–30. [CrossRef] [PubMed]
19. Hollenbeck, P.J.; Saxton, W.M. The axonal transport of mitochondria. *J. Cell Sci.* **2000**, *118*, 5411–5419. [CrossRef] [PubMed]
20. Erecinska, M.; Cherian, S.; Silver, I.A. Energy metabolism in mammalian brain during development. *Prog. Neurobiol.* **2004**, *73*, 397–445. [CrossRef] [PubMed]
21. Ogunleye, A.J.; Oduduga, A.A. The effect of riboflavin deficiency on cerebrum and cerebellum of developing rat brain. *J. Nutr. Sci. Vitaminol.* **1989**, *35*, 193–197. [CrossRef] [PubMed]
22. Wada, Y.; Kondo, H.; Itakura, C. Peripheral neuropathy of dietary riboflavin deficiency in racing pigeons. *J. Vet. Med. Sci.* **1996**, *58*, 161–163. [CrossRef] [PubMed]

23. Beard, S.E.; Spector, E.B.; Seltzer, W.K.; Frerman, F.E.; Goodman, S.I. Mutations in electron transfer flavoprotein: Ubiquinone oxidoreductase (ETF:QO) in glutaric acidemia type II (GA2). *Clin. Res.* **1993**, *41*, 271.

24. Frerman, F.E.; Goodman, S.I. Defects of electron transfer flavoprotein and electron transfer flavoprotein-ubiquinone oxidoreductase: Glutaric acidemia type II. In *The Metabolic and Molecular Bases of Inherited Disease*, 8th ed.; Scriver, C.R., Beaudet, A.L., Sly, W.S., Valle, D., Eds.; McGraw-Hill: New York, NY, USA, 2001; pp. 2357–2365.

25. Frerman, F.E.; Goodman, S.I. Deficiency of electron transfer flavoprotein or electron transfer flavoprotein: Ubiquinone oxidoreductase in glutaric academia type II fibroblasts. *Proc. Natl. Acad. Sci. USA* **1985**, *82*, 4517–4520. [CrossRef] [PubMed]

26. Ho, G.; Yonezawa, A.; Masuda, S.; Inui, K.; Sim, K.G.; Carpenter, K.; Olsen, R.K.; Mitchell, J.J.; Rhead, W.J.; Peters, G.; et al. Maternal riboflavin deficiency, resulting in transient neonatal-onset glutaric aciduria Type 2, is caused by a microdeletion in the riboflavin transporter gene GPR172B. *Hum. Mutat.* **2010**, *32*, 1976–1984. [CrossRef] [PubMed]

27. Schiff, M.; Veauville-Merllie, A.; Su, C.H.; Tzagoloff, A.; Rak, M.; Ogier de Baulny, H.; Boutron, A.; Smedts-Walters, H.; Romero, N.B.; Rigal, O.; et al. *SLC25A32* mutations and riboflavin-responsive exercise intolerance. *N. Engl. J. Med.* **2016**, *374*, 795–797. [CrossRef] [PubMed]

28. Olsen, R.K.; Konarikova, E.; Giancaspero, T.A.; Mosegaard, S.; Boczonadi, V.; Matakovic, L.; Veauville-Merllie, A.; Terrile, C.; Schwarzmayr, T.; Haack, T.B.; et al. Riboflavin-Responsive and Non-responsive Mutations in FAD Synthase Cause Multiple Acyl-CoA Dehydrogenase and Combined Respiratory-Chain Deficiency. *Am. J. Hum. Genet.* **2016**, *98*, 1130–1145. [CrossRef] [PubMed]

29. Triggs, W.J.; Roe, C.R.; Rhead, W.J.; Hanson, S.K.; Lin, S.N.; Willmore, L.J. Neuropsychiatric manifestations of defect in mitochondrial beta oxidation response to riboflavin. *J. Neurol. Neurosurg. Psychiatry* **1992**, *55*, 209–211. [CrossRef] [PubMed]

30. Cornelius, N.; Corydon, T.J.; Gregersen, N.; Olsen, R.K. Cellular consequences of oxidative stress in riboflavin responsive multiple acyl-CoA dehydrogenation deficiency patient fibroblasts. *Hum. Mol. Genet.* **2014**, *23*, 4285–4301. [CrossRef] [PubMed]

31. De Grandis, D.; Passadore, P.; Chinaglia, M.; Brazzo, F.; Ravenni, R.; Cudia, P. Clinical features and neurophysiological follow-up in a case of Brown-Vialetto-Van Laere syndrome. *Neuromuscul. Disorder.* **2005**, *15*, 565–568. [CrossRef] [PubMed]

32. Hawkins, S.A.; Nevin, N.C.; Harding, A.E. Pontobulbar palsy and neurosensory deafness (Brown-Vialetto-Van Laere syndrome) with possible autosomal dominant inheritance. *J. Med. Genet.* **1990**, *27*, 176–179. [CrossRef] [PubMed]

33. Sathasivam, S. Brown-Vialetto-Van Laere syndrome. *Orphanet J. Rare Dis.* **2008**, *3*, 9. [CrossRef] [PubMed]

34. Green, P.; Wiseman, M.; Crow, Y.J.; Houlden, H.; Riphagen, S.; Lin, J.P.; Raymon, F.L.; Childs, A.M.; Sheridan, E.; Edwards, S.; et al. Brown-Vialetto-Van Laere Syndrome, a Ponto-Bulbar Palsy with Deafness, Is Caused by Mutations in C20orf54. *Am. J. Hum. Genet.* **2010**, *86*, 485–489. [CrossRef] [PubMed]

35. Udhayabanu, T.; Subramanian, V.S.; Teafatiller, T.; Vykunta Raju, K.N.; Raghavan, V.S.; Varalakshmi, P.; Said, H.M.; Ashokkumar, B. *SLC52A2* [p.P141T] and *SLC52A3* [p.N21S] causing Brown-Vialetto-Van Laere Syndrome in an Indian patient: First genetically proven case with mutations in two riboflavin transporters. *Clin. Chim. Acta* **2016**, *462*, 210–214. [CrossRef] [PubMed]

36. Johnson, J.O.; Gibbs, J.R.; Megarbane, A.; Urtizberea, J.A.; Hernandez, D.G.; Foley, A.R.; Arepalli, S.; Pandraud, A.; Sanchez, J.S.; Clayton, P.; et al. Exome sequencing reveals riboflavin transporter mutations as a cause of motor neuron disease. *Brain* **2012**, *135*, 1–8. [CrossRef] [PubMed]

37. Cooper, J.M.; Mann, V.M.; Schapira, H. Analyses of mitochondrial respiratory chain function and mitochondrial DNA deletion in human skeletal muscle: Effect of ageing. *J. Neurol. Sci.* **1992**, *113*, 91–98. [CrossRef]

38. Swalwell, H.; Kirby, D.M.; Blakely, E.L.; Mitchell, A.; Salemi, R.; Sugiana, C.; Compton, A.G.; Tucker, E.J.; Ke, B.X.; Lamont, P.J.; et al. Respiratory chain complex I deficiency caused by mitochondrial DNA mutations. *Eur. J. Hum. Genet.* **2011**, *19*, 769–775. [CrossRef] [PubMed]

39. Grad, L.I.; Lemire, B.D. Riboflavin enhances the assembly of mitochondrial cytochrome c oxidase in C. *elegans* NADH-ubiquinone oxidoreductase mutants. *Biochim. Biophys. Acta Bioenerg.* **2006**, *1757*, 115–122. [CrossRef] [PubMed]

40. Griebel, V.; Krigeloh-Mann, I.; Ruitenbeek, W.; Trijbels, J.M.F.; Paulus, W. A mitochondrial myopathy in an infant with lactic acidosis. *Dev. Med. Child. Neurol.* **1990**, *32*, 528–531. [CrossRef] [PubMed]
41. Ogle, R.F.; Christodoulou, J.; Fagan, E.; Blok, R.B.; Kirby, D.M.; Seller, K.L.; Dahl, H.H.; Thorburn, D.R. Mitochondrial myopathy with tRNA Leu (UUR) mutation and complex I deficiency responsive to riboflavin. *J. Pediatr.* **1997**, *130*, 138–145. [CrossRef]
42. Feng, D.; Witkowski, A.; Smith, S. Down-regulation of mitochondrial acyl carrier protein in mammalian cells compromises protein lipoylation and respiratory complex I and results in cell death. *J. Biol. Chem.* **2009**, *284*, 11436–11445. [CrossRef] [PubMed]
43. Gerards, M.; van den Bosch, B.J.C.; Danhauser, K.; Serre, V.; van Weeghel, M.; Wanders, R.J.A.; Nicolaes, G.A.F.; Sluiter, W.; Schoonderwoerd, K.; Scholte, H.R.; et al. Riboflavin-responsive oxidative phosphorylation complex I deficiency caused by defective ACAD9: New function for an old gene. *Brain* **2011**, *134*, 210–219. [CrossRef] [PubMed]
44. Man, P.Y.W.; Turnbull, D.M.; Chinnery, P.F. Leber hereditary optic neuropathy. *J. Med. Genet.* **2002**, *39*, 162–169. [CrossRef]
45. Mashima, Y.; Kigasawa, K.; Wakakura, M.; Oguchi, Y. Do idebenone and vitamin therapy shorten the time to achieve visual recovery in Leber hereditary optic neuropathy? *J. Neuroophthalmol.* **2000**, *20*, 166–170. [CrossRef] [PubMed]
46. Zeviani, M.; Moraes, C.T.; DiMauro, S.; Nakase, H.; Bonilla, E.; Schon, E.A.; Rowland, L.P. Deletions of mitochondrial DNA in Kearns-Sayre syndrome. *Neurology* **1998**, *51*, 1525. [CrossRef] [PubMed]
47. Mita, S.; Schmidt, B.; Schon, E.A.; DiMauro, S.; Bonilla, E. Detection of "deleted" mitochondrial genomes in cytochrome-c oxidase-deficient muscle fibers of a patient with Kearns-Sayre syndrome. *Proc. Natl. Acad. Sci. USA* **1989**, *86*, 9509–9513. [CrossRef] [PubMed]
48. Rivner, M.H.; Shamsnia, M.; Swift, T.R.; Trefz, J.; Roesel, R.A.; Carter, A.L.; Yanamura, W.; Hommes, F.A. Kearns-Sayre syndrome and complex II deficiency. *Neurology* **1989**, *39*, 693–696. [CrossRef] [PubMed]
49. Nakagawa, E.; Osari, S.I.; Yamanouchi, H.; Matsuda, H.; Goto, Y.I.; Nonaka, I. Long-term therapy with cytochrome c, flavin mononucleotide and thiamine diphosphate for a patient with Kearns-Sayre syndrome. *Brain Dev.* **1996**, *18*, 68–70. [CrossRef]
50. Bugiani, M.; Lamantea, E.; Invernizzi, F.; Moroni, I.; Bizzi, A.; Zeviani, M.; Uziel, G. Effects of riboflavin in chicken with complex II deficiency. *Brain Dev.* **2006**, *28*, 576–581. [CrossRef] [PubMed]
51. Naviaux, R.K.; Nyhan, W.L.; Barshop, B.A.; Poulton, J.; Markusic, D.; Karpinski, N.C.; Haas, R.H. Mitochondrial DNA polymerase gamma deficiency and mtDNA depletion in a child with Alpers' syndrome. *Ann. Neurol.* **1999**, *45*, 54–58. [CrossRef]
52. Spelbrink, J.N.; Li, F.Y.; Tiranti, V.; Nikali, K.; Yuan, Q.P.; Wanrooij, S.; Garrido, N.; Comi, G.P.; Morandi, L.; Santoro, L.; et al. Human mitochondrial DNA deletions associated with mutations in the gene encoding Twinkle, a phage T7 gene 4-like protein localised in mitochondria. *Nat. Genet.* **2001**, *28*, 223–231. [CrossRef] [PubMed]
53. Hudson, G.; Deschauer, M.; Busse, K.; Zierz, S.; Chinnery, P.F. Sensory ataxic neuropathy due to a novel C10Orf2 mutation with probable germline mosaicism. *Neurology* **2005**, *64*, 371–373. [CrossRef] [PubMed]
54. Sadeghian, M.; Mastrolia, V.; Rezaei Haddad, A.; Mosley, A.; Mullali, G.; Schiza, D.; Sajic, M.; Hargreaves, I.; Heales, S.; Duchen, M.R.; et al. Mitochondrial dysfunction is an important cause of neurological deficits in an inflammatory model of multiple sclerosis. *Sci. Rep.* **2016**, *6*, 33249. [CrossRef] [PubMed]
55. Naghashpour, M.; Majdinasab, N.; Shakerinejad, G.; Kouchak, M.; Haghighizadeh, M.H.; Jarvandi, F.; Hajinajaf, S. Riboflavin supplementation to patients with multiple sclerosis does not improve disability status nor is riboflavin supplementation correlated to homocysteine. *Int. J. Vitam. Nutr. Res.* **2013**, *83*, 281–290. [CrossRef] [PubMed]
56. Naghashpour, M.; Amani, R.; Sarkaki, A.; Ghadiri, A.; Samarbafzadeh, A.; Jafarirad, S.; Saki Malehi, A. Brain-derived neurotrophic and immunologic factors: Beneficial effects of riboflavin on motor disability in murine model of multiple sclerosis. *Iran. J. Basic Med. Sci.* **2016**, *19*, 439–448. [PubMed]
57. Schapira, A.H.; Cooper, J.M.; Dexter, D.; Clark, J.B.; Jenner, P.; Marsden, C.D. Mitochondrial complex I deficiency in Parkinson's disease. *J. Neurochem.* **1990**, *54*, 823–827. [CrossRef] [PubMed]
58. Coimbra, C.G.; Junqueira, V.B.C. High doses of riboflavin and the elimination of dietary red meat promote the recovery of some motor functions in Parkinson's disease patients. *Braz. J. Med. Biol. Res.* **2003**, *36*, 1409–1417. [CrossRef] [PubMed]

59. Ray, A.; Martinez, B.A.; Berkowitz, L.A.; Caldwell, G.A.; Caldwell, K.A. Mitochondrial dysfunction, oxidative stress, and neurodegeneration elicited by a bacterial metabolite in a *C. elegans* Parkinson's model. *Cell Death Dis.* **2014**, *5*, e984. [CrossRef] [PubMed]

60. Bosetti, F.; Brizzi, F.; Barogi, S.; Mancuso, M.; Siciliano, G.; Tendi, E.A.; Murri, L.; Rapoport, S.I.; Solaini, G. Cytochrome c oxidase and mitochondrial F1F0-ATPase (ATP synthase) activities in platelets and brain from patients with Alzheimer's disease. *Neurobiol Aging.* **2002**, *23*, 371–376. [CrossRef]

61. Butterfield, D.A.; Boyd-Kimball, D. Amyloid β-peptide1–42 contributes to the oxidative stress and neurodegeneration found in Alzheimer disease brain. *Brain Pathol.* **2004**, *14*, 426–432. [CrossRef] [PubMed]

62. Shirendeb, U.P.; Calkins, M.J.; Manczak, M.; Anekonda, V.; Dufour, B.; McBride, J.L.; Mao, P.; Reddy, P.H. Mutant Huntingtin's interaction with mitochondrial protein Drp1 impairs mitochondrial biogenesis and causes defective axonal transport and synaptic degeneration in Huntington's disease. *Hum. Mol. Genet.* **2012**, *21*, 406–420. [CrossRef] [PubMed]

63. Seshadri, S.; Beiser, A.; Selhub, J.; Jacques, P.F.; Rosenberg, I.H.; D'Agostino, R.B.; Wilson, P.W.; Wolf, P.A. Plasma homocysteine as a risk factor for dementia and Alzheimer's disease. *N. Engl. J. Med.* **2002**, *346*, 476–483. [CrossRef] [PubMed]

64. Guenther, B.D.; Sheppard, C.A.; Tran, P.; Rozen, R.; Matthews, R.G.; Ludwig, M.L. The structure and properties of methylenetetrahydrofolate reductase from *Escherichia coli* suggest how folate ameliorates human hyperhomocysteinemia. *Nat. Struct. Biol.* **1999**, *6*, 359–365. [PubMed]

65. Kamat, P.K.; Vacek, J.C.; Kalani, A.; Tyagi, N. Homocysteine induced cerebrovascular dysfunction: A link to alzheimer's disease etiology. *Open Neurol. J.* **2015**, *9*, 9–14. [CrossRef] [PubMed]

Journal of
Clinical Medicine

MDPI

Review

Glutathione as a Redox Biomarker in Mitochondrial Disease—Implications for Therapy

Gregory M. Enns * and Tina M. Cowan

Departments of Pediatrics and Pathology, Stanford University, 300 Pasteur Drive, H-315, Stanford, CA 94005–5208, USA; tina.cowan@stanford.edu
* Correspondence: greg.enns@stanford.edu; Tel.: +1-650-498-5798; Fax: +1-650-498-4555

Academic Editor: Iain P. Hargreaves
Received: 17 February 2017; Accepted: 27 April 2017; Published: 3 May 2017

Abstract: Technical advances in the ability to measure mitochondrial dysfunction are providing new insights into mitochondrial disease pathogenesis, along with new tools to objectively evaluate the clinical status of mitochondrial disease patients. Glutathione (L-γ-glutamyl-L-cysteinylglycine) is the most abundant intracellular thiol, and the intracellular redox state, as reflected by levels of oxidized (GSSG) and reduced (GSH) glutathione, as well as the GSH/GSSG ratio, is considered to be an important indication of cellular health. The ability to quantify mitochondrial dysfunction in an affected patient will not only help with routine care, but also improve rational clinical trial design aimed at developing new therapies. Indeed, because multiple disorders have been associated with either primary or secondary deficiency of the mitochondrial electron transport chain and redox imbalance, developing mitochondrial therapies that have the potential to improve the intracellular glutathione status has been a focus of several clinical trials over the past few years. This review will also discuss potential therapies to increase intracellular glutathione with a focus on EPI-743 (α-tocotrienol quinone), a compound that appears to have the ability to modulate the activity of oxidoreductases, in particular NAD(P)H:quinone oxidoreductase 1.

Keywords: mitochondrial disease; glutathione; redox imbalance; EPI-743; *N*-acetylcysteine; RP103; cysteamine

1. Introduction

"Do you feel any better?" is a commonly asked question by a physician caring for a patient who has an underlying mitochondrial disorder during a clinic visit, typically after an interval of time following the start of various co-factors, vitamins, or supplements that may have a beneficial effect on mitochondrial function [1]. The lack of validated, widely available, and objective markers of mitochondrial function makes this state-of-the-art of mitochondrial medicine in the 21st century somewhat discouraging.

Nevertheless, there have been clear advances in our ability to determine clinical severity in mitochondrial disease patients. Clinical scoring tools, especially the Newcastle Mitochondrial Disease Adult Scale (NMDAS) and the Newcastle Paediatric Mitochondrial Disease Scale (NPMDS), provide the means to quantify the clinical burden of mitochondrial disease in individual patients [2,3]. Improvements in the resolution of traditional imaging techniques (e.g., magnetic resonance imaging and magnetic resonance spectroscopy) and the emergence of complementary new methods are also encouraging [4–7]. Finally, recent discoveries of minimally-invasive mitochondrial biomarkers, including blood creatine, FGF-21, and GDF-15, provide increased sensitivity and specificity compared to the standard practice of measuring lactate [8–11], although the link between these new biomarkers and mitochondrial disease pathophysiology is still unclear.

In contrast to these new analytes, which were detected by global metabolomic or transcriptomic profiling [9,12,13], there is a clear theoretical link between pathophysiology and biomarkers that are directly related to the biochemistry of redox imbalance. One important example is glutathione: mitochondrial dysfunction has been implicated in the generation of increased reactive oxygen and nitrogen species (RONS) leading, in turn, to increasing redox imbalance and decreased reduced glutathione (GSH) levels. Of course, the utility of this, and other, biomarkers still needs to be established in clinical practice. However, with improvements in both our understanding of mitochondrial disease pathogenesis and technologies to measure mitochondrial dysfunction, we may be approaching a time when clinicians will have quantitative, validated measures to gauge the clinical status of their mitochondrial disease patients. The ability to quantify mitochondrial dysfunction in an affected patient will not only help with routine care, but also improve rational clinical trial design aimed at developing new therapies.

2. Glutathione Levels in Mitochondrial Disorders

Glutathione (L-γ-glutamyl-L-cysteinylglycine) is the most abundant intracellular thiol, with intracellular concentrations ranging from about 0.5 to 10 μM. In its reduced form (GSH), it plays a key role in cellular free radical defense [14,15]. Glutathione synthesis occurs in the cytosol with 85% to 90% of GSH localized to the cytoplasm. The remainder is distributed between various organelles, including peroxisomes, the nuclear matrix, endoplasmic reticulum, and mitochondria [16–18].

The intracellular redox state, as reflected by levels of oxidized (GSSG) and reduced (GSH) glutathione, as well as the GSH/GSSG ratio, is an important indicator of cellular health [19,20]. By evaluating levels of GSH and GSSG, as well as the GSH/GSSG ratio in blood, one can get a glimpse into the degree of mitochondrial dysfunction at a tissue level as these compounds are leaked into the surrounding blood stream, urine or cerebrospinal fluid [8]. The GSSG/2GSH redox couple is representative of the redox environment in an individual, because the glutathione system plays a central role in maintaining the overall redox status of the body [19,21]. Dysfunction of the mitochondrial electron transport chain is associated with redox imbalance and abnormally low GSH levels in primary genetic mitochondrial disorders, as well as conditions associated with secondary mitochondrial impairment, such as organic acidemias, Friedreich ataxia, Alzheimer disease, Parkinson disease, amyotropic lateral sclerosis, and Rett syndrome [22–29]. Table 1 shows examples of glutathione determination by various methods in mitochondrial disorder and organic acidemia patients. Indeed, the documented redox abnormalities in patients who have either primary or secondary mitochondrial dysfunction have led to the development of mitochondrial therapies that have the potential to improve intracellular glutathione status [30].

While the glutathione redox couple represents an attractive mitochondrial biomarker, measured concentrations of GSH and GSSG have varied between different laboratories, likely because of instability of GSH during specimen handling and the use of various analytical methods, including high-performance liquid chromatography (HPLC), gas chromatography with mass spectrometry, capillary electrophoresis with ultraviolet absorbance or colormetric detection, and liquid chromatography-tandem mass spectrometry (LC-MS/MS) [31]. GSSG levels are especially influenced by oxidation during sample handling, and can appear elevated if conditions are not properly controlled [32–34]. Biological samples in which GSH, GSSG, and GSH/GSSG have been determined include whole blood, plasma, erythrocytes, leukocytes, urine, and skeletal muscle [24,31,35–44].

Table 1. Glutathione status in mitochondrial disorders and organic acidemias.

Conditions	Age	Analytical Method	Results	Reference
CPEO (n = 11)	30–70 years	HPLC	Plasma GSH: 8.64 ± 1.82 µM (controls 11 ± 3 µM; $p < 0.01$) RBC-GSH: 12.41 ± 2.37 nmol/mg protein (controls 18 ± 2.2 nmol/mg protein)	[43]
CPEO (n = 14) MELAS (n = 2) MERRF (n = 1)	19–86 years	GSH histo-chemistry	Muscle GSH: Induction of GSH in fibers with respiratory chain deficiency	[44]
Mitochondrial disorders (n = 24)	1 month–12 years	HPLC	Muscle GSH: Complexes I, II – III + IV (n = 7) 4.1 ± 0.98 nmol/mg protein ($p < 0.05$) Complex I (n = 1) 4.9 nmol/mg protein Complexes II – III + IV (n = 1) 3.0 nmol/mg protein Complex IV (n = 11) 9.7 ± 1.39 nmol/mg protein Complex I + IV (n = 7) 10.3 ± 1.80 nmol/mg protein Controls (n = 5) 12.3 ± 0.62 nmol/mg protein	[24]
Mitochondrial disorders (n = 20)	3–36 years	Hi-D FACS	Leukocyte GSH: Decreased iGSH in CD4 T cells ($p = 0.014$), CD8 T cells ($p = 0.005$), monocytes ($p = 0.016$), and neutrophils ($p = 0.044$)	[25]
Mitochondrial disorders (n = 10)	4–14 years	HPLC	Plasma GSH: 26.3 µM (controls 48.9 µM; $p = 0.031$) RBC-GSH: 6.4 ± 1.1 µmol/g protein (controls 6.7 ± 0.56 µmol/g protein)	[41]
Mitochondrial disorders (n = 58)	6 months–50 years	LC-MS/MS	Whole blood GSH: 808 ± 149 µM (controls 900 ± 140 µM; $p = 0.0008$) Whole blood GSSG: 2.23 ± 1.84 µM (controls 1.17 ± 0.43 µM; <0.0001) Whole blood GSH/GSSG: 596 ± 93 µM (controls 800 ± 370 µM; $p = 0.0002$) Whole blood redox potential: −251 ± 9.7 mV (controls −260 ± 6.4 mV)	[28,31]

Table 1. *Cont.*

Conditions	Age	Analytical Method	Results	Reference
Friedreich ataxia ($n = 14$)	8–22 years	HPLC	Whole blood GSH + GSSG: 0.55 nmol/mg hemoglobin (controls 8.4 ± 1.79 nmol/mg hemoglobin; $p < 0.001$) RBC hemoglobin-bound glutathione: 15 ± 1.5% (controls 8 ± 1.8%; $p < 0.05$)	[23]
Organic acidemias ($n = 9$)	1 week–6 years	Hi-D FACS	Leukocyte GSH: Decreased iGSH in CD4 T cells ($p = 0.008$), CD8 T cells ($p = 0.003$), monocytes ($p = 0.0008$), and neutrophils ($p = 0.0006$) in hospitalized patients Decreased iGSH in CD4 T cells (0.040) and CD8 T cells (0.045) in outpatients	[25]
Organic acidemias ($n = 11$)	1–16 years	HPLC	Plasma GSH: 32.9 ± 6.9 μM (controls 48.9 ± 25.7 μM)	[26]
Cobalamin C disease ($n = 18$)	1–14 years	HPLC	Lymphocyte total glutathione: 23 nmol/mg protein (95% CI 10.25–62.03) (controls 6.9 nmol/mg protein; 95% CI 21.96–60.42; $p < 0.05$) Lymphocyte GSH: 6.9 nmol/mg protein (95% CI 0.68–24.83) (controls 39.10 nmol/mg protein; 95% CI 19.31–54.55; $p < 0.001$) Lymphocyte GSSG: 7.9 nmol/mg protein (95% CI 1.87–24.78) (controls 2.94 nmol/mg protein; 95% CI 1.33–3.82; $p < 0.05$)	[42]

Note: The glutathione status as determined by different analytical methodologies in mitochondrial disease or organic acidemia patients is shown above. Statistical significance is shown where possible as provided by the individual references. CI = confidence interval; CPEO = chronic progressive external ophthalmoplegia; GSH = reduced glutathione; GSSG = glutathione disulfide; HiD-FACS = high-dimensional fluorescence-activated cell sorting; HPLC = high-performance liquid chromatography; iGSH = intracellular reduced glutathione; LC-MS/MS = liquid chromatography-tandem mass spectrometry; RBC = red blood cell.

Increased plasma lipid peroxidation and decreased plasma and erythrocyte GSH levels were detected using an HPLC method in 11 patients with chronic progressive external ophthalmoplegia (CPEO) and muscle biopsies with ragged-red fibers and scattered cytochrome *c* oxidase (COX) deficiency [43]. Skeletal muscle biopsies from 17 patients with CPEO, MELAS, or myoclonic epilepsy with ragged-red fibers (MERRF) showed induction of the antioxidant enzymes manganese and copper-zinc superoxide dismutase in fibers associated with ETC deficiency; GSH was found to be elevated in these fibers by histochemical analysis. Antioxidants were expressed in both ragged-red fibers and fibers with subsarcolemmal mitochondrial accumulations that were COX negative. The authors concluded that increased GSH represented the earliest defense against the toxic effects of ETC-produced hydrogen peroxide [44].

HPLC analysis of 24 skeletal muscle biopsies from mitochondrial disease patients with defined ETC defects showed a significant decrease in GSH concentration compared to 15 age-matched controls without evidence of mitochondrial ETC deficiency (7.7 $\pm$ 0.9 nmol/mg protein vs. 12.3 $\pm$ 0.6 nmol/mg protein). Furthermore, the most prominent GSH deficiency was noted in the patients who had multiple ETC defects in complexes I, II − III + IV [24]. The authors postulated that treatments designed to increase GSH levels, such as *N*-acetylcysteine or oxothiazolidine-4-carboxylate supplementation, may be beneficial to patients who have ETC deficiency associated with a GSH deficit [24].

Intracellular leukocyte GSH levels were evaluated in blood samples from patients with either mitochondrial diseases or organic acidemias using high-dimensional flow cytometry (Hi-D FACS). T lymphocyte subsets, monocytes, and neutrophils showed low GSH levels in both mitochondrial disease and organic acidemia patients, although levels were relatively normal in those patients who were taking antioxidants [25]. The Hi-D FACS results demonstrated redox imbalance in patients with either primary or secondary mitochondrial dysfunction, but the technique is semi-quantitative and not wholly amenable to the clinical setting, as samples have to be analyzed immediately and cannot be shipped.

A study of 10 mitochondrial disease patients with a variety of clinical presentations, including Leigh syndrome, Alpers syndrome, Kearns–Sayre syndrome, and multisystem disease analyzed plasma GSH and cysteine levels by HPLC. Plasma GSH levels were low in mitochondrial disease patients, mostly below the detection level of the method used, and reduced cysteine levels were also lower in mitochondrial disease patients compared to controls. Erythrocyte thiols and glutathione-related enzymes, such as glutathione peroxidase, glutathione reductase, and glutathione S-transferase, were also evaluated, but significant differences between patients and controls were not observed [41].

LC-MS/MS appears to be particularly promising as a methodology for analyzing glutathione samples. Whole blood samples can be deproteinized with sulfosalicylic acid and derivatized with *N*-ethylmaleimide (NEM) in order to prevent oxidation of GSH in a single step before being analyzed. Derivatized samples are stable for at least three years when stored at −80 °C, and underivatized samples for at least 24 h at room temperature, allowing potential implementation in clinical laboratories [31]. This LC-MS/MS method was initially used to study healthy individuals and mean $\pm$ SD glutathione levels were: GSH 900 $\pm$ 140 µM; GSSG 1.17 $\pm$ 0.43 µM; and GSH/GSSG 880 $\pm$ 370 [31].

A further study used LC-MS/MS to measure whole blood glutathione levels in mitochondrial disease patients with a variety of different clinical phenotypes, including Leigh syndrome, mitochondrial encephalomyopathy, lactic acidosis and stroke-like episodes (MELAS), mtDNA deletion syndrome, conditions associated with mtDNA depletion, and patients with a variety of electron transport disorders. A subset of patients was evaluated both while in times of relative good health and while hospitalized during a metabolic crisis. Compared to healthy controls, mitochondrial disease patients (n = 58), as a whole, showed significantly lower whole blood GSH levels (808 $\pm$ 149 µM vs. 900 $\pm$ 141 µM, p = 0.0008), GSH/GSSG ratio (881 $\pm$ 374 vs. 596 $\pm$ 424, p = 0.0002), and higher GSSG levels (2.23 $\pm$ 1.84 µM vs. 1.17 $\pm$ 0.43 µM, p < 0.0001) [28]. In addition to measuring absolute levels of GSH and GSSG, whole blood redox potential was calculated using the Nernst equation. Mitochondrial disease patients had significant redox imbalance, with an increased degree of oxidation of approximately 9 mV compared to controls (-251 ± 9.7 mV vs. -260 ± 6.4 mV). When subgroups of mitochondrial disease patients were evaluated, redox potential was significantly more oxidized in each group. Interestingly, the lowest GSH levels in relatively healthy patients were found in those with Leigh syndrome (735 ± 135 μM). Overall, patients who were hospitalized for treatment of a metabolic crisis showed the lowest GSH levels (550 ± 93 μM) and the greatest degree of redox imbalance (-242 ± 7.0 mV) [28].

In aggregate, there is clear evidence of GSH deficiency and redox imbalance in mitochondrial disease patients. However, further longitudinal studies are needed to determine the utility of using the glutathione system as a biomarker of disease severity and response to therapies.

3. Glutathione Levels in Other Disorders Associated with Mitochondrial Dysfunction

3.1. Organic Acidemias

Abnormal mitochondrial structure and function, as well as various abnormal measures of oxidative stress and redox imbalance, have been reported in animal models and patients affected by a variety of organic acidemias, including methylmalonic acidemia, cobalamin A disease, cobalamin C disease, cobalamin H/cobalamin D disease, propionic acidemia, isovaleric acidemia, 2-methyl-3-hydroxybutyric acidemia, 3-methylglutaconic acidemia types II and IV, D-2-hydroxyglutaric aciduria, L-2-hydroxyglutaric aciduria, and glutaric acidemia [42,45–70]. GSH levels have been found to be low in methylmalonic acidemia (MMA), propionic acidemia (PA), and isovaleric acidemia (IVA) [25,26,71]. A seven-year-old boy with *mut⁻* methylmalonic acidemia in metabolic crisis was found to have marked lactic acidemia, GSH deficiency, and 5-oxoprolinuria; treatment with ascorbate, in addition to other supportive management, resulted in an improvement in clinical status and resolution of the biochemical abnormalities [71]. Hi-D FACS analysis of peripheral blood leukocytes was performed in 13 patients with MMA, PA, or IVA. Organic acidemia patients who had samples collected during an illness severe enough to require hospitalization showed significantly lower intracellular GSH levels in CD4 T cells, CD8 T cells, monocytes, and neutrophils when compared to healthy controls. On the other hand, patients who had samples collected during routine outpatient visits had lower GSH levels detected only in CD4 and CD8 T cells [25]. A more recent study used an HPLC method to evaluate GSH levels in 11 MMA, PA, and IVA patients and showed that organic acidemia patients had lower plasma GSH levels than controls. Organic acidemia patients also had a greater fraction of GSH and cysteine in an oxidized state [41].

3.2. Friedreich Ataxia

Friedreich ataxia patients also have evidence of redox abnormalities and mitochondrial dysfunction [72–75]. A study of 14 unrelated Friedreich ataxia patients measured total and free GSH concentrations in erythrocytes by HPLC. Patients had a significant reduction of free glutathione levels, although total glutathione levels were comparable to controls. Friedreich ataxia patients were also found to have a significant increase in glutathione bound to hemoglobin in erythrocytes [23]. Glutathione homeostasis was, therefore, considered to be impaired in Friedreich ataxia, raising the possibility that free radicals play a role in disease pathophysiology [23].

3.3. Parkinson Disease and Other Neurodegenerative Disorders

Mitochondrial dysfunction and oxidative stress also appear to be central to disease pathogenesis in Parkinson disease and other neurodegenerative conditions, including Alzheimer disease, amyotropic lateral sclerosis, and Rett syndrome [27,76–81]. Not surprisingly, the glutathione axis has been found to be abnormal when evaluated in these conditions [20,27,82,83]. As reviewed elsewhere in this issue, children with autistic spectrum disorder have also been found to have biochemical abnormalities suggestive of an underlying impaired mitochondrial metabolism, including low levels of GSH and low GSH/GSSG ratio [84–86].

3.4. Genetic Syndromes

Finally, redox and GSH abnormalities have been identified in a number of genetic syndromes, including Down syndrome, Werner syndrome, fragile X syndrome, and Kindler syndrome [87–91]. Therefore, it is possible that mitochondrial dysfunction may play some role in the clinical findings associated with conditions that appear to have no or a minimal relationship to mitochondrial metabolism. Further studies are clearly needed in order to determine the significance of these initial reports.

4. Implications for Mitochondrial Disease Therapy

Since multiple disorders have been associated with either primary or secondary deficiency of the mitochondrial ETC and the resultant redox imbalance, developing mitochondrial therapies that have the potential to improve the intracellular glutathione status has been a focus of several clinical trials over the past few years [92–102]. For example, EPI-743 (α-tocotrienol quinone) is an investigational drug that is currently in clinical trials focusing on treatment of mitochondrial dysfunction related to primary genetic mitochondrial disease, including Leigh syndrome, Leber Hereditary Optic Neuropathy (LHON), and RARS2 deficiency. EPI-743 is also being used in other conditions that have redox imbalance linked to disease pathophysiology, including Friedreich ataxia, Parkinson disease, and Rett syndrome (Table 2).

Table 2. EPI-743 clinical trials.

Patient Population	Age	Trial Design	Duration	Outcomes	Reference
Mitochondrial disease ($n = 14$)	2–27 years	Open-label	98–444 days	11/12 survivors with clinical improvement; 3/11 partial relapse; 10/12 improvement in quality of life (NPMDS section IV); 2 deaths	[92]
LHON ($n = 5$)	8–52 years	Open-label	204–557 days	4/5 arrested disease progression and reversal of vision loss; 2/5 total recovery of visual acuity	[93]
Leigh syndrome ($n = 10$)	1–13 years	Open-label	6 months	Reversal of disease progression; Improvement in NPMDS, GMFM, PedsQL Neuromuscular Module ($p < 0.05$)	[94]
Leigh syndrome ($n = 35$)	9 months–14 years	Randomized, double-blind, placebo-controlled	36 months	Decreased rate of hospitalization and serious adverse events	[95]
RARS2 deficiency ($n = 5$)	5–13 years	Open-label	1 year	Improved neuromuscular function and redox state; Decreased seizure frequency with 2 patients showing resolution of status epilepticus	[96,97]
Friedreich ataxia ($n = 31$) [1]	18–66 years	Randomized, double-blind, placebo-controlled	28 days	Dose-dependent improvement in FARS score; No alteration in Disposition Index (measure of diabetic tendency)	[98]
Friedreich ataxia ($n = 63$)	19–43 years	Randomized, double-blind, placebo-controlled	2 years	Dose-dependent improvement in FARS score	[99]
Freidreich ataxia (point mutations) ($n = 4$)	21–63 years	Open-label	18 months	Improvement in FARS	[100]
Rett syndrome ($n = 24$)	2.5–8 years	Open-label	6 months	Primary endpoint of improvement in Rett syndrome disease severity score not met; Increase in head circumference ($p = 0.05$); Improved oxygenation, hand function and disease biomarkers in subgroup with greatest degree of head growth	[101]
Parkinson disease ($n = 10$)	43–69 years	Open-label	6 months	Improvement in UPDRS Parts II/III; Decrease in brain glutamine/glutamate levels; Improvement of retinal function on electroretinogram	[102]

[1] EPI-A0001 was used in this study, not EPI-743. EPI-A0001 is an α-tocopheryl quinone drug with a chemical structure similar to EPI-743. FARS = Friedreich Ataxia Rating Scale; GMFM = Gross Motor Function Measure; NPMDS = Newcastle Paediatric Mitochondrial Disease Scale; PedsQL = Pediatrics Quality of Life Inventory; UPDRS = Unified Parkinson Disease Rating Scale.

5. EPI-743

EPI-743 is a *para*-benzoquinone analog that is approximately one thousand- to ten thousand-fold more potent than coenzyme Q_{10} or idebenone in protecting mitochondrial patient fibroblasts when a strong oxidant stress is applied [103]. This beneficial effect is considered to be related to the ability of

EPI-743 to modulate the activity of oxidoreductases, in particular NAD(P)H:quinone oxidoreductase 1, resulting in increased cellular GSH concentration and improvement in redox status [30,92,103]. EPI-743 may also affect antioxidant gene expression, as pre-treatment of fibroblasts derived from a polymerase γ deficiency patient with EPI-743 before exposure of cells to oxidative stress resulted in a blunting of antioxidant response element (ARE) gene expression in genes under direct control of nuclear factor-erythroid 2 p45-related factor (Nrf2), including genes related to GSH synthesis [92].

5.1. Mitochondrial Disorders

The initial experience with EPI-743 in mitochondrial disease was reported in 13 children and one adult enrolled in a 13-week emergency treatment protocol for patients who were considered to be at risk for progressing to end-of-life care within 90 days by experienced clinicians. Surviving patients were then placed in an extension protocol. Twelve of the 14 patients survived during the period of observation; 11 of the survivors demonstrated clinical improvement, with three showing partial relapse. NPMDS scores were not significantly different for sections I, II, or III (sections related to clinical status) when scores from before and after EPI-743 treatment were compared, whereas 10 patients showed improvement in section IV (quality of life) [92].

Twelve of the 14 patients also underwent serial brain imaging using technetium-99m-hexamethylpropyleneamine oxime (HMPAO) SPECT. HMPAO is a lipophilic radionuclide tracer that is sensitive to intracellular redox status that is used to measure cerebral blood flow, as well as intracellular GSH and reduced protein thiols. This tracer is retained inside cells, locked in a hydrophilic state, in the presence of adequate reducing equivalents generated by functional mitochondria [104]. Before the administration of EPI-743, all 12 individuals had decreased brain HMPAO uptake compared to normal controls. After three months of EPI-743 therapy, there was a significant increase in whole brain HMPAO uptake [92]. A further study of 22 patients enrolled in the EPI-743 emergency treatment protocol demonstrated an increase in HMPAO uptake in the cerebellum in all patients. Furthermore, there was a significant correlation between increased cerebellar uptake and improved Newcastle score ($r = 0.623$; $p = 0.00161$). The subgroup of five patients with MELAS showed a significant relationship between whole brain HMPAO uptake and Newcastle score improvement ($r = 0.917$; $p = 0.028$) [7].

5.2. Leber Hereditary Optic Neuropathy

EPI-743 has also been used to treat LHON patients. An open-label trial using EPI-743 was performed in five LHON patients, including a child harboring the m.14484T > C variant (associated with spontaneous recovery in some cases), three patients with the m.11778G > A variant and one with the m.3460G > A variant. EPI-743 arrested disease progression and reversed vision loss in all but one of these consecutively-treated patients [93].

An open-label study of EPI-743 therapy in ten children with genetically-confirmed Leigh syndrome showed stabilization and even reversal of disease progression. A significant improvement was noted for each of the primary outcome endpoints, which included the NPMDS, and measures of gross motor function, and quality of life ($p < 0.05$) [94]. Enrolled subjects also had total, reduced, and protein-bound glutathione levels measured in lymphocytes before and after treatment with EPI-743. At baseline, the Leigh syndrome subjects had decreased total and reduced glutathione levels, as well as high levels of oxidized glutathione. Following treatment with EPI-743 a marked increase in reduced glutathione ($p < 0.001$) and a 96% decrease in the ratio of oxidized-to-reduced glutathione ($p < 0.001$) was observed [30].

5.3. Leigh Syndrome

EPI-743 is currently being used in a randomized, double blind, placebo-controlled clinical trial in children with Leigh syndrome (NCT01721733; NCT02352896). Clinical trial design included a six-month placebo-controlled phase, followed by a 30-month extension phase to assess long-term drug safety and impact on disease morbidity. In the initial six-month phase, treatment with EPI-743 was associated with fewer subjects requiring hospitalization or experiencing serious adverse events as compared with those subjects who received a placebo (11.8% vs. 42.8%). Further follow-up of enrolled subjects indicated that there was a progressive decline in hospitalizations and serious adverse events from the first six months of EPI-743 treatment to months 19 to 24 [95].

5.4. RARS2 Deficiency

Autosomal recessive pathogenic variants in *RARS2*, the gene encoding mitochondrial arginyl-transfer RNA synthetase, have been associated with an early-onset mitochondrial encephalopathy characterized by microcephaly, profound developmental delay, intractable seizures, dystonia, and pontocerebellar hypoplasia [96]. In an open-label study, five children with RARS2 deficiency were given EPI-743 over a 12-month treatment phase, followed by an extension phase that is still ongoing. All subjects demonstrated an improvement in clinical status regardless of the severity of baseline disease. Status epilepticus resolved in two children, and the other three children demonstrated a reduction in seizure frequency and duration [97].

5.5. Friedreich Ataxia

In a study of EPI-A0001, an α-tocopheryl quinone structurally related to EPI-743, 31 adults with Friedreich ataxia were evaluated using a measure of diabetic tendency as the primary clinical trial outcome measure (NCT01035671). The Friedreich Ataxia Rating Scale (FARS) was used as a secondary neurological outcome measure [105]. No significant difference was observed in the measure of diabetic tendency between treated subjects and controls after four weeks of therapy. However, a dose-dependent improvement in the FARS score was observed, indicating that this compound potentially has an effect on the central nervous system [98].

In a phase 2 double-blind placebo-controlled trial of EPI-743 in adults with Friedreich ataxia (NCT01728064), EPI-743 treatment resulted in a significant improvement in neurological function and disease progression when compared to controls as measured by the FARS. The improvement in FARS score was dose-dependent; subjects who received EPI-743 at the highest dosage for the entire 24-month study period registered the greatest degree of improvement [99].

Friedreich ataxia may rarely be caused by a point mutation in *FXN* on one allele in combination with a typical GAA trinucleotide repeat expansion on the other allele. Four Friedreich ataxia patients who harbor a point mutation in one *FXN* allele were treated with EPI-743 in an open-label study (NCT01962363). The patients showed clinical improvement as assessed by FARS score over 18 months of therapy [100].

5.6. Rett Syndrome

In a six-month, randomized, double-blind, placebo-controlled trial involving 24 Rett syndrome patients aged 2.5–8 years (NCT01822249), those who were treated with EPI-743 showed a significant increase in head circumference relative to placebo subjects ($p = 0.05$). In a subgroup of children with the greatest degree of head growth, improvements in oxygenation, hand function, and disease biomarkers were also observed [101].

5.7. Parkinson Disease

A phase 2a open-label pilot study was performed to determine if EPI-743 might improve the treatment of Parkinson disease (NCT01923584). The Unified Parkinson Disease Rating Scale (UPDRS) [106], electroretinography, and brain metabolite levels as measured by magnetic resonance spectroscopy (MRS) were used as clinical outcome measures. Six of seven patients with follow-up MRS studies showed a decrease in glutamine/glutamate levels in the basal ganglia opposite the side most severely affected by Parkinson disease. In addition, improvement in retinal function was noted on evaluation by electroretinogram. Subjects also demonstrated an improvement in UPDRS scores that approached statistical significance [102].

J. Clin. Med. **2017**, *6*, 50

6. *N*-Acetylcysteine and Cysteamine

Other compounds that have the potential to increase intracellular GSH include *N*-acetylcysteine (NAC) and cysteamine [107,108]. NAC is a drug that is best known for its therapeutic effects in acetaminophen-induced liver failure [107,109,110], but has also shown promise in acute liver failure in the absence of acetaminophen overdose [110,111]. Since oxidative and nitrosative stress play a role in the pathogenesis of liver failure and the subsequent CNS effects of hepatic encephalopathy [112,113], the replenishment of intracellular GSH by NAC may be beneficial [107]. NAC has also been used as a mucolytic for the treatment of cystic fibrosis, and to treat chronic obstructive pulmonary disease, diabetes mellitus, and patients infected with human immunodeficiency virus [107].

NAC has been shown to improve markers of oxidative stress in an animal model of Huntington disease and cell lines derived from patients with Huntington disease and mitochondrial respiratory chain disorders [114–116]. Although there have been case reports using NAC to treat primary mitochondrial disorders, for example, in mitochondrial disease patients who have liver dysfunction [117], to our knowledge there have not yet been controlled clinical trials that explore the efficacy of NAC in these conditions. On the other hand, NAC has been used in controlled trials in several conditions with likely secondary mitochondrial involvement, including Alzheimer disease, amyotropic lateral sclerosis, and autism [118]. Improvement in some measures of cognitive ability was observed in Alzheimer disease patients, but no improvement in survival or disease progression was noted in those with amyotropic lateral sclerosis. Autistic patients have shown improvement in some aberrant behaviors, especially irritability, following treatment with NAC [118].

NAC has also been used in combination with metronidazole to treat ethylmalonic encephalopathy, a disorder caused by mutations in *ETHE1* that result in secondary inhibition of cytochrome *c* oxidase and other enzymes. Treated *Ethe1*-deficient mice had a prolonged lifespan, and five ethylmalonic encephalopathy patients demonstrated marked clinical improvement following combined therapy [119].

Cysteamine, an established therapy for cystinosis, serves to decrease the abnormal lysosomal storage of cystine. Cysteamine also appears to promote the transport of cysteine into cells, which could increase intracellular glutathione levels [108]. Cysteamine has been shown to improve symptoms in mouse models of Huntington disease, and has been used in a small open-label clinical trial in Huntington disease patients [120]. Clinical efficacy was not demonstrated, although the trial established a safe cysteamine dosage regimen in Huntington disease patients [120,121]. Cysteamine bitartrate delayed-release (RP103) is a microsphere formulation associated with decreased gastrointestinal symptoms [122]. RP103 is currently being used in a randomized, controlled, double-blind multicenter trial for Huntington disease (NCT02101957) and an open-label study in children with mitochondrial disease (NCT02023866), but results have not yet been published.

In summary, the glutathione system continues to be evaluated as a potentially valuable biomarker of mitochondrial dysfunction across multiple diseases. There have been clear advances in the field of mitochondrial redox biomarker analysis, with glutathione levels being able to be measured by improved analytical techniques in virtually any tissue sample, as well as by using relatively non-invasive brain imaging techniques, such as HMPAO SPECT. Measuring of glutathione metabolites has provided investigators with unique insights into the redox imbalance present in patients who have mitochondrial dysfunction, which has led to clinical trials designed to address this issue. In the near future, clinicians caring for individuals affected by mitochondrial disease may not only have improved therapies to offer their patients, but may also be able to monitor individuals by blood and imaging biomarkers. By doing so, physicians may be able to both predict and understand in advance the answer to the question, "Do you feel any better?"

Acknowledgments: The authors are grateful to our patients and their families, who continue to provide inspiration for the pursuit of improved techniques to diagnose, monitor and treat mitochondrial disease.

Author Contributions: G.M.E. and T.M.C. wrote the manuscript.

Conflicts of Interest: G.M.E. reports receiving funding for being an investigator in clinical trials related to EPI-743 (Edison Pharmaceuticals, Inc.) and RP-103 (Raptor Pharmaceuticals, Inc.), and has received unrestricted gift research funds from Edison Pharmaceuticals, Inc. T.M.C. received research funds from Raptor Pharmaceuticals, Inc.

References

1. Enns, G.M. Treatment of Mitochondrial Disorders: Antioxidants and Beyond. *J. Child Neurol.* **2014**, *29*, 1235–1240. [CrossRef] [PubMed]
2. Schaefer, A.M.; Phoenix, C.; Elson, J.L.; McFarland, R.; Chinnery, P.F.; Turnbull, D.M. Mitochondrial Disease in Adults: A Scale to Monitor Progression and Treatment. *Neurology* **2006**, *66*, 1932–1934. [CrossRef] [PubMed]
3. Phoenix, C.; Schaefer, A.M.; Elson, J.L.; Morava, E.; Bugiani, M.; Uziel, G.; Smeitink, J.A.; Turnbull, D.M.; McFarland, R. A Scale to Monitor Progression and Treatment of Mitochondrial Disease in Children. *Neuromuscul. Disord.* **2006**, *16*, 814–820. [CrossRef] [PubMed]
4. Bianchi, M.C.; Sgandurra, G.; Tosetti, M.; Battini, R.; Cioni, G. Brain Magnetic Resonance in the Diagnostic Evaluation of Mitochondrial Encephalopathies. *Biosci. Rep.* **2007**, *27*, 69–85. [CrossRef] [PubMed]
5. Mitochondrial Medicine Society's Committee on Diagnosi; Haas, R.H.; Parikh, S.; Falk, M.J.; Saneto, R.P.; Wolf, N.I.; Darin, N.; Wong, L.J.; Cohen, B.H.; Naviaux, R.K. The in-Depth Evaluation of Suspected Mitochondrial Disease. *Mol. Genet. Metab.* **2008**, *94*, 16–37. [CrossRef] [PubMed]
6. Mancuso, M.; Orsucci, D.; Coppede, F.; Nesti, C.; Choub, A.; Siciliano, G. Diagnostic Approach to Mitochondrial Disorders: The Need for a Reliable Biomarker. *Curr. Mol. Med.* **2009**, *9*, 1095–1107. [CrossRef] [PubMed]
7. Blankenberg, F.G.; Kinsman, S.L.; Cohen, B.H.; Goris, M.L.; Spicer, K.M.; Perlman, S.L.; Krane, E.J.; Kheifets, V.; Thoolen, M.; Miller, G.; et al. Brain Uptake of Tc99m-Hmpao Correlates with Clinical Response to the Novel Redox Modulating Agent Epi-743 in Patients with Mitochondrial Disease. *Mol. Genet. Metab.* **2012**, *107*, 690–699. [CrossRef] [PubMed]
8. Suomalainen, A. Biomarkers for Mitochondrial Respiratory Chain Disorders. *J. Inherit. Metab. Dis.* **2011**, *34*, 277–282. [CrossRef] [PubMed]
9. Shaham, O.; Slate, N.G.; Goldberger, O.; Xu, Q.; Ramanathan, A.; Souza, A.L.; Clish, C.B.; Sims, K.B.; Mootha, V.K. A Plasma Signature of Human Mitochondrial Disease Revealed through Metabolic Profiling of Spent Media from Cultured Muscle Cells. *Proc. Natl. Acad. Sci. USA* **2010**, *107*, 1571–1575. [CrossRef] [PubMed]
10. Suomalainen, A.; Elo, J.M.; Pietilainen, K.H.; Hakonen, A.H.; Sevastianova, K.; Korpela, M.; Isohanni, P.; Marjavaara, S.K.; Tyni, T.; Kiuru-Enari, S.; et al. Fgf-21 as a Biomarker for Muscle-Manifesting Mitochondrial Respiratory Chain Deficiencies: A Diagnostic Study. *Lancet Neurol.* **2011**, *10*, 806–818. [CrossRef]
11. Yatsuga, S.; Fujita, Y.; Ishii, A.; Fukumoto, Y.; Arahata, H.; Kakuma, T.; Kojima, T.; Ito, M.; Tanaka, M.; Saiki, R.; et al. Growth Differentiation Factor 15 as a Useful Biomarker for Mitochondrial Disorders. *Ann. Neurol.* **2015**, *78*, 814–823. [CrossRef] [PubMed]
12. Tyynismaa, H.; Carroll, C.J.; Raimundo, N.; Ahola-Erkkila, S.; Wenz, T.; Ruhanen, H.; Guse, K.; Hemminki, A.; Peltola-Mjosund, K.E.; Tulkki, V.; et al. Mitochondrial Myopathy Induces a Starvation-Like Response. *Hum. Mol. Genet.* **2010**, *19*, 3948–3958. [CrossRef] [PubMed]
13. Kalko, S.G.; Paco, S.; Jou, C.; Rodriguez, M.A.; Meznaric, M.; Rogac, M.; Jekovec-Vrhovsek, M.; Sciacco, M.; Moggio, M.; Fagiolari, G.; et al. Transcriptomic Profiling of Tk2 Deficient Human Skeletal Muscle Suggests a Role for the P53 Signalling Pathway and Identifies Growth and Differentiation Factor-15 as a Potential Novel Biomarker for Mitochondrial Myopathies. *BMC Genomics* **2014**, *15*, 91. [CrossRef] [PubMed]
14. Cadenas, E. Mitochondrial Free Radical Production and Cell Signaling. *Mol. Asp. Med.* **2004**, *25*, 17–26. [CrossRef] [PubMed]
15. Lash, L.H. Mitochondrial Glutathione Transport: Physiological, Pathological and Toxicological Implications. *Chem. Biol. Interact.* **2006**, *163*, 54–67. [CrossRef] [PubMed]
16. Lu, S.C. Regulation of Glutathione Synthesis. *Curr. Top. Cell Regul.* **2000**, *36*, 95–116. [PubMed]
17. Wu, G.; Fang, Y.Z.; Yang, S.; Lupton, J.R.; Turner, N.D. Glutathione Metabolism and Its Implications for Health. *J. Nutr.* **2004**, *134*, 489–492. [PubMed]
18. Mari, M.; Morales, A.; Colell, A.; Garcia-Ruiz, C.; Kaplowitz, N.; Fernandez-Checa, J.C. Mitochondrial Glutathione: Features, Regulation and Role in Disease. *Biochim. Biophys. Acta* **2013**, *1830*, 3317–3328. [CrossRef] [PubMed]
19. Jones, D.P. Redefining Oxidative Stress. *Antioxid. Redox Signal.* **2006**, *8*, 1865–1879. [CrossRef] [PubMed]

20. Ballatori, N.; Krance, S.M.; Notenboom, S.; Shi, S.; Tieu, K.; Hammond, C.L. Glutathione Dysregulation and the Etiology and Progression of Human Diseases. *Biol. Chem.* **2009**, *390*, 191–214. [CrossRef] [PubMed]

21. Schafer, F.Q.; Buettner, G.R. Redox Environment of the Cell as Viewed through the Redox State of the Glutathione Disulfide/Glutathione Couple. *Free Radic. Biol. Med.* **2001**, *30*, 1191–1212. [CrossRef]

22. Merad-Boudia, M.; Nicole, A.; Santiard-Baron, D.; Saille, C.; Ceballos-Picot, I. Mitochondrial Impairment as an Early Event in the Process of Apoptosis Induced by Glutathione Depletion in Neuronal Cells: Relevance to Parkinson's Disease. *Biochem. Pharmacol.* **1998**, *56*, 645–655. [CrossRef]

23. Piemonte, F.; Pastore, A.; Tozzi, G.; Tagliacozzi, D.; Santorelli, F.M.; Carrozzo, R.; Casali, C.; Damiano, M.; Federici, G.; Bertini, E. Glutathione in Blood of Patients with Friedreich's Ataxia. *Eur. J. Clin. Investig.* **2001**, *31*, 1007–1011. [CrossRef] [PubMed]

24. Hargreaves, I.P.; Sheena, Y.; Land, J.M.; Heales, S.J. Glutathione Deficiency in Patients with Mitochondrial Disease: Implications for Pathogenesis and Treatment. *J. Inherit. Metab. Dis.* **2005**, *28*, 81–88. [CrossRef] [PubMed]

25. Atkuri, K.R.; Cowan, T.M.; Kwan, T.; Ng, A.; Herzenberg, L.A.; Herzenberg, L.A.; Enns, G.M. Inherited Disorders Affecting Mitochondrial Function Are Associated with Glutathione Deficiency and Hypocitrullinemia. *Proc. Natl. Acad. Sci. USA* **2009**, *106*, 3941–3945. [CrossRef] [PubMed]

26. Salmi, H.; Leonard, J.V.; Lapatto, R. Patients with Organic Acidaemias Have an Altered Thiol Status. *Acta Paediatr.* **2012**, *101*, e505–e508. [CrossRef] [PubMed]

27. Signorini, C.; Leoncini, S.; De Felice, C.; Pecorelli, A.; Meloni, I.; Ariani, F.; Mari, F.; Amabile, S.; Paccagnini, E.; Gentile, M.; et al. Redox Imbalance and Morphological Changes in Skin Fibroblasts in Typical Rett Syndrome. *Oxid. Med. Cell Longev.* **2014**, *2014*, 195935. [CrossRef] [PubMed]

28. Enns, G.M.; Moore, T.; Le, A.; Atkuri, K.; Shah, M.K.; Cusmano-Ozog, K.; Niemi, A.K.; Cowan, T.M. Degree of Glutathione Deficiency and Redox Imbalance Depend on Subtype of Mitochondrial Disease and Clinical Status. *PLoS One* **2014**, *9*, e100001. [CrossRef] [PubMed]

29. Gu, F.; Chauhan, V.; Chauhan, A. Glutathione Redox Imbalance in Brain Disorders. *Curr. Opin. Clin. Nutr. Metab. Care* **2015**, *18*, 89–95. [CrossRef] [PubMed]

30. Pastore, A.; Petrillo, S.; Tozzi, G.; Carrozzo, R.; Martinelli, D.; Dionisi-Vici, C.; Di Giovamberardino, G.; Ceravolo, F.; Klein, M.B.; Miller, G.; et al. Glutathione: A Redox Signature in Monitoring Epi-743 Therapy in Children with Mitochondrial Encephalomyopathies. *Mol. Genet. Metab.* **2013**, *109*, 208–214. [CrossRef] [PubMed]

31. Moore, T.; Le, A.; Niemi, A.K.; Kwan, T.; Cusmano-Ozog, K.; Enns, G.M.; Cowan, T.M. A New Lc-Ms/Ms Method for the Clinical Determination of Reduced and Oxidized Glutathione from Whole Blood. *J. Chromatogr. B Anal. Technol. Biomed. Life Sci.* **2013**, *929*, 51–55. [CrossRef] [PubMed]

32. Rossi, R.; Milzani, A.; Dalle-Donne, I.; Giustarini, D.; Lusini, L.; Colombo, R.; Di Simplicio, P. Blood Glutathione Disulfide: In Vivo Factor or in Vitro Artifact? *Clin. Chem.* **2002**, *48*, 742–753. [PubMed]

33. Rossi, R.; Dalle-Donne, I.; Milzani, A.; Giustarini, D. Oxidized Forms of Glutathione in Peripheral Blood as Biomarkers of Oxidative Stress. *Clin. Chem.* **2006**, *52*, 1406–1414. [CrossRef] [PubMed]

34. Giustarini, D.; Dalle-Donne, I.; Milzani, A.; Rossi, R. Detection of Glutathione in Whole Blood after Stabilization with *N*-Ethylmaleimide. *Anal. Biochem.* **2011**, *415*, 81–83. [CrossRef] [PubMed]

35. Steghens, J.P.; Flourie, F.; Arab, K.; Collombel, C. Fast Liquid Chromatography-Mass Spectrometry Glutathione Measurement in Whole Blood: Micromolar Gssg Is a Sample Preparation Artifact. *J. Chromatogr. B Anal. Technol. Biomed. Life Sci.* **2003**, *798*, 343–349. [CrossRef] [PubMed]

36. Monostori, P.; Wittmann, G.; Karg, E.; Turi, S. Determination of Glutathione and Glutathione Disulfide in Biological Samples: An in-Depth Review. *J. Chromatogr. B Anal. Technol. Biomed. Life Sci.* **2009**, *877*, 3331–3346. [CrossRef] [PubMed]

37. Iwasaki, Y.; Saito, Y.; Nakano, Y.; Mochizuki, K.; Sakata, O.; Ito, R.; Saito, K.; Nakazawa, H. Chromatographic and Mass Spectrometric Analysis of Glutathione in Biological Samples. *J. Chromatogr. B Anal. Technol. Biomed. Life Sci.* **2009**, *877*, 3309–3317. [CrossRef] [PubMed]

38. Harwood, D.T.; Kettle, A.J.; Brennan, S.; Winterbourn, C.C. Simultaneous Determination of Reduced Glutathione, Glutathione Disulphide and Glutathione Sulphonamide in Cells and Physiological Fluids by Isotope Dilution Liquid Chromatography-Tandem Mass Spectrometry. *J. Chromatogr. B Anal. Technol. Biomed. Life Sci.* **2009**, *877*, 3393–3399. [CrossRef] [PubMed]

39. Forman, H.J.; Zhang, H.; Rinna, A. Glutathione: Overview of Its Protective Roles, Measurement, and Biosynthesis. *Mol. Asp. Med.* **2009**, *30*, 1–12. [CrossRef] [PubMed]

40.   Yap, L.P.; Sancheti, H.; Ybanez, M.D.; Garcia, J.; Cadenas, E.; Han, D. Determination of Gsh, Gssg, and Gsno Using Hplc with Electrochemical Detection. *Methods Enzymol.* **2010**, *473*, 137–147. [PubMed]
41.   Salmi, H.; Leonard, J.V.; Rahman, S.; Lapatto, R. Plasma Thiol Status Is Altered in Children with Mitochondrial Diseases. *Scand. J. Clin. Lab. Investig.* **2012**, *72*, 152–157. [CrossRef] [PubMed]
42.   Pastore, A.; Martinelli, D.; Piemonte, F.; Tozzi, G.; Boenzi, S.; Di Giovamberardino, G.; Petrillo, S.; Bertini, E.; Dionisi-Vici, C. Glutathione Metabolism in Cobalamin Deficiency Type C (Cblc). *J. Inherit. Metab. Dis.* **2014**, *37*, 125–129. [CrossRef] [PubMed]
43.   Piccolo, G.; Banfi, P.; Azan, G.; Rizzuto, R.; Bisson, R.; Sandona, D.; Bellomo, G. Biological Markers of Oxidative Stress in Mitochondrial Myopathies with Progressive External Ophthalmoplegia. *J. Neurol. Sci.* **1991**, *105*, 57–60. [CrossRef]
44.   Filosto, M.; Tonin, P.; Vattemi, G.; Spagnolo, M.; Rizzuto, N.; Tomelleri, G. Antioxidant Agents Have a Different Expression Pattern in Muscle Fibers of Patients with Mitochondrial Diseases. *Acta Neuropathol.* **2002**, *103*, 215–220. [CrossRef] [PubMed]
45.   Hayasaka, K.; Metoki, K.; Satoh, T.; Narisawa, K.; Tada, K.; Kawakami, T.; Matsuo, N.; Aoki, T. Comparison of Cytosolic and Mitochondrial Enzyme Alterations in the Livers of Propionic or Methylmalonic Acidemia: A Reduction of Cytochrome Oxidase Activity. *Tohoku J. Exp. Med.* **1982**, *137*, 329–334. [CrossRef] [PubMed]
46.   Krahenbuhl, S.; Ray, D.B.; Stabler, S.P.; Allen, R.H.; Brass, E.P. Increased Hepatic Mitochondrial Capacity in Rats with Hydroxy-Cobalamin[C-Lactam]-Induced Methylmalonic Aciduria. *J. Clin. Investig.* **1990**, *86*, 2054–2061. [CrossRef] [PubMed]
47.   Krahenbuhl, S.; Chang, M.; Brass, E.P.; Hoppel, C.L. Decreased Activities of Ubiquinol:Ferricytochrome C Oxidoreductase (Complex Iii) and Ferrocytochrome C:Oxygen Oxidoreductase (Complex Iv) in Liver Mitochondria from Rats with Hydroxycobalamin[C-Lactam]-Induced Methylmalonic Aciduria. *J. Biol. Chem.* **1991**, *266*, 20998–21003. [PubMed]
48.   Tandler, B.; Krahenbuhl, S.; Brass, E.P. Unusual Mitochondria in the Hepatocytes of Rats Treated with a Vitamin B12 Analogue. *Anat. Rec.* **1991**, *231*, 1–6 [CrossRef] [PubMed]
49.   Wajner, M.; Dutra, J.C.; Cardoso, S.E.; Wannmacher, C.M.; Motta, E.R. Effect of Methylmalonate on in Vitro Lactate Release and Carbon Dioxide Production by Brain of Suckling Rats. *J. Inherit. Metab. Dis.* **1992**, *15*, 92–96. [CrossRef] [PubMed]
50.   Dutra, J.C.; Dutra-Filho, C.S.; Cardozo, S.E.; Wannmacher, C.M.; Sarkis, J.J.; Wajner, M. Inhibition of Succinate Dehydrogenase and Beta-Hydroxybutyrate Dehydrogenase Activities by Methylmalonate in Brain and Liver of Developing Rats. *J. Inherit. Metab. Dis.* **1993**, *16*, 147–153. [CrossRef] [PubMed]
51.   McLaughlin, B.A.; Nelson, D.; Silver, I.A.; Erecinska, M.; Chesselet, M.F. Methylmalonate Toxicity in Primary Neuronal Cultures. *Neuroscience* **1998**, *86*, 279–290. [CrossRef]
52.   Brusque, A.M.; Borba Rosa, R.; Schuck, P.F.; Dalcin, K.B.; Ribeiro, C.A.; Silva, C.G.; Wannmacher, C.M.; Dutra-Filho, C.S.; Wyse, A.T.; Briones, P.; et al. Inhibition of the Mitochondrial Respiratory Chain Complex Activities in Rat Cerebral Cortex by Methylmalonic Acid. *Neurochem. Int.* **2002**, *40*, 593–601. [CrossRef]
53.   Okun, J.G.; Horster, F.; Farkas, L.M.; Feyh, P.; Hinz, A.; Sauer, S.; Hoffmann, G.F.; Unsicker, K.; Mayatepek, E.; Kolker, S. Neurodegeneration in Methylmalonic Aciduria Involves Inhibition of Complex Ii and the Tricarboxylic Acid Cycle, and Synergistically Acting Excitotoxicity. *J. Biol. Chem.* **2002**, *277*, 14674–14680. [PubMed]
54.   Chandler, R.J.; Zerfas, P.M.; Shanske, S.; Sloan, J.; Hoffmann, V.; DiMauro, S.; Venditti, C.P. Mitochondrial Dysfunction in Mut Methylmalonic Acidemia. *FASEB J.* **2009**, *23*, 1252–1261. [CrossRef] [PubMed]
55.   Murphy, G.E.; Lowekamp, B.C.; Zerfas, P.M.; Chandler, R.J.; Narasimha, R.; Venditti, C.P.; Subramaniam, S. Ion-Abrasion Scanning Electron Microscopy Reveals Distorted Liver Mitochondrial Morphology in Murine Methylmalonic Acidemia. *J. Struct. Biol.* **2010**, *171*, 125–132. [CrossRef] [PubMed]
56.   Fernandes, C.G.; Borges, C.G.; Seminotti, B.; Amaral, A.U.; Knebel, L.A.; Eichler, P.; de Oliveira, A.B.; Leipnitz, G.; Wajner, M. Experimental Evidence That Methylmalonic Acid Provokes Oxidative Damage and Compromises Antioxidant Defenses in Nerve Terminal and Striatum of Young Rats. *Cell Mol. Neurobiol.* **2011**, *31*, 775–785. [CrossRef] [PubMed]
57.   Valayannopoulos, V.; Hubert, L.; Benoist, J.F.; Romano, S.; Arnoux, J.B.; Chretien, D.; Kaplan, J.; Fakhouri, F.; Rabier, D.; Rotig, A.; et al. Multiple Oxphos Deficiency in the Liver of a Patient with Cbla Methylmalonic Aciduria Sensitive to Vitamin B(12). *J. Inherit. Metab. Dis.* **2009**, *32*, 159–162. [CrossRef] [PubMed]

58. Richard, E.; Jorge-Finnigan, A.; Garcia-Villoria, J.; Merinero, B.; Desviat, L.R.; Gort, L.; Briones, P.; Leal, F.; Perez-Cerda, C.; Ribes, A.; et al. Genetic and Cellular Studies of Oxidative Stress in Methylmalonic Aciduria (Mma) Cobalamin Deficiency Type C (Cblc) with Homocystinuria (Mmachc). *Hum. Mutat.* **2009**, *30*, 1558–1566. [CrossRef] [PubMed]

59. Melo, D.R.; Kowaltowski, A.J.; Wajner, M.; Castilho, R.F. Mitochondrial Energy Metabolism in Neurodegeneration Associated with Methylmalonic Acidemia. *J. Bioenerg. Biomembr.* **2011**, *43*, 39–46. [CrossRef] [PubMed]

60. Richard, E.; Monteoliva, L.; Juarez, S.; Perez, B.; Desviat, L.R.; Ugarte, M.; Albar, J.P. Quantitative Analysis of Mitochondrial Protein Expression in Methylmalonic Acidemia by Two-Dimensional Difference Gel Electrophoresis. *J. Proteome Res.* **2006**, *5*, 1602–1610. [CrossRef] [PubMed]

61. Fontella, F.U.; Pulrolnik, V.; Gassen, E.; Wannmacher, C.M.; Klein, A.B.; Wajner, M.; Dutra-Filho, C.S. Propionic and L-Methylmalonic Acids Induce Oxidative Stress in Brain of Young Rats. *Neuroreport* **2000**, *11*, 541–544. [CrossRef] [PubMed]

62. Brusque, A.M.; Rotta, L.N.; Tavares, R.G.; Emanuelli, T.; Schwarzbold, C.V.; Dutra-Filho, C.S.; de Souza Wyse, A.T.; Duval Wannmacher, C.M.; Gomes de Souza, D.O.; Wajner, M. Effects of Methylmalonic and Propionic Acids on Glutamate Uptake by Synaptosomes and Synaptic Vesicles and on Glutamate Release by Synaptosomes from Cerebral Cortex of Rats. *Brain Res.* **2011**, *920*, 194–201. [CrossRef]

63. Mardach, R.; Verity, M.A.; Cederbaum, S.D. Clinical, Pathological, and Biochemical Studies in a Patient with Propionic Acidemia and Fatal Cardiomyopathy. *Mol. Genet. Metab.* **2005**, *85*, 286–290. [CrossRef] [PubMed]

64. Schwab, M.A.; Sauer, S.W.; Okun, J.G.; Nijtmans, L.G.; Rodenburg, R.J.; van den Heuvel, L.P.; Drose, S.; Brandt, U.; Hoffmann, G.F.; Ter Laak, H.; et al. Secondary Mitochondrial Dysfunction in Propionic Aciduria: A Pathogenic Role for Endogenous Mitochondrial Toxins. *Biochem. J.* **2006**, *398*, 107–112. [CrossRef] [PubMed]

65. De Keyzer, Y.; Valayannopoulos, V.; Benoist, J.F.; Batteux, F.; Lacaille, F.; Hubert, L.; Chretien, D.; Chadefeaux-Vekemans, B.; Niaudet, P.; Touati, G.; et al. Multiple Oxphos Deficiency in the Liver, Kidney, Heart, and Skeletal Muscle of Patients with Methylmalonic Aciduria and Propionic Aciduria. *Pediatr. Res.* **2009**, *66*, 91–95. [CrossRef] [PubMed]

66. Ribeiro, C.A.; Balestro, F.; Grando, V.; Wajner, M. Isovaleric Acid Reduces Na^+, K^+-Atpase Activity in Synaptic Membranes from Cerebral Cortex of Young Rats. *Cell Mol. Neurobiol.* **2007**, *27*, 529–540. [CrossRef] [PubMed]

67. Solano, A.F.; Leipnitz, G.; De Bortoli, G.M.; Seminotti, B.; Amaral, A.U.; Fernandes, C.G.; Latini, A.S.; Dutra-Filho, C.S.; Wajner, M. Induction of Oxidative Stress by the Metabolites Accumulating in Isovaleric Acidemia in Brain Cortex of Young Rats. *Free Radic. Res.* **2008**, *42*, 707–715. [CrossRef] [PubMed]

68. Lehnert, W.; Sass, J.O. Glutaconyl-Coa Is the Main Toxic Agent in Glutaryl-Coa Dehydrogenase Deficiency (Glutaric Aciduria Type I). *Med. Hypotheses.* **2005**, *65*, 330–333. [CrossRef] [PubMed]

69. Magni, D.V.; Furian, A.F.; Oliveira, M.S.; Souza, M.A.; Lunardi, F.; Ferreira, J.; Mello, C.F.; Royes, L.F.; Fighera, M.R. Kinetic Characterization of L-[(3)H]Glutamate Uptake Inhibition and Increase Oxidative Damage Induced by Glutaric Acid in Striatal Synaptosomes of Rats. *Int. J. Dev. Neurosci.* **2009**, *27*, 65–72. [CrossRef] [PubMed]

70. Wajner, M.; Goodman, S.I. Disruption of Mitochondrial Homeostasis in Organic Acidurias: Insights from Human and Animal Studies. *J. Bioenerg. Biomembr.* **2011**, *43*, 31–38. [CrossRef] [PubMed]

71. Treacy, E.; Arbour, L.; Chessex, P.; Graham, G.; Kasprzak, L.; Casey, K.; Bell, L.; Mamer, O.; Scriver, C.R. Glutathione Deficiency as a Complication of Methylmalonic Acidemia: Response to High Doses of Ascorbate. *J. Pediatr.* **1996**, *129*, 445–448. [CrossRef]

72. Seznec, H.; Simon, D.; Bouton, C.; Reutenauer, L.; Hertzog, A.; Golik, P.; Procaccio, V.; Patel, M.; Drapier, J.C.; Koenig, M.; et al. Friedreich Ataxia: The Oxidative Stress Paradox. *Hum. Mol. Genet.* **2005**, *14*, 463–474. [CrossRef] [PubMed]

73. Santos, R.; Lefevre, S.; Sliwa, D.; Seguin, A.; Camadro, J.M.; Lesuisse, E. Friedreich Ataxia: Molecular Mechanisms, Redox Considerations, and Therapeutic Opportunities. *Antioxid. Redox. Signal.* **2010**, *13*, 651–690. [CrossRef] [PubMed]

74. Armstrong, J.S.; Khdour, O.; Hecht, S.M. Does Oxidative Stress Contribute to the Pathology of Friedreich's Ataxia? A Radical Question. *FASEB J.* **2010**, *24*, 2152–2163. [CrossRef] [PubMed]

75. Cotticelli, M.G.; Crabbe, A.M.; Wilson, R.B.; Shchepinov, M.S. Insights into the Role of Oxidative Stress in the Pathology of Friedreich Ataxia Using Peroxidation Resistant Polyunsaturated Fatty Acids. *Redox. Biol.* **2013**, *1*, 398–404. [CrossRef] [PubMed]

76. Murray, J.; Taylor, S.W.; Zhang, B.; Ghosh, S.S.; Capaldi, R.A. Oxidative Damage to Mitochondrial Complex I Due to Peroxynitrite: Identification of Reactive Tyrosines by Mass Spectrometry. *J. Biol. Chem.* **2003**, *278*, 37223–37230. [CrossRef] [PubMed]

77. Hauser, D.N.; Hastings, T.G. Mitochondrial Dysfunction and Oxidative Stress in Parkinson's Disease and Monogenic Parkinsonism. *Neurobiol. Dis.* **2013**, *51*, 35–42. [CrossRef] [PubMed]

78. Ansari, M.A.; Scheff, S.W. Oxidative Stress in the Progression of Alzheimer Disease in the Frontal Cortex. *J. Neuropathol. Exp. Neurol.* **2010**, *69*, 155–167. [CrossRef] [PubMed]

79. Sultana, R.; Mecocci, P.; Mangialasche, F.; Cecchetti, R.; Baglioni, M.; Butterfield, D.A. Increased Protein and Lipid Oxidative Damage in Mitochondria Isolated from Lymphocytes from Patients with Alzheimer's Disease: Insights into the Role of Oxidative Stress in Alzheimer's Disease and Initial Investigations into a Potential Biomarker for This Dementing Disorder. *J. Alzheimers. Dis.* **2011**, *24*, 77–84. [PubMed]

80. Filosa, S.; Pecorelli, A.; D'Esposito, M.; Valacchi, G.; Hajek, J. Exploring the Possible Link between Mecp2 and Oxidative Stress in Rett Syndrome. *Free Radic. Biol. Med.* **2015**, *88*, 81–90. [CrossRef] [PubMed]

81. Murata, T.; Ohtsuka, C.; Terayama, Y. Increased Mitochondrial Oxidative Damage in Patients with Sporadic Amyotrophic Lateral Sclerosis. *J. Neurol. Sci.* **2008**, *267*, 66–69. [CrossRef] [PubMed]

82. Zeevalk, G.D.; Razmpour, R.; Bernard, L.P. Glutathione and Parkinson's Disease: Is This the Elephant in the Room? *Biomed. Pharmacother.* **2008**, *62*, 236–249. [CrossRef] [PubMed]

83. Martin, H.L.; Teismann, P. Glutathione—A Review on Its Role and Significance in Parkinson's Disease. *FASEB J.* **2009**, *23*, 3263–3272. [CrossRef] [PubMed]

84. Rossignol, D.A.; Frye, R.E. Mitochondrial Dysfunction in Autism Spectrum Disorders: A Systematic Review and Meta-Analysis. *Mol. Psychiatry* **2012**, *17*, 290–314. [CrossRef] [PubMed]

85. Frye, R.E.; Melnyk, S.; Macfabe, D.F. Unique Acyl-Carnitine Profiles Are Potential Biomarkers for Acquired Mitochondrial Disease in Autism Spectrum Disorder. *Transl. Psychiatry* **2013**, *3*, e220. [CrossRef] [PubMed]

86. Frye, R.E.; Delatorre, R.; Taylor, H.; Slattery, J.; Melnyk, S.; Chowdhury, N.; James, S.J. Redox Metabolism Abnormalities in Autistic Children Associated with Mitochondrial Disease. *Transl. Psychiatry* **2013**, *3*, e273. [CrossRef] [PubMed]

87. Pastore, A.; Tozzi, G.; Gaeta, L.M.; Giannotti, A.; Bertini, E.; Federici, G.; Digilio, M.C.; Piemonte, F. Glutathione Metabolism and Antioxidant Enzymes in Children with Down Syndrome. *J. Pediatr.* **2003**, *142*, 583–585. [CrossRef] [PubMed]

88. Infantino, V.; Castegna, A.; Iacobazzi, F.; Spera, I.; Scala, I.; Andria, G.; Iacobazzi, V. Impairment of Methyl Cycle Affects Mitochondrial Methyl Availability and Glutathione Level in Down's Syndrome. *Mol. Genet. Metab.* **2011**, *102*, 378–382. [CrossRef] [PubMed]

89. Pagano, G.; Zatterale, A.; Degan, P.; d'Ischia, M.; Kelly, F.J.; Pallardo, F.V.; Calzone, R.; Castello, G.; Dunster, C.; Giudice, A.; et al. In Vivo Prooxidant State in Werner Syndrome (Ws): Results from Three Ws Patients and Two Ws Heterozygotes. *Free Radic. Res.* **2005**, *39*, 529–533. [CrossRef] [PubMed]

90. Song, G.; Napoli, E.; Wong, S.; Hagerman, R.; Liu, S.; Tassone, F.; Giulivi, C. Altered Redox Mitochondrial Biology in the Neurodegenerative Disorder Fragile X-Tremor/Ataxia Syndrome: Use of Antioxidants in Precision Medicine. *Mol. Med.* **2016**, *22*, 548–559. [CrossRef] [PubMed]

91. Zapatero-Solana, E.; Garcia-Gimenez, J.L.; Guerrero-Aspizua, S.; Garcia, M.; Toll, A.; Baselga, E.; Duran-Moreno, M.; Markovic, J.; Garcia-Verdugo, J.M.; Conti, C.J.; et al. Oxidative Stress and Mitochondrial Dysfunction in Kindler Syndrome. *Orphanet. J. Rare Dis.* **2014**, *9*, 211. [CrossRef] [PubMed]

92. Enns, G.M.; Kinsman, S.L.; Perlman, S.L.; Spicer, K.M.; Abdenur, J.E.; Cohen, B.H.; Amagata, A.; Barnes, A.; Kheifets, V.; Shrader, W.D.; et al. Initial Experience in the Treatment of Inherited Mitochondrial Disease with Epi-743. *Mol. Genet. Metab.* **2012**, *105*, 91–102. [CrossRef] [PubMed]

93. Sadun, A.A.; Chicani, C.F.; Ross-Cisneros, F.N.; Barboni, P.; Thoolen, M.; Shrader, W.D.; Kubis, K.; Carelli, V.; Miller, G. Effect of Epi-743 on the Clinical Course of the Mitochondrial Disease Leber Hereditary Optic Neuropathy. *Arch. Neurol.* **2012**, *69*, 331–338. [CrossRef] [PubMed]

94. Martinelli, D.; Catteruccia, M.; Piemonte, F.; Pastore, A.; Tozzi, G.; Dionisi-Vici, C.; Pontrelli, G.; Corsetti, T.; Livadiotti, S.; Kheifets, V.; et al. Epi-743 Reverses the Progression of the Pediatric Mitochondrial Disease—Genetically Defined Leigh Syndrome. *Mol. Genet. Metab.* **2012**, *107*, 383–388. [CrossRef] [PubMed]

95. Klein, M. Personal communication; manuscript in preparation. In Presented in part at the United Mitochondrial Disease Foundation Annual Meeting, Seattle, WA, USA, 15–18 June 2016.

96. Luhl, S.; Bode, H.; Schlotzer, W.; Bartsakoulia, M.; Horvath, R.; Abicht, A.; Stenzel, M.; Kirschner, J.; Grunert, S.C. Novel Homozygous Rars2 Mutation in Two Siblings without Pontocerebellar Hypoplasia—Further Expansion of the Phenotypic Spectrum. *Orphanet. J. Rare Dis.* **2016**, *11*, 140. [CrossRef] [PubMed]

97. Martinelli, D.; Catteruccia, M.; Klein, M.; Bevivino, E.; Thoolen, M.; Piemonte, F.; Pastore, A.; Tozzi, G.; Pontrelli, G.; Bertini, E.; et al. Epi-743 Reduces Seizure Frequency in Rars2 Defect Syndrome. *J. Inherit. Metab. Dis.* **2013**, *36*, S110.

98. Lynch, D.R.; Willi, S.M.; Wilson, R.B.; Cotticelli, M.G.; Brigatti, K.W.; Deutsch, E.C.; Kucheruk, O.; Shrader, W.; Rioux, P.; Miller, G.; et al. A0001 in Friedreich Ataxia: Biochemical Characterization and Effects in a Clinical Trial. *Mov. Disord.* **2012**, *27*, 1026–1033. [CrossRef] [PubMed]

99. Klein, M. Personal communication; manuscript in preparation. In Presented at the Friedreich's Ataxia Research Alliance Annual Meeting, Tampa, FL, USA, June 2016.

100. Sullivan, K.; Freeman, M.; Shaw, J.; Gooch, C.; Huang, Y.; Klein, M.; Miller, G.; Zesiewicz, T. Epi-743 for Friedreichs Ataxia Patients with Point Mutations. *Neurology* **2016**, *86*. Supplement P5.388.

101. Hayek, J. Epi-743 in Rett Syndrome: Improved Head Growth in a Randomized Double-Blind Placebo-Controlled Trial. In Proceedings of the 4th European Congress on Rett Syndrome, Rome, Italy, 30 October–1 November 2015.

102. Zesiewicz, T.; Allison, K.; Jahan, I.; Shaw, J.; Murtagh, F.; Jones, T.; Gooch, C.; Salemi, J.; Klein, M.; Miller, G.; et al. Epi-743 Improves Motor Function and Cns Biomarkers in Pd: Results from a Phase 2a Pilot Trial. *Neurology* **2016**, *86*. Supplement I1.012.

103. Shrader, W.D.; Amagata, A.; Barnes, A.; Enns, G.M.; Hinman, A.; Jankowski, O.; Kheifets, V.; Komatsuzaki, R.; Lee, E.; Mollard, P.; et al. Alpha-Tocotrienol Quinone Modulates Oxidative Stress Response and the Biochemistry of Aging. *Bioorg. Med. Chem. Lett.* **2011**, *21*, 3693–3698. [CrossRef] [PubMed]

104. Jacquier-Sarlin, M.R.; Polla, B.S.; Slosman, D.O. Oxido-Reductive State: The Major Determinant for Cellular Retention of Technetium-99m-Hmpao. *J. Nucl. Med.* **1996**, *37*, 1413–1416. [PubMed]

105. Subramony, S.H.; May, W.; Lynch, D.; Gomez, C.; Fischbeck, K.; Hallett, M.; Taylor, P.; Wilson, R.; Ashizawa, T.; Group Cooperative Ataxia. Measuring Friedreich Ataxia: Interrater Reliability of a Neurologic Rating Scale. *Neurology* **2005**, *64*, 1261–1262. [CrossRef] [PubMed]

106. Movement Disorder Society Task Force on Rating Scales for Parkinson's Disease. The Unified Parkinson's Disease Rating Scale (Updrs): Status and Recommendations. *Mov. Disord.* **2003**, *18*, 738–750.

107. Atkuri, K.R.; Mantovani, J.J.; Herzenberg, L.A.; Herzenberg, L.A. N-Acetylcysteine—A Safe Antidote for Cysteine/Glutathione Deficiency. *Curr. Opin. Pharmacol.* **2007**, *7*, 355–359. [CrossRef] [PubMed]

108. Besouw, M.; Masereeuw, R.; van den Heuvel, L.; Levtchenko, E. Cysteamine: An Old Drug with New Potential. *Drug. Discov. Today* **2013**, *18*, 785–792. [CrossRef] [PubMed]

109. Harrison, P.M.; Keays, R.; Bray, G.P.; Alexander, G.J.; Williams, R. Improved Outcome of Paracetamol-Induced Fulminant Hepatic Failure by Late Administration of Acetylcysteine. *Lancet* **1990**, *335*, 1572–1573. [CrossRef]

110. Harrison, P.M.; Wendon, J.A.; Gimson, A.E.; Alexander, G.J.; Williams, R. Improvement by Acetylcysteine of Hemodynamics and Oxygen Transport in Fulminant Hepatic Failure. *N. Engl. J. Med.* **1991**, *324*, 1852–1857. [CrossRef] [PubMed]

111. Stravitz, R.T.; Sanyal, A.J.; Reisch, J.; Bajaj, J.S.; Mirshahi, F.; Cheng, J.; Lee, W.M.; Group Acute Liver Failure Study. Effects of N-Acetylcysteine on Cytokines in Non-Acetaminophen Acute Liver Failure: Potential Mechanism of Improvement in Transplant-Free Survival. *Liver Int.* **2013**, *33*, 1324–1331. [CrossRef] [PubMed]

112. Hilgier, W.; Wegrzynowicz, M.; Ruszkiewicz, J.; Oja, S.S.; Saransaari, P.; Albrecht, J. Direct Exposure to Ammonia and Hyperammonemia Increase the Extracellular Accumulation and Degradation of Astroglia-Derived Glutathione in the Rat Prefrontal Cortex. *Toxicol. Sci.* **2010**, *117*, 163–168. [CrossRef] [PubMed]

113. Gimenez-Garzo, C.; Urios, A.; Agusti, A.; Gonzalez-Lopez, O.; Escudero-Garcia, D.; Escudero-Sanchis, A.; Serra, M.A.; Giner-Duran, R.; Montoliu, C.; Felipo, V. Is Cognitive Impairment in Cirrhotic Patients Due to Increased Peroxynitrite and Oxidative Stress? *Antioxid. Redox. Signal.* **2015**, *22*, 871–877. [CrossRef] [PubMed]

114. Moreira, P.I.; Harris, P.L.; Zhu, X.; Santos, M.S.; Oliveira, C.R.; Smith, M.A.; Perry, G. Lipoic Acid and N-Acetyl Cysteine Decrease Mitochondrial-Related Oxidative Stress in Alzheimer Disease Patient Fibroblasts. *J. Alzheimers Dis.* **2007**, *12*, 195–206. [CrossRef] [PubMed]

115. Wright, D.J.; Renoir, T.; Smith, Z.M.; Frazier, A.E.; Francis, P.S.; Thorburn, D.R.; McGee, S.L.; Hannan, A.J.; Gray, L.J. N-Acetylcysteine Improves Mitochondrial Function and Ameliorates Behavioral Deficits in the R6/1 Mouse Model of Huntington's Disease. *Transl. Psychiatry* **2015**, *5*, e492. [CrossRef] [PubMed]

116. Douiev, L.; Soiferman, D.; Alban, C.; Saada, A. The Effects of Ascorbate, *N*-Acetylcysteine, and Resveratrol on Fibroblasts from Patients with Mitochondrial Disorders. *J. Clin. Med.* **2017**, *6*, 1. [CrossRef] [PubMed]

117. Niemi, A.K.; Enns, G.M. The Role of N-Acetylcysteine in Treating Mitochondrial Liver Disease. *ASHG Annual Meeting Program Guide.* 2012, p. 140. Available online: http://www.ashg.org/2012meeting/abstracts/fulltext/f120121092.htm (accessed on 3 May 2017).

118. Deepmala; Slattery, J.; Kumar, N.; Delhey, L.; Berk, M.; Dean, O.; Spielholz, C.; Frye, R. Clinical Trials of N-Acetylcysteine in Psychiatry and Neurology: A Systematic Review. *Neurosci. Biobehav. Rev.* **2015**, *55*, 294–321.

119. Viscomi, C.; Burlina, A.B.; Dweikat, I.; Savoiardo, M.; Lamperti, C.; Hildebrandt, T.; Tiranti, V.; Zeviani, M. Combined Treatment with Oral Metronidazole and *N*-Acetylcysteine Is Effective in Ethylmalonic Encephalopathy. *Nat. Med.* **2010**, *16*, 869–871. [CrossRef] [PubMed]

120. Gibrat, C.; Cicchetti, F. Potential of Cystamine and Cysteamine in the Treatment of Neurodegenerative Diseases. *Prog. Neuropsychopharmacol. Biol. Psychiatry* **2011**, *35*, 380–389. [CrossRef] [PubMed]

121. Dubinsky, R.; Gray, C. Cyte-I-Hd: Phase I Dose Finding and Tolerability Study of Cysteamine (Cystagon) in Huntington's Disease. *Mov. Disord.* **2006**, *21*, 530–533. [CrossRef] [PubMed]

122. Dohil, R.; Rioux, P. Pharmacokinetic Studies of Cysteamine Bitartrate Delayed-Release. *Clin. Pharmacol. Drug Dev.* **2013**, *2*, 178–185. [CrossRef] [PubMed]

Journal of
Clinical Medicine

MDPI

Review

The Value of Coenzyme Q$_{10}$ Determination in Mitochondrial Patients

Delia Yubero [1], George Allen [2], Rafael Artuch [1,*] and Raquel Montero [1]

[1] Clinical Biochemistry and Molecular Medicine Department, Institut de Recerca Sant Joan de Déu and
    CIBERER-ISCIII, Passeig Sant Joan de Déu, 2, 08950 Esplugues, Barcelona, Spain;
    dyubero@hsjdbcn.org (D.Y.); rmontero@hsjdbcn.org (R.M.)
[2] Department of Blood Sciences, Royal Devon and Exeter NHS Foundation Trust, Exeter EX2 5DW, UK;
    george.allen@nhs.net
* Correspondence: rartuch@hsjdbcn.org; Tel.: +34-93-280-6169

Academic Editor: Iain P. Hargreaves
Received: 28 February 2017; Accepted: 17 March 2017; Published: 24 March 2017

Abstract: Coenzyme Q$_{10}$ (CoQ) is a lipid that is ubiquitously synthesized in tissues and has a key role in mitochondrial oxidative phosphorylation. Its biochemical determination provides insight into the CoQ status of tissues and may detect CoQ deficiency that can result from either an inherited primary deficiency of CoQ metabolism or may be secondary to different genetic and environmental conditions. Rapid identification of CoQ deficiency can also allow potentially beneficial treatment to be initiated as early as possible. CoQ may be measured in different specimens, including plasma, blood mononuclear cells, platelets, urine, muscle, and cultured skin fibroblasts. Blood and urinary CoQ also have good utility for CoQ treatment monitoring.

Keywords: coenzyme Q$_{10}$ deficiency; mitochondrial diseases; treatment monitoring

1. Introduction

Coenzyme Q$_{10}$ (CoQ) is a lipid that acts in the mitochondrial respiratory chain (MRC) as the electron transporter from Enzymatic Complexes I and II to Complex III. Recognized biological functions of CoQ include an essential role in energy biosynthesis in the form of ATP, free radical detoxification, stabilization of mitochondrial enzymatic complexes, binding to the permeability transition pore, and the function of mitochondrial uncoupling proteins [1]. Almost all cells have the capacity for CoQ, and at least 13 genes have been shown to be required for endogenous production of CoQ. Furthermore, additional genes also influence CoQ availability such as those related with acetyl-CoA metabolism or those that can cause secondary reduction in CoQ biosynthesis or increase its degradation [2,3]. CoQ deficiency has been associated with different clinical phenotypes and genetic conditions [4] and environmental factors can also influence CoQ availability [5]. Regardless of the cause, the impairment of CoQ status can result in profound deficits to mitochondrial function. Treatment with CoQ supplementation can result in clinical improvement in CoQ deficiency, and early measurement of CoQ status is therefore of fundamental importance to allow the rapid initiation of treatment. Unfortunately, response to CoQ supplementation in trials with other mitochondrial disorders has been disappointing [6,7]. However, the lack of efficacy could potentially relate to delayed treatment; consequently, further studies are needed to ascertain whether early identification of CoQ deficiency in mitochondrial patients may help identify those in whom CoQ supplementation may yet prove beneficial [7].

2. Diagnostic Issues of CoQ Deficiency Syndromes

Currently, it is known that 8 of the 13 genes related to CoQ biosynthesis (*COQ* genes) can cause human disease [4], but these primary conditions are extremely rare. However, secondary CoQ deficiency is a common feature in a range of diseases. This susceptibility may be due to the intricate mechanisms and biological functions in which CoQ participates. Secondary deficiencies can occur in mitochondrial oxidative phosphorylation (OXPHOS) disorders [2,6,7] and in a broad spectrum of non-OXPHOS disorders [3]. Interestingly, the prevalence of muscle CoQ deficiency was demonstrated to be similar for both OXPHOS and non-OXPHOS disease patients [3]. Furthermore, it has been suggested that CoQ status may be an accurate predictor of deficient activity of MRC components [8,9], so routine CoQ measurement within the diagnostic workflow of OXPHOS disease seems advisable, especially for muscle biopsies.

Primary CoQ deficiency is considered a rare mitochondrial disorder associated with a heterogeneous clinical phenotype [4]. Nevertheless, clinical identification of potential cases is of paramount importance to initiate investigations that may provide early diagnosis and initiation of specific treatment, especially as some CoQ-deficient patients respond well to high oral doses of CoQ [10]. The clinical picture in primary CoQ deficiency can include ataxia with cerebellar involvement (the most common phenotype of CoQ deficiency syndromes), multiple organ failure in neonatal-onset forms, kidney disease, deafness, or muscular involvement [4], amongst others. The biochemical findings consist of a variable degree of CoQ deficiency in tissues (muscle/fibroblasts), which in turn may cause a reduction in the activity of the CoQ-dependent mitochondrial respiratory chain (Complexes I + III and Complexes II + III in muscle, Complexes glycerol-3-phosphate (G3P) + III and Complexes II + III in fibroblasts). However, it is not possible to biochemically distinguish between primary and secondary CoQ deficiencies nor to identify candidate genes for mutational analysis [11].

The initial stage of laboratory analysis is the biochemical identification of CoQ deficiency. Reduced activities of CoQ-dependent enzymes are indicative of CoQ deficiency, suggesting a decrease in electron transfer related to the quinone pool. This is supported by the restoration of Complex II + III activity after incubation with exogenous ubiquinone [12–14]. Nevertheless, direct quantitative measurement of CoQ levels is the most reliable test for diagnosis [15]. Essential to this is the choice of tissue for analysis. This may often be a balance between obtaining the most reliable sample for CoQ measurement and minimizing invasive procedures. However, some specimens such as plasma may not be suitable for the diagnosis of primary CoQ deficiency since misleading partial restoration of CoQ values from dietary sources of CoQ can occur in plasma.

After establishing the biochemical diagnosis, the next step is to identify the specific genetic defect. Next-generation sequencing has largely replaced the need to serially sequence individual *COQ* genes and other genes associated with secondary deficiency and thus has profoundly changed the diagnostic process [11]. Nevertheless, biochemical measurements still play an important role in the diagnostic pathway by providing rapid and reliable demonstration of CoQ deficiency that allows early treatment initiation.

In this chapter, we will review the state of the art in CoQ measurement, utilizing different biological specimens for the investigation of mitochondrial disorders for both diagnosis and therapeutic follow-up. Additionally, we will highlight the advantages and pitfalls of CoQ determination in such specimens.

3. CoQ Determination in Biological Samples. What Can We Expect?

CoQ is ubiquitously synthesized and found in almost all human cells, with higher CoQ concentrations found in organs with high-energy demand and metabolic rate. The measurement of both reduced and oxidized forms of CoQ allows for the determination of total CoQ, and this provides an optimal measure to detect CoQ deficiencies. CoQ levels in a range of specimen types from patients with mitochondrial diseases have been demonstrated to be lower than control values [6,7,16,17]. However, the particular CoQ distribution in distinct cellular fractions and the complexity of biological matrices makes the biological sample choice and preparation a critical step in the CoQ quantification process [18]. Additionally, since CoQ deficiency may be tissue-specific [19], invasive procedures are frequently needed in order to assess endogenous CoQ in the target organ, especially in muscle. Thus, it can be of value to analyze CoQ status in a full range of sample types, as a deficiency may remain undetected if the appropriate specimen is not chosen. Table 1 summarizes the different biological specimens and technical approaches for accurate CoQ determination.

Table 1. Advantages and limitations for the CoQ analysis in different biological specimens.

Tissue	Advantages	Limitations
Plasma	Minimally invasive Identification of secondary CoQ deficiencies CoQ treatment monitoring	Low diagnostic yield for CoQ deficiency in mitochondrial disorders CoQ values modified by external sources
Leukocytes Platelets	Minimally invasive Correlation with CoQ tissue levels CoQ treatment monitoring	Fresh preparation Time-consuming Few reported experiences in mitochondrial disorders.
Muscle	Good diagnostic yield for CoQ deficiency Other mitochondrial studies can be performed	Invasive No treatment monitoring
Fibroblasts	Good diagnostic yield for some CoQ deficiencies Functional studies can be performed (CoQ biosynthesis) Unlimited biological material for further studies	False negative results in some cases
Urine	Non-invasive Easily detectable CoQ values Treatment monitoring purposes	Correlation with kidney CoQ status remains to be established

Note: Coenzyme Q_{10} (CoQ).

3.1. Blood Plasma

Plasma CoQ is influenced by both dietary intake and hepatic biosynthesis [20]. Exogenous CoQ is absorbed through the gut by a complex process that can involve both active and passive mechanisms [21]. CoQ is then redistributed via the blood linked to cholesterol transporter lipoproteins [22] that act as the major carrier of CoQ in circulation [23]. The first step of patient CoQ estimation may be based on plasma measurement. However, CoQ status in plasma can be affected by both dietary supply and lipoprotein concentration. It is noteworthy that dietary sources of CoQ can significantly influence plasma CoQ concentrations, contributing up to 25% of the total amount [23]. For this reason, it has been suggested that plasma CoQ evaluation is not reliable for the diagnosis of primary CoQ deficiencies [24] as partial correction of CoQ levels may occur due to dietary consumption of CoQ or increases in cholesterol availability. Indeed, in most patients with primary CoQ deficiencies, plasma CoQ values are normal. Conversely, a reduction of plasma CoQ is not frequently observed in the general population, and although there is no demonstrated correlation of plasma and tissue CoQ values, decreased levels may reliably indicate secondary CoQ deficiencies associated with diseases such as phenylketonuria and lysosomal storage diseases [25–30]. Whether plasma CoQ can be used to indicate deficient tissue CoQ status in patients with mitochondrial disorders remains unknown at present.

While the usefulness of plasma CoQ analysis for the diagnosis of CoQ deficiency remains to be established, plasma CoQ determination has a critical role for CoQ treatment monitoring [31]. CoQ therapy is commonly used for the treatment of mitochondrial disorders and the follow-up of these patients should include regular plasma CoQ quantification. This allows for informed adjustment of the oral CoQ dose, the control of treatment compliance and confirmation of adequate CoQ intestinal absorption. The degree of distribution of this supplemented plasma CoQ from blood to affected tissues remains to be demonstrated and is still a matter of debate.

3.2. Blood Cells

Since analysis of plasma CoQ has limitations for diagnosis and avoidance of invasive first-step diagnostic procedures for the investigation of mitochondrial patients is desirable, a range of studies have tested the reliability of CoQ determination in different blood cell types. Remarkably, leukocyte CoQ levels correlate well with that of skeletal muscle and therefore represent a good alternative to evaluate endogenous CoQ [20]. Moreover, these cells can also display changes in cellular CoQ status upon CoQ supplementation [32]. This approach should be applicable for the identification of some patients with primary CoQ deficiencies. However, there is a lack of experience with these cells in most specialist clinical chemistry laboratories. Reference values have been reported [33], but no large-scale studies of mitochondrial diseases have been published. Unfortunately, the utility of this specimen currently remains constrained by technical limitations, including difficulties concerning sample collection and processing.

Similarly, platelet CoQ evaluation seems to be a good indicator of mitochondrial electron transport chain function [34,35]. Some studies have reported platelets as being a useful material for the determination of cellular CoQ content and of great utility for clinical monitoring CoQ treatment [36,37]. Although no reference values have been established, it is of note that platelet CoQ measurement may be advantageous compared to plasma during CoQ supplementation by providing a more representative measurement of cellular uptake and steady-state conditions [37]. However, even though detailed information about sample preparation and methods of detection are available, no studies of a large patient series have been published in relation to platelet CoQ evaluation for diagnosis or for follow-up in mitochondrial diseases.

Another possibility is to employ lymphoblastoid cell lines [11]. These cells combine the advantages and disadvantages of mononuclear cells and fibroblasts (see below) in that they do not require invasive procedures and allow for functional measurements. However, experience with these cells in clinical laboratories is very limited and the immortalization procedure may cause artifacts. In particular, immortalized cells tend to compensate the CoQ deficiency by overexpressing components of the CoQ biosynthetic machinery, so this may mask partial deficiencies.

3.3. Muscle

Muscle biopsy material continues to be the best current option for investigations of CoQ status in mitochondrial disorders, with a biochemical diagnosis of CoQ deficiency indicated from the measurement of total CoQ content in muscle homogenates. The main advantage of this biological specimen is that other biochemical and histological studies may be conducted in parallel to evaluate the functional condition of the mitochondria, including measurement of mitochondrial respiratory chain enzyme activities, expression and assembly of mitochondrial complexes, and an assessment of the characteristic histopathological features of mitochondrial diseases [8]. Typically, a part of the muscle biopsy is processed in fresh conditions for histopathological and functional studies whilst the remaining sample is frozen immediately at $-80\,^{\circ}$C until analysis. The frozen samples are useful for CoQ determination and some other mitochondrial studies, with a minimum of 20–40 mg of muscle biopsy required for CoQ measurement [8].

Muscle CoQ deficiency is relatively common in patients with mitochondrial disorders [2–8]. Biochemically, it may be concomitantly present with decreased Complexes I + III and Complexes II + III activities, although other MRC complexes may be affected [8]. However, not all patients with muscle CoQ deficiency show MRC abnormalities, supporting the role of this molecule in other essential biological processes beyond energy production [2,17]. Some investigators have proposed CoQ redox status as a biomarker for oxidative stress [8,38]. However, this requires extensive and complicated investigations and is unnecessary to establish the diagnosis of CoQ deficiency.

In the last few years, a large list of mutated genes in patients who display secondary muscle CoQ deficiency has been reported [2–7]. All of these authors concluded that muscle CoQ deficiency is a frequent finding in mitochondrial disorders in general, with primary CoQ deficiency a much rarer condition. Clearly, some of these patients may benefit from CoQ supplementation aimed at restoring CoQ values and thus improving clinical outcomes [39]. Recently, it has been demonstrated that muscle CoQ values are a good predictor of MRC enzyme function with a utility at least equal to citrate synthase activity [9]. This provides added value for muscle CoQ analysis in the investigation of mitochondrial disorders.

3.4. Fibroblasts

Skin fibroblasts are also a good specimen for demonstrating CoQ deficiency [40]. CoQ content and MRC enzyme activities can be measured in fibroblasts alongside different in vitro studies for assessing CoQ synthesis and other metabolic pathways [41,42]. Fibroblast CoQ deficiency may be accompanied by decreased activities of Complexes G3P + III and Complexes II + III that may or may not be associated with deficiency of other MRC complexes [17].

Fibroblasts are usually obtained from minimally invasive skin punch biopsies and after culturing in standard medium (Dulbecco's Modified Eagle Medium (DMEM) containing 10% Fetal Bovine Serum (FBS) and 1% penicillin-streptomycin) [40], these cells are amenable to biobank storage, making them a very valuable material for future investigations. To analyze CoQ content, cultured skin fibroblasts are homogenized and lipids are extracted before CoQ quantification [43]. In addition, in vitro investigation of CoQ biosynthetic capacity can be performed in cultured fibroblasts. These assays are used to analyze the incorporation of radiolabeled substrates into CoQ, such as [3H]-mevalonate and [14C]-4-hydroxybenzoate, or that of stable isotopes with the measurement of synthesized CoQ by high pressure liquid chromatography (HPLC) with radiometric or tandem mass spectometry (MS–MS) detection [41,44,45]. These studies have been demonstrated to be very useful in discriminating between a primary CoQ deficiency, where the CoQ biosynthesis downstream of the provided substrates is impaired, and secondary CoQ deficiencies, where other mechanisms leading to CoQ deficiency are expected [41]. However, some limitations inherent to fibroblasts have been reported. For example, patients with muscle CoQ deficiency may show normal CoQ values in fibroblasts [46], and secondary fibroblast CoQ deficiency has been described in patients with other non-mitochondrial diseases [41].

Other investigations in cultured skin fibroblast have provided insights about key pathophysiological aspects implicated in CoQ deficiency, such as impairment of ATP synthesis or increased free radical damage [47,48]. Other pathophysiological mechanisms proposed include the implication of CoQ in pyrimidine biosynthesis (demonstrated in *COQ2*-mutant fibroblasts) [49] and the potential induction of mitophagy in response to CoQ deficiency [50]. Cotan et al. [51] reported that MELAS fibroblasts show a significant reduction of the mitochondrial membrane potential associated with secondary CoQ deficiency, which triggers mitochondrial degradation by mitophagy. Fragaki et al. [52] also observed a secondary mitochondrial dysfunction in ganglioside GM3 synthase (EC:2.4.99.9) deficient patients (including decrease in Complexes I + III and Complexes II + III in the liver among other MRC abnormalities in fibroblasts), leading to the impairment of normal mitochondrial electron flow and proton pumping, including a drop in mitochondrial membrane potential and an increase in apoptosis.

3.5. Urine

Mitochondrial diseases can be associated with renal involvement. Interestingly, primary CoQ deficiency patients can present with a nephrotic syndrome either in isolation or in combination with other clinical signs [53,54]. This is perhaps not surprising given that renal tubules and glomerular podocytes are rich in mitochondria, allowing them to satisfy a high metabolic demand. Moreover, dramatic clinical improvement of patients with renal disease and CoQ deficiency has been observed following CoQ supplementation [10]. Urinary tract CoQ analysis could be an appropriate approach to assessing kidney CoQ status and may help fulfill the critical need for less invasive procedures to determine tissue CoQ status. Recently, a new methodology for the measurement of CoQ in urine has been standardized, including the establishment of reference values for a pediatric control population [55]. This new evaluation of urinary tract CoQ is a noninvasive procedure that might be useful for estimating CoQ kidney status for diagnosis and especially for CoQ treatment monitoring.

3.6. Other Biological Samples

CoQ is present in all tissues with its abundance in individual tissues associated with energy requirements or metabolic activity. Consequently, CoQ content displays a great variability between different organs and even between cells within the same organ [33]. Several authors have reported reduced CoQ levels in other tissues such as liver or kidney in patients with mitochondrial diseases [6,56,57]. However, these procedures require invasive biopsy procedures that can only be justified if other options for investigations of CoQ status have been exhausted. Experimental measurement of CoQ has also been reported in cardiac tissue obtained from patients undergoing heart transplantation [58]. This study found moderate decreases in CoQ in patients with heart failure in comparison to those without heart failure.

CoQ deficiency can present with profound neurological features, so measurement of CoQ in cerebrospinal fluid (CSF) has the potential to provide important clinical insight as an indicator of brain CoQ status. The concentration of CSF CoQ is in the low nanomolar range and therefore requires specialist analysis by tandem mass spectrometry [59]. Reference ranges have been established for CoQ in CSF, and a suggested application for this technique is in the identification of cerebral CoQ deficiency, although pathological samples were not reported [59].

Very recently, a new approach for CoQ determination has been described using non-invasive mouth swab collection of buccal mucosa cells for CoQ measurement by micro-HPLC. This technique may prove to be valuable for monitoring pediatric patients [60].

4. CoQ Quantification: Technical Aspects

The gold standard procedure for biochemical diagnosis of CoQ deficiency is the analysis of CoQ concentration by HPLC with ultraviolet or electrochemical detection [11]. Recently, new procedures for CoQ determination have been developed based on liquid chromatography-tandem mass spectrometry [41], allowing not only CoQ quantification but also an estimation of the CoQ biosynthetic rate in fibroblast cell cultures incubated with adequate CoQ precursors. Methodological approaches for CoQ measurement are reviewed in another chapter of this issue. Typical CoQ chromatograms from serum, urine, muscle, and cultured skin fibroblasts are depicted in Figure 1.

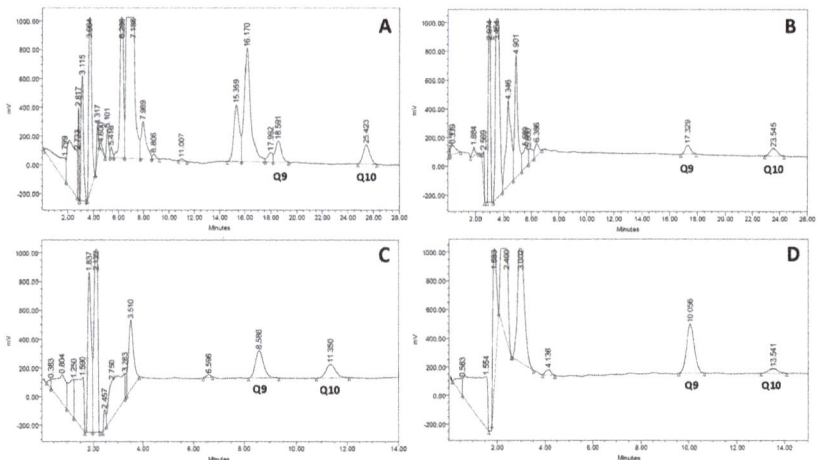

Figure 1. Normal Coenzime Q_{10} (CoQ) chromatograms of different biological specimens. (**A**) serum; (**B**) urine; (**C**) muscle; (**D**) cultured skin fibroblasts. In each specimen, type Q_9 and Q_{10} have a different retention time that is related to differences in sample matrices and the high-pressure liquid chromatography (HPLC) column length required for separation.

5. Conclusions

CoQ is a molecule involved in multiple essential biological functions mainly within the mitochondria. The intricate metabolic pathways related to CoQ biosynthesis and metabolism underlie a vulnerability to frequent reductions in the concentration of this molecule as a consequence of different disease states, but are especially relevant in mitochondrial disorders. Because of this, the measurement of CoQ status in different biological specimens can be considered an essential part of the diagnostic and research workflows for patients with mitochondrial disorders and for CoQ treatment monitoring. This is particularly crucial as CoQ deficiency can be a treatable condition in some cases, so early recognition of the CoQ-deficient status is important to allow for the commencement of CoQ therapy as soon as possible.

Acknowledgments: This work was supported by grants from the Ministerio de Economia y Competitividad de España (FIS PI14/00005, PI14/00028) and Federación Española de Enfermedades Raras (FEDER) Funding Program from the European Union. The "Centro de Investigación Biomédica en Red de Enfermedades Raras (CIBERER)" is an initiative of the Instituto de Salud Carlos III.

Author Contributions: D.Y., G.A., R.A., and R.M. all contributed to the manuscript and have critically reviewed it.

Conflicts of Interest: The authors declare no conflict of interest.

References

1. Genova, M.L.; Lenaz, G. New developments on the functions of coenzyme Q in mitochondria. *Biofactors* **2011**, *37*, 330–354. [CrossRef] [PubMed]
2. Emmanuele, V.; López, L.C.; Berardo, A.; Naini, A.; Tadesse, S.; Wen, B.; D'Agostino, E.; Solomon, M.; DiMauro, S.; Quinzii, C.; et al. Heterogeneity of coenzyme Q10 deficiency: Patient study and literature review. *Arch. Neurol.* **2012**, *69*, 978–983. [CrossRef] [PubMed]
3. Yubero, D.; Montero, R.; Martín, M.A.; Montoya, J.; Ribes, A.; Grazina, M.; Trevisson, E.; Rodriguez-Aguilera, J.C.; Hargreaves, I.P.; Salviati, L.; et al. CoQ deficiency study group. Secondary coenzyme Q10 deficiencies in oxidative phosphorylation (OXPHOS) and non-OXPHOS disorders. *Mitochondrion* **2016**, *30*, 51–58. [CrossRef] [PubMed]
4. Desbats, M.A.; Lunardi, G.; Doimo, M.; Trevisson, E.; Salviati, L. Genetic bases and clinical manifestations of coenzyme Q10 (CoQ 10) deficiency. *J. Inherit. Metab. Dis.* **2015**, *38*, 145–156. [CrossRef] [PubMed]
5. Miles, M.O.; Putnam, P.E.; Miles, L.; Tang, P.H.; De Grauw, A.J.; Wong, B.L.; Horn, P.S.; Foote, H.L.; Rothenberg, M.E. Acquired coenzyme Q10 deficiency in children with recurrent food intolerance and allergies. *Mitochondrion* **2011**, *11*, 127–135. [CrossRef] [PubMed]
6. Sacconi, S.; Trevisson, E.; Salviati, L.; Aymé, S.; Rigal, O.; Redondo, A.G.; Mancuso, M.; Siciliano, G.; Tonin, P.; Angelini, C.; et al. Coenzyme Q10 is frequently reduced in muscle of patients with mitochondrial myopathy. *Neuromuscul. Disord.* **2010**, *20*, 44–48. [CrossRef] [PubMed]
7. Montero, R.; Grazina, M.; López-Gallardo, E.; Montoya, J.; Briones, P.; Navarro-Sastre, A.; Land, J.M.; Hargreaves, I.P.; Artuch, R.; Coenzyme Q10 Deficiency Study Group. Coenzyme Q10 deficiency in mitochondrial DNA depletion syndromes. *Mitochondrion* **2013**, *13*, 337–341. [CrossRef] [PubMed]
8. Miles, M.V.; Miles, L.; Tang, P.H.; Horn, P.S.; Steele, P.E.; DeGrauw, A.J.; Wong, B.L.; Bove, K.E. Systematic evaluation of muscle coenzyme Q10 content in children with mitochondrial respiratory chain enzyme deficiencies. *Mitochondrion* **2008**, *8*, 170–180. [CrossRef] [PubMed]
9. Yubero, D.; Adin, A.; Montero, R.; Jou, C.; Jiménez-Mallebrera, C.; García-Cazorla, A.; Nascimento, A.; O'Callaghan, M.M.; Montoya, J.; Gort, L.; et al. A statistical algorithm showing coenzyme Q10 and citrate synthase as biomarkers for mitochondrial respiratory chain enzyme activities. *Sci. Rep.* **2016**, *6*, 15. [CrossRef] [PubMed]
10. Montini, G.; Malaventura, C.; Salviati, L. Early coenzyme Q10 supplementation in primary coenzyme Q10 deficiency. *N. Engl. J. Med.* **2008**, *358*, 2849–2850. [CrossRef] [PubMed]
11. Yubero, D.; Montero, R.; Armstrong, J.; Espinós, C.; Palau, F.; Santos-Ocaña, C.; Salviati, L.; Navas, P.; Artuch, R. Molecular diagnosis of coenzyme Q10 deficiency. *Expert Rev. Mol. Diagn.* **2015**, *15*, 1049–1059. [CrossRef] [PubMed]

12. Leshinsky-Silver, E.; Levine, A.; Nissenkorn, A.; Barash, V.; Perach, M.; Buzhaker, E.; Shahmurov, M.; Polak-Charcon, S.; Lev, D.; Lerman-Sagie, T. Neonatal liver failure and Leigh syndrome possibly due to CoQ-responsive OXPHOS deficiency. *Mol. Genet. Metab.* **2003**, *79*, 288–293. [CrossRef]

13. Lerman-Sagie, T.; Rustin, P.; Lev, D.; Yanoov, M.; Leshinsky-Silver, E.; Sagie, A.; Ben-Gal, T.; Munnich, A. Dramatic improvement in mitochondrial cardiomyopathy following treatment with idebenone. *J. Inherit. Metab. Dis.* **2001**, *24*, 28–34. [CrossRef] [PubMed]

14. Fragaki, K.; Cano, A.; Benoist, J.F.; Rigal, O.; Chaussenot, A.; Rouzier, C.; Bannwarth, S.; Caruba, C.; Chabrol, B.; Paquis-Flucklinger, V. Fatal heart failure associated with CoQ10 and multiple OXPHOS deficiency in a child with propionic acidaemia. *Mitochondrion* **2011**, *11*, 533–536. [CrossRef] [PubMed]

15. Rustin, P.; Munnich, A.; Rotig, A. Mitochondrial respiratory chain dysfunction caused by coenzyme Q deficiency. *Methods Enzymol.* **2004**, *382*, 81–88. [PubMed]

16. Quinzii, C.M.; Kattah, A.G.; Naini, A.; Akman, H.O.; Mootha, V.K.; DiMauro, S.; Hirano, M. Coenzyme Q deficiency and cerebellar ataxia associated with an aprataxin mutation. *Neurology* **2005**, *64*, 539–541. [CrossRef] [PubMed]

17. Fragaki, K.; Chaussenot, A.; Benoist, J.F.; Ait-El-Mkadem, S.; Bannwarth, S.; Rouzier, C.; Cochaud, C.; Paquis-Flucklinger, V. Coenzyme Q10 defects may be associated with a deficiency of Q10-independent mitochondrial respiratory chain complexes. *Biol. Res.* **2016**, *49*, 4. [CrossRef] [PubMed]

18. Turkowicz, M.J.; Karpińska, J. Analytical problems with the determination of coenzyme Q10 in biological samples. *Biofactors* **2013**, *39*, 176–185. [CrossRef] [PubMed]

19. Ogasahara, S.; Engel, A.G.; Frens, D.; Mack, D. Muscle coenzyme Q deficiency in familial mitochondrial encephalomyopathy. *Proc. Natl. Acad. Sci. USA* **1989**, *86*, 2379–2382. [CrossRef] [PubMed]

20. Duncan, A.J.; Heales, S.J.; Mills, K.; Eaton, S.; Land, J.M.; Hargreaves, I.P. Determination of coenzyme Q10 status in blood mononuclear cells, skeletal muscle, and plasma by HPLC with di-propoxy-coenzyme Q10 as an internal standard. *Clin. Chem.* **2005**, *51*, 2380–2382. [CrossRef] [PubMed]

21. Palamakula, A.; Soliman, M.; Khan, M.M. Regional permeability of coenzyme Q10 in isolated rat gastrointestinal tracts. *Pharmazie* **2005**, *60*, 212–214. [PubMed]

22. Miles, M.V. The uptake and distribution of coenzyme Q10. *Mitochondrion* **2007**, *7*, S72–S77. [CrossRef] [PubMed]

23. Weber, C.; Bysted, A.; Holmer, G. Coenzyme Q10 in the diet–daily intake and relative bioavailability. *Mol. Aspects Med.* **1997**, *18*, S251–S254. [CrossRef]

24. Salviati, L.; Sacconi, S.; Murer, L.; Zacchello, G.; Franceschini, L.; Laverda, A.M.; Basso, G.; Quinzii, C.; Angelini, C.; Hirano, M.; et al. Infantile encephalomyopathy and nephropathy with CoQ10 deficiency: A CoQ10-responsive condition. *Neurology* **2005**, *65*, 606–608. [CrossRef] [PubMed]

25. Artuch, R.; Vilaseca, M.A.; Moreno, J.; Lambruschini, N.; Cambra, F.J.; Campistol, J. Decreased serum ubiquinone-10 concentrations in phenylketonuria. *Am. J. Clin. Nutr.* **1999**, *70*, 892–895. [PubMed]

26. Hübner, C.; Hoffmann, G.F.; Charpentier, C.; Gibson, K.M.; Finckh, B.; Puhl, H.; Lehr, H.A.; Kohlschütter, A. Decreased plasma ubiquinone-10 concentration in patients with mevalonate kinase deficiency. *Pediatr. Res.* **1993**, *34*, 129–133. [CrossRef] [PubMed]

27. Delgadillo, V.; O'Callaghan, M.M.; Artuch, R.; Montero, R.; Pineda, M. Genistein supplementation in patients affected by Sanfilippo disease. *J. Inherit. Metab. Dis.* **2011**, *34*, 1039–1044. [CrossRef] [PubMed]

28. Fu, R.; Yanjanin, N.M.; Bianconi, S.; Pavan, W.J.; Porter, F.D. Oxidative stress in Niemann-Pick disease, type C. *Mol. Genet. Metab.* **2010**, *101*, 214–218. [CrossRef] [PubMed]

29. Cooper, J.M.; Korlipara, L.V.; Hart, P.E.; Bradley, J.L.; Schapira, A.H. Coenzyme Q10 and vitamin E deficiency in Friedreich's ataxia: Predictor of efficacy of vitamin E and coenzyme Q10 therapy. *Eur. J. Neurol.* **2008**, *15*, 1371–1379. [CrossRef] [PubMed]

30. Haas, D.; Niklowitz, P.; Hoffmann, G.F.; Andler, W.; Menke, T. Plasma and thrombocyte levels of coenzyme Q10 in children with Smith-Lemli-Opitz syndrome (SLOS) and the influence of HMG-CoA reductase inhibitors. *Biofactors* **2008**, *32*, 191–197. [CrossRef] [PubMed]

31. Bhagavan, H.N.; Chopra, R.K. Plasma coenzyme Q10 response to oral ingestion of coenzyme Q10 formulations. *Mitochondrion* **2007**, *7*, S78–88. [CrossRef] [PubMed]

32. Turunen, M.; Olson, J.; Dallner, G. Metabolism and function of coenzyme Q. *Biochim. Biophys. Acta* **2004**, *1660*, 171–199. [CrossRef] [PubMed]

33. Arias, A.; García-Villoria, J.; Rojo, A.; Buján, N.; Briones, P.; Ribes, A. Analysis of coenzyme Q(10) in lymphocytes by HPLC-MS/MS. *J. Chromatogr. B Anal. Technol. Biomed. Life Sci.* **2012**, *908*, 23–26. [CrossRef] [PubMed]

34. Barshop, B.A.; Gangoiti, J.A. Analysis of coenzyme Q in human blood and tissues. *Mitochondrion* **2007**, *7*, S89–S93. [CrossRef] [PubMed]

35. Crane, F.L. Discovery of ubiquinone (coenzyme Q) and an overview of function. *Mitochondrion* **2007**, *7*, S2–S7. [CrossRef] [PubMed]

36. Niklowitz, P.; Menke, T.; Andler, W.; Okun, J.G. Simultaneous analysis of coenzyme Q10 in plasma, erythrocytes and platelets: Comparison of the antioxidant level in blood cells and their environment in healthy children and after oral supplementation in adults. *Clin. Chim. Acta* **2004**, *342*, 219–226. [CrossRef] [PubMed]

37. Miles, M.V.; Tang, P.H.; Miles, L.; Steele, P.E.; Moye, M.J.; Horn, P.S. Validation and application of an HPLC-EC method for analysis of coenzyme Q10 in blood platelets. *Biomed. Chromatogr.* **2008**, *22*, 1403–1408. [CrossRef] [PubMed]

38. Galinier, A.; Carrière, A.; Fernandez, Y.; Bessac, A.M.; Caspar-Bauguil, S.; Periquet, B.; Comtat, M.; Thouvenot, J.P.; Casteilla, L. Biological validation of coenzyme Q redox state by HPLC-EC measurement: Relationship between coenzyme Q redox state and coenzyme Q content in rat tissues. *FEBS Lett.* **2004**, *578*, 53–57. [CrossRef] [PubMed]

39. Trevisson, E.; DiMauro, S.; Navas, P.; Salviati, L. Coenzyme Q deficiency in muscle. *Curr. Opin. Neurol.* **2011**, *24*, 449–456. [CrossRef] [PubMed]

40. Montero, R.; Sánchez-Alcázar, J.A.; Briones, P.; Hernández, A.R.; Cordero, M.D.; Trevisson, E.; Salviati, L.; Pineda, M.; García-Cazorla, A.; Navas, P.; et al. Analysis of coenzyme Q10 in muscle and fibroblasts for the diagnosis of CoQ10 deficiency syndromes. *Clin. Biochem.* **2008**, *41*, 697–700. [CrossRef] [PubMed]

41. Buján, N.; Arias, A.; Montero, R.; García-Villoria, J.; Lissens, W.; Seneca, S.; Espinós, C.; Navas, P.; De Meirleir, L.; Artuch, R.; et al. Characterization of CoQ10 biosynthesis in fibroblasts of patients with primary and secondary CoQ10deficiency. *J. Inherit. Metab. Dis.* **2014**, *37*, 53–62. [CrossRef] [PubMed]

42. López, L.C.; Quinzii, C.M.; Area, E.; Naini, A.; Rahman, S.; Schuelke, M.; Salviati, L.; Dimauro, S.; Hirano, M. Treatment of CoQ(10) deficient fibroblasts with ubiquinone, CoQ analogs, and vitamin C: Time-and compound-dependent effects. *PLoS ONE* **2010**, *5*, e11897. [CrossRef] [PubMed]

43. Duncan, A.J.; Bitner-Glindzicz, M.; Meunier, B.; Costello, H.; Hargreaves, I.P.; López, L.C.; Hirano, M.; Quinzii, C.M.; Sadowski, M.I.; Hardy, J.; et al. A nonsense mutation in COQ9 causes autosomal-recessive neonatal-onset primary coenzyme Q10 deficiency: A potentially treatable form of mitochondrial disease. *Am. J. Hum. Genet.* **2009**, *84*, 558–566. [CrossRef] [PubMed]

44. López, L.C.; Luna-Sánchez, M.; García-Corzo, L.; Quinzii, C.M.; Hirano, M. Pathomechanisms in coenzyme q10-deficient human fibroblasts. *Mol. Syndromol.* **2014**, *5*, 163–169. [CrossRef] [PubMed]

45. Quinzii, C.; Naini, A.; Salviati, L.; Trevisson, E.; Navas, P.; Dimauro, S.; Hirano, M. A mutation in para-hydroxybenzoate-polyprenyl transferase (COQ2) causes primary coenzyme Q10 deficiency. *Am. J. Hum. Genet.* **2006**, *78*, 345–349. [CrossRef] [PubMed]

46. Lagier-Tourenne, C.; Tazir, M.; Lopez, L.C.; Quinzii, C.M.; Assoum, M.; Drouot, N.; Busso, C.; Makri, S.; Ali-Pacha, L.; Benhassine, T.; et al. ADCK3, an ancestral kinase, is mutated in a form of recessive ataxia associated with coenzyme Q10 deficiency. *Am. J. Hum. Genet.* **2008**, *82*, 661–672. [CrossRef] [PubMed]

47. Geromel, V.; Kadhom, N.; Cebalos-Picot, I.; Ouari, O.; Polidori, A.; Munnich, A.; Rötig, A.; Rustin, P. Superoxide-induced massive apoptosis in cultured skin fibroblasts harboring the neurogenic ataxia retinitis pigmentosa (NARP) mutation in the ATPase-6 gene of the mitochondrial DNA. *Hum. Mol. Genet.* **2001**, *10*, 1221–1228. [CrossRef] [PubMed]

48. Quinzii, C.M.; López, L.C.; Gilkerson, R.W.; Dorado, B.; Coku, J.; Naini, A.B.; Lagier-Tourenne, C.; Schuelke, M.; Salviati, L.; Carrozzo, R.; et al. Reactive oxygen species, oxidative stress, and cell death correlate with level of CoQ10 deficiency. *FASEB J.* **2010**, *24*, 3733–3743. [CrossRef] [PubMed]

49. López-Martín, J.M.; Salviati, L.; Trevisson, E.; Montini, G.; DiMauro, S.; Quinzii, C.; Hirano, M.; Rodriguez-Hernandez, A.; Cordero, M.D.; Sánchez-Alcázar, J.A.; et al. Missense mutation of the COQ2 gene causes defects of bioenergetics and de novo pyrimidine synthesis. *Hum. Mol. Genet.* **2007**, *16*, 1091–1097. [CrossRef] [PubMed]

50. Rodríguez-Hernández, A.; Cordero, M.D.; Salviati, L.; Artuch, R.; Pineda, M.; Briones, P.; Gómez Izquierdo, L.; Cotán, D.; Navas, P.; Sánchez-Alcázar, J.A. Coenzyme Q deficiency triggers mitochondria degradation by mitophagy. *Autophagy* **2009**, *5*, 19–32. [CrossRef] [PubMed]

51. Cotan, D.; Cordero, M.D.; Garrido-Maraver, J.; Oropesa-Avila, M.; Rodriquez-Hernandez, A.; Gomez Izquierdo, L.; De la Mata, M.; De Miquel, M.; Lorite, J.B.; Infante, E.R.; et al. Secondary coenzyme Q10 deficiency triggers mitocondria degradation by mitophagy in MELAS fibroblasts. *FABEB J.* **2011**, *25*, 2669–2687.

52. Fragaki, K.; Ait-El-Mkadem, S.; Chaussenot, A.; Gire, C.; Menqual, R.; Bonesso, L.; Beneteau, M.; Ricci, J.E.; Desquiret-Dumas, V.; Procaccio, V.; et al. Refractory epilepsy and mitocondrial dysfunction due to GM3 synthase deficiency. *Eur. J. Hum. Genet.* **2013**, *21*, 528–534. [CrossRef] [PubMed]

53. Heeringa, S.F.; Chernin, G.; Chaki, M.; Zhou, W.; Sloan, A.J.; Ji, Z.; Xie, L.X.; Salviati, L.; Hurd, T.W.; Vega-Warner, V.; et al. COQ6 mutations in human patients produce nephrotic syndrome with sensorineural deafness. *J. Clin. Investig.* **2011**, *121*, 2013–2024. [CrossRef] [PubMed]

54. Ashraf, S.; Gee, H.Y.; Woerner, S.; Xie, L.X.; Vega-Warner, V.; Lovric, S.; Fang, H.; Song, X.; Cattran, D.C.; Avila-Casado, C.; et al. ADCK4 mutations promote steroid-resistant nephrotic syndrome through CoQ10 biosynthesis disruption. *J. Clin. Investig.* **2013**, *123*, 5179–5189. [CrossRef] [PubMed]

55. Yubero, D.; Montero, R.; Ramos, M.; Neergheen, V.; Navas, P.; Artuch, R.; Hargreaves, I. Determination of urinary coenzyme Q10 by HPLC with electrochemical detection: Reference values for a paediatric population. *Biofactors* **2015**, *41*, 424–230. [CrossRef] [PubMed]

56. Emma, F.; Montini, G.; Parikh, S.M.; Salviati, L. Mitochondrial dysfunction in inherited renal disease and acute kidney injury. *Nat. Rev. Nephrol.* **2016**, *12*, 267–280. [CrossRef] [PubMed]

57. Diomedi-Camassei, F.; Di Giandomenico, S.; Santorelli, F.M.; Cardi, G.; Piemonte, F.; Montini, G.; Ghiggeri, G.M.; Murer, L.; Barisoni, L.; Pastore, A.; et al. COQ2 nephropathy: A newly described inherited mitochondriopathy with primary renal involvement. *J. Am. Soc. Nephrol.* **2007**, *18*, 2773–2780. [CrossRef] [PubMed]

58. Sheeran, F.L.; Pepe, S. Posttranslational modifications and dysfunction of mitochondrial enzymes in human heart failure. *Am. J. Physiol. Endocrinol. Metab.* **2016**, *311*, E449–E460. [CrossRef] [PubMed]

59. Duberley, K.E.; Hargreaves, I.P.; Chaiwatanasirikul, K.A.; Heales, S.J.; Land, J.M.; Rahman, S.; Mills, K.; Eaton, S. Coenzyme Q10 quantification in muscle, fibroblasts and cerebrospinal fluid by liquid chromatography/tandem mass spectrometry using a novel deuterated internal standard. *Rapid Commun. Mass Spectrom.* **2013**, *27*, 924–930. [CrossRef] [PubMed]

60. Martinefski, M.; Samassa, P.; Lucangioli, S.; Tripodi, V. A novel non-invasive sampling method using buccal mucosa cells for determination of coenzyme Q10. *Anal. Bioanal. Chem.* **2015**, *407*, 5529–5533. [CrossRef] [PubMed]

Journal of
Clinical Medicine

MDPI

Review

Biochemical Assessment of Coenzyme Q_{10} Deficiency

Juan Carlos Rodríguez-Aguilera [1,2], **Ana Belén Cortés** [1,2], **Daniel J. M. Fernández-Ayala** [2,3] **and Plácido Navas** [2,3,*]

1   Laboratorio de Fisiopatología Celular y Bioenergética, 41013 Sevilla, Spain; jcrodagu@upo.es (J.C.R.-A.); abcorrod@upo.es (A.B.C.)
2   Centro de Investigación Biomédica en Red de Enfermedades Raras, Instituto de Salud Carlos III, Universidad Pablo de Olavide-CISC, 41013 Sevilla, Spain; dmorfer@upo.es
3   Centro Andaluz de Biología del Desarrollo, 41013 Sevilla, Spain
*   Correspondence: pnavas@upo.es; Tel.: +34-954-349-385

Academic Editor: Iain P. Hargreaves
Received: 18 January 2017; Accepted: 28 February 2017; Published: 5 March 2017

Abstract: Coenzyme Q_{10} (CoQ$_{10}$) deficiency syndrome includes clinically heterogeneous mitochondrial diseases that show a variety of severe and debilitating symptoms. A multiprotein complex encoded by nuclear genes carries out CoQ$_{10}$ biosynthesis. Mutations in any of these genes are responsible for the primary CoQ$_{10}$ deficiency, but there are also different conditions that induce secondary CoQ$_{10}$ deficiency including mitochondrial DNA (mtDNA) depletion and mutations in genes involved in the fatty acid β-oxidation pathway. The diagnosis of CoQ$_{10}$ deficiencies is determined by the decrease of its content in skeletal muscle and/or dermal skin fibroblasts. Dietary CoQ$_{10}$ supplementation is the only available treatment for these deficiencies that require a rapid and distinct diagnosis. Here we review methods for determining CoQ$_{10}$ content by HPLC separation and identification using alternative approaches including electrochemical detection and mass spectrometry. Also, we review procedures to determine the CoQ$_{10}$ biosynthesis rate using labeled precursors.

Keywords: coenzyme Q_{10}; CoQ$_{10}$ deficiency syndrome; CoQ$_{10}$ biosynthesis; mitochondria diseases

---

1. Introduction

The mitochondrial respiratory chain (MRC) generates most of the cellular ATP and is comprised of five multi-subunit enzyme complexes. Both the mitochondrial DNA (mtDNA) and the nuclear DNA (nDNA) encode for polypeptides of these complexes and also proteins involved in mitochondrial function. Besides MRC enzyme complexes, two electron carriers, coenzyme Q (CoQ) and cytochrome c, are vital for mitochondrial synthesis of ATP. Mutations in genes of either genome may cause mitochondrial diseases, which are common among inherited metabolic and neurological disorders [1].

CoQ is a lipid-soluble component of virtually all cell membranes. It is composed of a benzoquinone ring with a polyprenyl side chain, the number of isoprene units being a characteristic of given specie, e.g., 10 in humans (CoQ$_{10}$). CoQ$_{10}$ transports electrons from MRC Complexes I and II to Complex III. These electrons come from either NADH or succinate [2] although CoQ$_{10}$ can be alternatively reduced with electrons provided by different redox reactions in mitochondria [3]. Consequently, CoQ$_{10}$ is essential for ATP production inside mitochondria, although it is also an indispensible antioxidant in extramitochondrial membranes and a key factor for pyrimidine nucleotide synthesis [4].

CoQ biosynthesis depends on a pathway that involves at least 11 genes (*COQ* genes), showing a high degree of conservation among species, and is carried out by a putative multi-subunit enzyme complex [5]. Most of the information about the CoQ biosynthesis pathway comes from yeast, and maintains a high homology with mammal gene components (Table 1) [6]. The CoQ$_{10}$ biosynthesis pathway is highly regulated by transcription factors PPARα and NFκB [7–9]. HuR and hnRNP C1/C2 binding proteins stabilize *COQ7* mRNA as another CoQ$_{10}$ biosynthesis regulatory mechanism [10].

Table 1. Yeast *COQ* genes and their characterized human homologues.

Yeast	Human	Function
COQ1	*PDSS1 */PDSS2 *	Synthesis of polyprenyl-diphosphate
COQ2	*COQ2 *	*p*HB-prenyl-transferase
COQ3	*COQ3 *	Methyltransferase
COQ4	*COQ4 *	Organization of the multi-enzyme complex
COQ5	*COQ5*	Methyltransferase
COQ6	*COQ6 *	Mono-oxygenase
COQ7	*COQ7 *	Hydroxylase
COQ8	*ADCK3 */ADCK4 *	Unorthodox kinase (regulatory)
COQ9	*COQ9 *	Lipid binding protein
COQ10	*COQ10A/COQ10B*	CoQ chaperone
PTC7	*PPTC7*	Phosphatase (regulatory)

* These genes were mutated in human causing primary CoQ$_{10}$ deficiency.

Coq7p is post-translationally regulated in yeast that involves mitochondrial phosphatase Ptc7 [11,12]. Ptc7 human orthologue (*PPTC7*) is related to cellular bioenergetics and stress resistance [13]. Coq7p activity is a key regulator of the CoQ biosynthesis complex [6,14], which may depend on the interaction with *Coq9p* contributing to the stabilization of the biosynthesis complex [15–18]. The level of CoQ is highly regulated inside cells and tissues but its concentration is different in each tissue and organ, and depends on dietary conditions and age [19,20]. CoQ also varies greatly in human diseases such as Alzheimer's disease, cardiomyopathy, Niemann-Pick and diabetes.

2. CoQ$_{10}$ Deficiency Syndrome

CoQ$_{10}$ deficiency syndrome includes diverse inherited pathological diseases defined by the decrease of CoQ$_{10}$ content in muscle and/or cultured skin fibroblasts. CoQ$_{10}$ deficiency impairs oxidative phosphorylation and causes clinically heterogeneous mitochondrial diseases [21,22]. When the decrease in CoQ$_{10}$ content is due to mutations in genes encoding proteins of the CoQ biosynthesis pathway or its regulation (COQ genes), it causes primary CoQ$_{10}$ deficiency [23,24]. Secondary CoQ$_{10}$ deficiencies may be due to defects in genes unrelated to the CoQ$_{10}$ biosynthetic pathway. Secondary CoQ$_{10}$ deficiency is a common finding in oxidative phosphorylation (OXPHOS) and non-OXPHOS disorders [25]. A low mitochondrial CoQ$_{10}$ content is described in mtDNA depletion [26], mutations in the DNA repairing aprataxin [27], mutations of the enzyme *ETFDH* of the β-oxidation of fatty acids [28], recurrent food intolerance and allergies [29], methylmalonic aciduria [30], myalgic encephalomyelitis chronic fatigue syndrome [31], and propionic acidemia [32]. We propose that cases of secondary CoQ$_{10}$ deficiency associated with OXPHOS defects could be adaptive mechanisms to maintain a balanced OXPHOS which is required to keep cells alive, although the mechanisms explaining these deficiencies and the pathophysiological role in the disease are unknown.

The clinical phenotypes of primary CoQ$_{10}$-deficient patients are broader than initially reported in 1989 [33], including (i) a multisystem disorder with steroid-resistant nephrotic syndrome as the main clinical manifestation (*COQ1-PDSS2*) [34], (*COQ2*) [35], (*COQ6*) [36] and (*ADCK4*) [37]; (ii) a multisystem disorder without nephrotic syndrome (*COQ1-PDSS1*) [38], (*COQ9*) [39] and (*COQ7*) [40]; (iii) cerebellar ataxia (*COQ8-ADCK3*) [41–47]; and (iv) myopathy and encephalopathy (*COQ4*) [48–50].

3. Primary CoQ$_{10}$ Deficiency Therapy

Primary CoQ$_{10}$ deficiency is unique among mitochondrial diseases because an effective therapy is available for patients, which is the supplementation of CoQ$_{10}$. Ubiquinol, the reduced form of CoQ$_{10}$, was recently approved as an orphan drug for primary CoQ$_{10}$ deficiency [51]

While this approach is quite successful in some patients, with a clear improvement of the pathological phenotype [52], some cases do not show any clinical relief as would be expected [53], probably because they are suffering secondary CoQ_{10} deficiency. High-dose oral CoQ_{10} supplementation can stop the progression of the encephalopathy and allows the recovery of renal damage [52]. High-dose CoQ_{10} supplementation was also able to prevent the onset of renal symptoms in *PDSS2*-deficient mice [54]. Furthermore, CoQ_{10} but not other quinones can restore mitochondrial function in deficient human fibroblasts [55]. Due to the therapeutic possibility of CoQ_{10} supplementation for these patients, a rapid and unequivocal diagnosis of the deficiency is essential.

4. CoQ_{10} Determination in Cells and Tissues

Content of CoQ_{10} has been determined in plasma, white blood cells, skin fibroblasts and skeletal muscle biopsies to assess a deficiency diagnosis [56–58], and recently useful determination in the urine of pediatric patients was demonstrated [59]. Although CoQ can be measured in plasma and white blood cells, you cannot use it for the diagnosis of mitochondrial diseases since CoQ content in plasma and white blood cells is often not decreased in these conditions.

CoQ_{10} content is mainly analyzed by the injection of lipid extracts in HPLC and detected by either electrochemical and/or UV-vis detectors, or mass spectrometry. Electrochemical detection has significant advantages compared to UV-vis detection; these include higher sensitivity and also the ability to measure oxidized and reduced forms of CoQ, either separately or combined, according to differential positioning of the conditioning cell (before or after the injector valve, respectively).

CoQ_{10} extraction from biological samples (0.5 mg protein) requires the disruption of hydrophobic elements (lipid bilayers and lipoproteins) by adding SDS (1% final concentration). Lipids are dispersed with an alcohol cocktail (2-propanol 5% in ethanol) mixed with the disrupted biological sample (ratio 1:2 v/v), and they undergo subsequent triplicated hexane extraction (dispersed sample:hexane ratio 3:5 v/v). Hexane fractions are mixed and dried under vacuum, and then reconstituted in ethanol prior to HPLC analysis. To estimate CoQ_{10} recovery, 100 pmol CoQ_9 was included in the alcohol cocktail (2-propanol 5% in ethanol). Trace amounts of CoQ9 may have eventually been found in human tissues (probably from dietary uptake), but this does not interfere with the significant amount of internal standard added.

For convenience in high-throughput analysis, volumes are scaled down for extraction and vortex in 1.5 mL polypropylene tubes or 2 mL cryo vials.

Separation in C18 RP-HPLC columns (5 μm, 150 × 4.6 mm) requires 20 mM $AcNH_4$ pH 4.4 in methanol (solvent A) and 20 mM $AcNH_4$ pH 4.4 in propanol (solvent B). A gradient method with a 85:15 solvent mixture (A:B ratio), and a flow rate of 1.2 mL/min, is regularly used as the starting conditions. The mobile phase turns to a 50:50 A:B ratio starting in minute 6 and completed in minute 8, as the flow rate decreases to 1.1 mL/min. After 20 min (run time) at 40 °C, the columns are re-equilibrated to the initial conditions for three additional minutes.

The detection of total CoQ_{10} can be achieved either by UV-vis (set to 275 nm) or electrochemical (ECD) detectors (channel 1 set to −700 mV and channel 2 set to +500 mV, conditioning guard cell after injection valve). For complex samples including many peaks, the CoQ_{10} peak is confirmed by spectral information (UV-vis) or by the redox area ratio (ECD detector, −700/+500 area ratio), compared to pure CoQ_{10}. Figure 1 illustrates two chromatograms that correspond to normal age-matched human dermal fibroblasts (black plot) compared to patient dermal fibroblasts with CoQ_{10} deficiency (red plot).

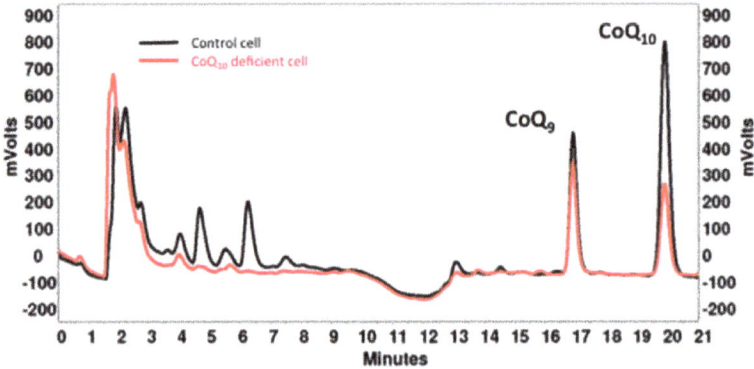

Figure 1. HPLC elution profile of lipid extracts from human skeletal muscular tissue. Patient pathological profile (red plot) shows that CoQ$_{10}$ is clearly diminished compared to healthy control volunteers (black plot). CoQ$_9$ is used as internal standard for normalization.

5. Analysis of CoQ$_{10}$ Biosynthesis

Another important approach to assess CoQ$_{10}$ deficiency in cells is to determine the rate of biosynthesis by the level of incorporation of labeled of CoQ$_{10}$ precursors such as *para*-hydroxybenzoate (*p*-HB) labeled with either ^{13}C-*p*-HB or ^{14}C-*p*-HB, which is the precursor of the benzoquinone ring, or ^{2}H-mevalonate, which is the precursor of the isoprenyl side chain [10,60].

Polyprenyl-*p*HB transferase activity was assayed by measuring the incorporation of ^{14}C-*p*-HB into nonaprenyl-4-hydroxybenzoate [35]. Isolated mitochondria (0.1–1 mg protein) were mixed with assay buffer (50 mM phosphate buffer, pH 7.5, 10 mM MgCl$_2$, 5 mM EGTA containing 1 mM PMSF, 20 µg/mL each of the protease inhibitors chymostatin, leupeptin, antipain, and pepstatin A, 5 µM solanesyl pyrophosphate solubilized in detergent solution (1% in water), and 10^5 DPM of ^{14}C-*p*-HB). A sufficient volume of a 10% detergent stock solution was also added to the reaction medium to achieve a final detergent concentration of 1%. The following detergents were tested: Triton X-100, Chaps, sodium cholate, sodium deoxycholate, lysophosphatidyl choline, and octylglucoside. After incubation for 30 min at 37 °C with gentle stirring, the reaction was stopped by chilling samples to 4 °C. Prenylated ^{14}C-*p*-HB was separated by organic extraction with hexane and then measured using a liquid scintillation counter. Specific activity was expressed as disintegrations per minute (DPM) min^{-1}·mg·protein^{-1}.

Biosynthesis of ^{14}C-CoQ$_{10}$ has been quantified in any type of cell culture, such as cancer cells, human skin fibroblasts, and murine embryonic fibroblast and stem cells [10,61]. Previously, cultures were incubated with 4.5 nM ^{14}C-*p*-HB for one to three days, depending on the cell-specific rate of growth. The ^{14}C-*p*-HB was chemically synthesized in our laboratory from ^{14}C-thyrosine [61]. Labeled-CoQ$_{10}$ content is analyzed by lipid extract injection in HPLC and detected by the radio-flow detector LB 509 with a solid cell YG 150 Al-U4D (Berthold Technologies, Bad Wildbad, Germany) in parallel with either electrochemical or UV-vis detectors. Lipid extraction is done as we described above for CoQ$_{10}$ determination in cells and tissues, but isocratic HPLC analysis lipid separation is performed with methanol:propanol (65:35) plus 20 mM AcNH$_4$ pH 4.4 at a constant flow rate of 1 mL/min (Figure 2).

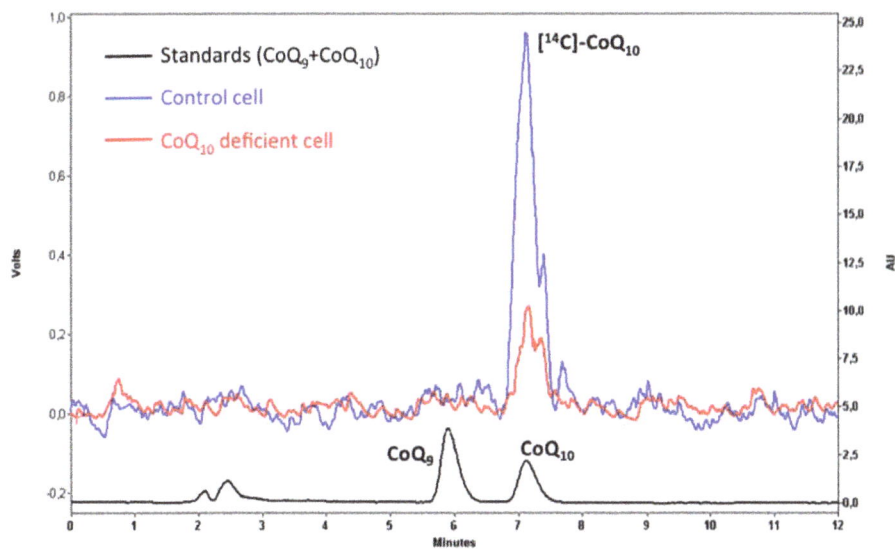

Figure 2. HPLC elution profile of lipid extracts from human fibroblasts cultured with the radiolabeled precursor ^{14}C-*p*-HB. Patient pathological profile (red plot) shows that CoQ_{10} is clearly diminished compared to control cells from healthy humans (blue plot). Left Y-axis shows the radio-flow detector scale (volts). Right Y-axis shows the UV-detector scale (absorbance units) for a standard pool of CoQ_{10} and CoQ_9 (black plot). Notice that the only peak detected in this analysis corresponded with CoQ_{10}.

Alternatively, a non-radioactive protocol to analyze CoQ_{10} biosynthesis was developed using either ^{2}H-mevalonate or ^{13}C-phydroxybenzoate as CoQ_{10} precursors as described by Buján et al. (2014) [60]. Human fibroblasts at 60%–70% were incubated with these precursors for 24–72 h at different concentrations. After incubation, cells were trypsinized and washed twice with isotonic buffer. Pelleted cells were resuspended with 300 µL of a buffer solution containing 0.25 mmol/L sucrose, 2 mmol/L EDTA, 10 mmol/L Tris and 100 UI/mL heparin, pH 7.4, and sonicated twice for 5 s. These homogenates were used to determine CoQ_{10} biosynthesis measuring by HPLC-MS/MS, as described in Arias et al. (2012) [62]. Briefly, HPLC separation was as indicated above and extracted peaks were analyzed by MS/MS in a Micromass Quattro micro™ (Waters/Micromass, Manchester, UK). The MS/MS was operated in the electrospray positive ion mode with a cone voltage (CV), and collision energy (CE) of 15 V and 20 eV, respectively. The following multiple-reaction monitoring transitions were selected: m/z 900 > 203 and 897>197 for ^{13}C-CoQ_{10} or ^{2}H-CoQ_{10}, respectively, 894 > 197 for the physiological CoQ_{10} and 826 > 197 for CoQ_9 (internal standard). The dwell time for each transition was 200 ms and the run-time was 16 min. Nitrogen (at a flow rate of 50 L/h) and argon (adjusted to obtain a vacuum of $3°$—10^{-3} bar) were used as the nebulizing and collision gas, respectively.

6. Concluding Remarks

Coenzyme Q_{10} deficiency syndrome includes a group of mitochondrial diseases showing diverse inherited pathological phenotypes. The common aspect of them is the lower content of CoQ_{10} in tissues and organs. Primary deficiency is caused by defects in proteins encoded by *COQ* genes, which are components of the biosynthesis pathway or its regulation. CoQ_{10} supplementation is the current treatment of primary CoQ_{10} deficiency, which highly improves symptoms. A rapid and distinct characterization of the deficiency is important, and it is mainly determined in skeletal muscle and/or skin dermal fibroblasts. The main approach is to analyze the total content of CoQ_{10} in lipid extracts by

HPLC and UV and/or electrochemical detection. Alternatively, the CoQ_{10} biosynthesis rate in cultured cells can be determined by incubation with radiolabeled precursors.

Acknowledgments: This work has been funded by the Instituto de Salud Carlos III FIS PI14-01962 grant. Authors were also funded by the Junta de Andalucía BIO177 research group.

Author Contributions: J.C.R.-A., A.B.C., D.J.M.F.-A., and P.N. have contributed to writing and editing the review and figures.

Conflicts of Interest: The authors declare no conflict of interest.

References

1. Gorman, G.S.; Chinnery, P.F.; DiMauro, S.; Hirano, M.; Koga, Y.; McFarland, R.; Suomalainen, A.; Thorburn, D.R.; Zeviani, M.; Turnbull, D.M. Mitochondrial diseases, Nature reviews. *Dis. Prim.* **2016**, *2*, 16080. [CrossRef] [PubMed]

2. Lapuente-Brun, E.; Moreno-Loshuertos, R.; Acin-Perez, R.; Latorre-Pellicer, A.; Colas, C.; Balsa, E.; Perales-Clemente, E.; Quiros, P.M.; Calvo, E.; Rodriguez-Hernandez, M.A.; et al. Supercomplex assembly determines electron flux in the mitochondrial electron transport chain. *Science* **2013**, *340*, 1567–1570. [CrossRef] [PubMed]

3. Alcazar-Fabra, M.; Navas, P.; Brea-Calvo, G. Coenzyme Q biosynthesis and its role in the respiratory chain structure. *Biochim. Biophys. Acta* **2016**, *1857*, 1073–1078. [CrossRef] [PubMed]

4. Lopez-Lluch, G.; Rodriguez-Aguilera, J.C.; Santos-Ocana, C.; Navas, P. Is coenzyme Q a key factor in aging? *Mech. Ageing Dev.* **2010**, *131*, 225–235. [CrossRef] [PubMed]

5. Bentinger, M.; Tekle, M.; Dallner, G. Coenzyme Q—Biosynthesis and functions. *Biochem. Biophys. Res. Commun.* **2010**, *396*, 74–79. [CrossRef] [PubMed]

6. Gonzalez-Mariscal, I.; Garcia-Teston, E.; Padilla, S.; Martin-Montalvo, A.; Viciana, T.P.; Vazquez-Fonseca, L.; Dominguez, P.G.; Santos-Ocana, C. The regulation of coenzyme q biosynthesis in eukaryotic cells: All that yeast can tell us. *Mol. Syndromol.* **2014**, *5*, 107–118. [CrossRef] [PubMed]

7. Turunen, M.; Peters, J.M.; Gonzalez, F.J.; Schedin, S.; Dallner, G. Influence of peroxisome proliferator-activated receptor alpha on ubiquinone biosynthesis. *J. Mol. Biol.* **2000**, *297*, 607–614. [CrossRef] [PubMed]

8. Bentinger, M.; Tekle, M.; Brismar, K.; Chojnacki, T.; Swiezewska, E.; Dallner, G. Stimulation of coenzyme Q synthesis. *Biofactors* **2008**, *32*, 99–111. [CrossRef] [PubMed]

9. Brea-Calvo, G.; Siendones, E.; Sanchez-Alcazar, J.A.; de Cabo, R.; Navas, P. Cell survival from chemotherapy depends on NF-kappaB transcriptional up-regulation of coenzyme Q biosynthesis. *PLoS ONE* **2009**, *4*, e5301. [CrossRef] [PubMed]

10. Cascajo, M.V.; Abdelmohsen, K.; Noh, J.H.; Fernandez-Ayala, D.J.; Willers, I.M.; Brea, G.; Lopez-Lluch, G.; Valenzuela-Villatoro, M.; Cuezva, J.M.; Gorospe, M.; et al. RNA-binding proteins regulate cell respiration and coenzyme Q biosynthesis by post-transcriptional regulation of COQ7. *RNA Biol.* **2016**, *13*, 622–634. [CrossRef] [PubMed]

11. Martin-Montalvo, A.; Gonzalez-Mariscal, I.; Padilla, S.; Ballesteros, M.; Brautigan, D.L.; Navas, P.; Santos-Ocana, C. Respiratory-induced coenzyme Q biosynthesis is regulated by a phosphorylation cycle of Cat5p/Coq7p. *Biochem. J.* **2011**, *440*, 107–114. [CrossRef] [PubMed]

12. Martín-Montalvo, A.; González-Mariscal, I.; Pomares-Viciana, T.; Padilla-López, S.; Ballesteros, M.; Vazquez-Fonseca, L.; Gandolfo, P.; Brautigan, D.L.; Navas, P.; Santos-Ocaña, C. The phosphatase Ptc7 induces coenzyme Q biosynthesis by activating the hydroxylase Coq7 in yeast. *J. Biol. Chem.* **2013**, *288*, 28126–28137. [CrossRef] [PubMed]

13. Lanning, N.J.; Looyenga, B.D.; Kauffman, A.L.; Niemi, N.M.; Sudderth, J.; DeBerardinis, R.J.; MacKeigan, J.P. A mitochondrial RNAi screen defines cellular bioenergetic determinants and identifies an adenylate kinase as a key regulator of ATP levels. *Cell Rep.* **2014**, *7*, 907–917. [CrossRef] [PubMed]

14. Gonzalez-Mariscal, I.; Garcia-Teston, E.; Padilla, S.; Martin-Montalvo, A.; Pomares-Viciana, T.; Vazquez-Fonseca, L.; Gandolfo-Dominguez, P.; Santos-Ocana, C. Regulation of coenzyme Q biosynthesis in yeast: A new complex in the block. *IUBMB Life* **2014**, *66*, 63–70. [CrossRef] [PubMed]

15. Padilla, S.; Tran, U.C.; Jimenez-Hidalgo, M.; Lopez-Martin, J.M.; Martin-Montalvo, A.; Clarke, C.F.; Navas, P.; Santos-Ocana, C. Hydroxylation of demethoxy-Q6 constitutes a control point in yeast coenzyme Q6 biosynthesis. *Cell. Mol. Life Sci.* **2009**, *66*, 173–186. [CrossRef] [PubMed]
16. Lohman, D.C.; Forouhar, F.; Beebe, E.T.; Stefely, M.S.; Minogue, C.E.; Ulbrich, A.; Stefely, J.A.; Sukumar, S.; Luna-Sanchez, M.; Jochem, A.; et al. Mitochondrial COQ9 is a lipid-binding protein that associates with COQ7 to enable coenzyme Q biosynthesis. *Proc. Natl. Acad. Sci. USA* **2014**, *111*, E4697–E4705. [CrossRef] [PubMed]
17. Stefely, J.A.; Licitra, F.; Laredj, L.; Reidenbach, A.G.; Kemmerer, Z.A.; Grangeray, A.; Jaeg-Ehret, T.; Minogue, C.E.; Ulbrich, A.; Hutchins, P.D.; et al. Cerebellar Ataxia and Coenzyme Q Deficiency through Loss of Unorthodox Kinase Activity. *Mol. Cell* **2016**, *63*, 608–620. [CrossRef] [PubMed]
18. Stefely, J.A.; Reidenbach, A.G.; Ulbrich, A.; Oruganty, K.; Floyd, B.J.; Jochem, A.; Saunders, J.M.; Johnson, I.E.; Minogue, C.E.; Wrobel, R.L.; et al. Mitochondrial ADCK3 Employs an Atypical Protein Kinase-like Fold to Enable Coenzyme Q Biosynthesis. *Mol. Cell* **2015**, *57*, 83–94. [CrossRef] [PubMed]
19. Parrado-Fernandez, C.; Lopez-Lluch, G.; Rodriguez-Bies, E.; Santa-Cruz, S.; Navas, P.; Ramsey, J.J.; Villalba, J.M. Calorie restriction modifies ubiquinone and COQ transcript levels in mouse tissues. *Free Radic. Biol. Med.* **2011**, *50*, 1728–1736. [CrossRef] [PubMed]
20. Lopez-Lluch, G.; Santos-Ocana, C.; Sanchez-Alcazar, J.A.; Fernandez-Ayala, D.J.; Asencio-Salcedo, C.; Rodriguez-Aguilera, J.C.; Navas, P. Mitochondrial responsibility in ageing process: Innocent, suspect or guilty. *Biogerontology* **2015**, *16*, 599–620. [CrossRef] [PubMed]
21. Schon, E.A.; DiMauro, S.; Hirano, M.; Gilkerson, R.W. Therapeutic prospects for mitochondrial disease. *Trends Mol. Med.* **2010**, *16*, 268–276. [CrossRef] [PubMed]
22. Quinzii, C.M.; Emmanuele, V.; Hirano, M. Clinical presentations of coenzyme q10 deficiency syndrome. *Mol. Syndromol.* **2014**, *5*, 141–146. [CrossRef] [PubMed]
23. Doimo, M.; Desbats, M.A.; Cerqua, C.; Cassina, M.; Trevisson, E.; Salviati, L. Genetics of coenzyme q10 deficiency. *Mol. Syndromol.* **2014**, *5*, 156–162. [CrossRef] [PubMed]
24. Desbats, M.A.; Lunardi, G.; Doimo, M.; Trevisson, E.; Salviati, L. Genetic bases and clinical manifestations of coenzyme Q10 (CoQ 10) deficiency. *J. Inherit. Metab. Dis.* **2015**, *38*, 145–156. [CrossRef] [PubMed]
25. Yubero, D.; Montero, R.; Martin, M.A.; Montoya, J.; Ribes, A.; Grazina, M.; Trevisson, E.; Rodriguez-Aguilera, J.C.; Hargreaves, I.P.; Salviati, L.; et al. Secondary coenzyme Q10 deficiencies in oxidative phosphorylation (OXPHOS) and non-OXPHOS disorders. *Mitochondrion* **2016**, *30*, 51–58. [CrossRef] [PubMed]
26. Montero, R.; Sanchez-Alcazar, J.A.; Briones, P.; Navarro-Sastre, A.; Gallardo, E.; Bornstein, B.; Herrero-Martin, D.; Rivera, H.; Martin, M.A.; Marti, R.; et al. Coenzyme Q10 deficiency associated with a mitochondrial DNA depletion syndrome: A case report. *Clin. Biochem.* **2009**, *42*, 742–745. [CrossRef] [PubMed]
27. Quinzii, C.M.; Kattah, A.G.; Naini, A.; Akman, H.O.; Mootha, V.K.; DiMauro, S.; Hirano, M. Coenzyme Q deficiency and cerebellar ataxia associated with an aprataxin mutation. *Neurology* **2005**, *64*, 539–541. [CrossRef] [PubMed]
28. Gempel, K.; Topaloglu, H.; Talim, B.; Schneiderat, P.; Schoser, B.G.; Hans, V.H.; Palmafy, B.; Kale, G.; Tokatli, A.; Quinzii, C.; et al. The myopathic form of coenzyme Q10 deficiency is caused by mutations in the electron-transferring-flavoprotein dehydrogenase (ETFDH) gene. *Brain* **2007**, *130*, 2037–2044. [CrossRef] [PubMed]
29. Miles, M.V.; Putnam, P.E.; Miles, L.; Tang, P.H.; DeGrauw, A.J.; Wong, B.L.; Horn, P.S.; Foote, H.L.; Rothenberg, M.E. Acquired coenzyme Q10 deficiency in children with recurrent food intolerance and allergies. *Mitochondrion* **2011**, *11*, 127–135. [CrossRef] [PubMed]
30. Haas, D.; Niklowitz, P.; Horster, F.; Baumgartner, E.R.; Prasad, C.; Rodenburg, R.J.; Hoffmann, G.F.; Menke, T.; Okun, J.G. Coenzyme Q(10) is decreased in fibroblasts of patients with methylmalonic aciduria but not in mevalonic aciduria. *J. Inherit. Metab. Dis.* **2009**, *32*, 570–575. [CrossRef] [PubMed]
31. Maes, M.; Mihaylova, I.; Kubera, M.; Uytterhoeven, M.; Vrydags, N.; Bosmans, E. Coenzyme Q10 deficiency in myalgic encephalomyelitis/chronic fatigue syndrome (ME/CFS) is related to fatigue, autonomic and neurocognitive symptoms and is another risk factor explaining the early mortality in ME/CFS due to cardiovascular disorder. *Neuro Endocrinol. Lett.* **2009**, *30*, 470–476. [PubMed]

32. Fragaki, K.; Cano, A.; Benoist, J.F.; Rigal, O.; Chaussenot, A.; Rouzier, C.; Bannwarth, S.; Caruba, C.; Chabrol, B.; Paquis-Flucklinger, V. Fatal heart failure associated with CoQ10 and multiple OXPHOS deficiency in a child with propionic acidemia. *Mitochondrion* **2011**, *11*, 533–536. [CrossRef] [PubMed]

33. Ogasahara, S.; Engel, A.G.; Frens, D.; Mack, D. Muscle coenzyme Q deficiency in familial mitochondrial encephalomyopathy. *Proc. Natl. Acad. Sci. USA* **1989**, *86*, 2379–2382. [CrossRef] [PubMed]

34. Lopez, L.C.; Schuelke, M.; Quinzii, C.M.; Kanki, T.; Rodenburg, R.J.; Naini, A.; Dimauro, S.; Hirano, M. Leigh syndrome with nephropathy and CoQ10 deficiency due to decaprenyl diphosphate synthase subunit 2 (PDSS2) mutations. *Am. J. Hum. Genet.* **2006**, *79*, 1125–1129. [CrossRef] [PubMed]

35. Quinzii, C.; Naini, A.; Salviati, L.; Trevisson, E.; Navas, P.; Dimauro, S.; Hirano, M. A mutation in para-hydroxybenzoate-polyprenyl transferase (COQ2) causes primary coenzyme Q10 deficiency. *Am. J. Hum. Genet.* **2006**, *78*, 345–349. [CrossRef] [PubMed]

36. Heeringa, S.F.; Chernin, G.; Chaki, M.; Zhou, W.; Sloan, A.J.; Ji, Z.; Xie, L.X.; Salviati, L.; Hurd, T.W.; Vega-Warner, V.; et al. COQ6 mutations in human patients produce nephrotic syndrome with sensorineural deafness. *J. Clin. Investig.* **2011**, *121*, 2013–2024. [CrossRef] [PubMed]

37. Ashraf, S.; Gee, H.Y.; Woerner, S.; Xie, L.X.; Vega-Warner, V.; Lovric, S.; Fang, H.; Song, X.; Cattran, D.C.; Avila-Casado, C.; et al. ADCK4 mutations promote steroid-resistant nephrotic syndrome through CoQ10 biosynthesis disruption. *J. Clin. Investig.* **2013**, *123*, 5179–5189. [CrossRef] [PubMed]

38. Mollet, J.; Giurgea, I.; Schlemmer, D.; Dallner, G.; Chretien, D.; Delahodde, A.; Bacq, D.; de Lonlay, P.; Munnich, A.; Rotig, A. Prenyldiphosphate synthase, subunit 1 (PDSS1) and OH-benzoate polyprenyltransferase (COQ2) mutations in ubiquinone deficiency and oxidative phosphorylation disorders. *J. Clin. Investig.* **2007**, *117*, 765–772. [CrossRef] [PubMed]

39. Duncan, A.J.; Bitner-Glindzicz, M.; Meunier, B.; Costello, H.; Hargreaves, I.P.; Lopez, L.C.; Hirano, M.; Quinzii, C.M.; Sadowski, M.I.; Hardy, J.; et al. A nonsense mutation in COQ9 causes autosomal-recessive neonatal-onset primary coenzyme Q10 deficiency: A potentially treatable form of mitochondrial disease. *Am. J. Hum. Genet.* **2009**, *84*, 558–566. [CrossRef] [PubMed]

40. Freyer, C.; Stranneheim, H.; Naess, K.; Mourier, A.; Felser, A.; Maffezzini, C.; Lesko, N.; Bruhn, H.; Engvall, M.; Wibom, R.; et al. Rescue of primary ubiquinone deficiency due to a novel COQ7 defect using 2,4-dihydroxybensoic acid. *J. Med. Genet.* **2015**, *52*, 779–783. [CrossRef]

41. Mollet, J.; Delahodde, A.; Serre, V.; Chretien, D.; Schlemmer, D.; Lombes, A.; Boddaert, N.; Desguerre, I.; de Lonlay, P.; de Baulny, H.O.; et al. CABC1 gene mutations cause ubiquinone deficiency with cerebellar ataxia and seizures. *Am. J. Hum. Genet.* **2008**, *82*, 623–630. [CrossRef] [PubMed]

42. Lagier-Tourenne, C.; Tazir, M.; Lopez, L.C.; Quinzii, C.M.; Assoum, M.; Drouot, N.; Busso, C.; Makri, S.; Ali-Pacha, L.; Benhassine, T.; et al. ADCK3, an ancestral kinase, is mutated in a form of recessive ataxia associated with coenzyme Q10 deficiency. *Am. J. Hum. Genet.* **2008**, *82*, 661–672. [CrossRef] [PubMed]

43. Horvath, R.; Czermin, B.; Gulati, S.; Demuth, S.; Houge, G.; Pyle, A.; Dineiger, C.; Blakely, E.L.; Hassani, A.; Foley, C.; et al. Adult-onset cerebellar ataxia due to mutations in CABC1/ADCK3. *J. Neurol. Neurosurg. Psychiatry* **2012**, *83*, 174–178. [CrossRef] [PubMed]

44. Mignot, C.; Apartis, E.; Durr, A.; Lourenco, C.M.; Charles, P.; Devos, D.; Moreau, C.; de Lonlay, P.; Drouot, N.; Burglen, L.; et al. Phenotypic variability in ARCA2 and identification of a core ataxic phenotype with slow progression. *Orphanet J. Rare Dis.* **2013**, *8*, 173. [CrossRef]

45. Liu, Y.T.; Hersheson, J.; Plagnol, V.; Fawcett, K.; Duberley, K.E.; Preza, E.; Hargreaves, I.P.; Chalasani, A.; Laura, M.; Wood, N.W.; et al. Autosomal-recessive cerebellar ataxia caused by a novel ADCK3 mutation that elongates the protein: Clinical, genetic and biochemical characterisation. *J. Neurol. Neurosurg. Psychiatry* **2014**, *85*, 493–498. [CrossRef]

46. Blumkin, L.; Leshinsky-Silver, E.; Zerem, A.; Yosovich, K.; Lerman-Sagie, T.; Lev, D. Heterozygous Mutations in the ADCK3 Gene in Siblings with Cerebellar Atrophy and Extreme Phenotypic Variability. *JIMD Rep.* **2014**, *12*, 103–107. [PubMed]

47. Hikmat, O.; Tzoulis, C.; Knappskog, P.M.; Johansson, S.; Boman, H.; Sztromwasser, P.; Lien, E.; Brodtkorb, E.; Ghezzi, D.; Bindoff, L.A. ADCK3 mutations with epilepsy, stroke-like episodes and ataxia: A POLG mimic? *Eur. J. Neurol.* **2016**, *23*, 1188–1194. [CrossRef] [PubMed]

48. Salviati, L.; Trevisson, E.; Hernandez, M.A.R.; Casarin, A.; Pertegato, V.; Doimo, M.; Cassina, M.; Agosto, C.; Desbats, M.A.; Sartori, G.; et al. Haploinsufficiency of COQ4 causes coenzyme Q10 deficiency. *J. Med. Genet.* **2012**, *49*, 187–191. [CrossRef] [PubMed]

49. Brea-Calvo, G.; Haack, T.B.; Karall, D.; Ohtake, A.; Invernizzi, F.; Carrozzo, R.; Kremer, L.; Dusi, S.; Fauth, C.; Scholl-Burgi, S.; et al. COQ4 Mutations Cause a Broad Spectrum of Mitochondrial Disorders Associated with CoQ10 Deficiency. *Am. J. Hum. Genet.* **2015**, *96*, 309–317. [CrossRef] [PubMed]

50. Chung, W.K.; Martin, K.; Jalas, C.; Braddock, S.R.; Juusola, J.; Monaghan, K.G.; Warner, B.; Franks, S.; Yudkoff, M.; Lulis, L.; et al. Mutations in COQ4, an essential component of coenzyme Q biosynthesis, cause lethal neonatal mitochondrial encephalomyopathy. *J. Med. Genet.* **2015**, *52*, 627–635. [CrossRef] [PubMed]

51. Public Health. Community Register of Orphan Medicinal Products. Available online: http://ec.europa.eu/health/documents/community-register/html/o1765.htm (accessed on 4 March 2017).

52. Montini, G.; Malaventura, C.; Salviati, L. Early coenzyme Q10 supplementation in primary coenzyme Q10 deficiency. *N. Engl. J. Med.* **2008**, *358*, 2849–2850. [CrossRef] [PubMed]

53. Pineda, M.; Montero, R.; Aracil, A.; O'Callaghan, M.M.; Mas, A.; Espinos, C.; Martinez-Rubio, D.; Palau, F.; Navas, P.; Briones, P.; et al. Coenzyme Q(10)-responsive ataxia: 2-year-treatment follow-up. *Mov. Disord.* **2010**, *25*, 1262–1268. [CrossRef] [PubMed]

54. Saiki, R.; Lunceford, A.L.; Shi, Y.; Marbois, B.; King, R.; Pachuski, J.; Kawamukai, M.; Gasser, D.L.; Clarke, C.F. Coenzyme Q10 supplementation rescues renal disease in Pdss2kd/kd mice with mutations in prenyl diphosphate synthase subunit 2, American journal of physiology. *Ren. Physiol.* **2008**, *295*, F1535–F1544. [CrossRef] [PubMed]

55. Lopez, L.C.; Quinzii, C.M.; Area, E.; Naini, A.; Rahman, S.; Schuelke, M.; Salviati, L.; Dimauro, S.; Hirano, M. Treatment of CoQ(10) deficient fibroblasts with ubiquinone, CoQ analogs, and vitamin C: Time- and compound-dependent effects. *PLoS ONE* **2010**, *5*, e11897. [CrossRef] [PubMed]

56. Yubero, D.; Montero, R.; Armstrong, J.; Espinos, C.; Palau, F.; Santos-Ocana, C.; Salviati, L.; Navas, P.; Artuch, R. Molecular diagnosis of coenzyme Q10 deficiency. *Expert Rev. Mol. Diagn.* **2015**, *15*, 1049–1059. [CrossRef] [PubMed]

57. Yubero, D.; Montero, R.; Artuch, R.; Land, J.M.; Heales, S.J.; Hargreaves, I.P. Biochemical diagnosis of coenzyme q10 deficiency. *Mol. Syndromol.* **2014**, *5*, 147–155.

58. Trevisson, E.; DiMauro, S.; Navas, P.; Salviati, L. Coenzyme Q deficiency in muscle. *Curr. Opin. Neurol.* **2011**, *24*, 449–456. [CrossRef] [PubMed]

59. Yubero, D.; Montero, R.; Ramos, M.; Neergheen, V.; Navas, P.; Artuch, R.; Hargreaves, I. Determination of urinary coenzyme Q10 by HPLC with electrochemical detection: Reference values for a paediatric population. *Biofactors* **2015**, *41*, 424–430. [CrossRef] [PubMed]

60. Bujan, N.; Arias, A.; Montero, R.; Garcia-Villoria, J.; Lissens, W.; Seneca, S.; Espinos, C.; Navas, P.; de Meirleir, L.; Artuch, R.; et al. Characterization of CoQ(1)(0) biosynthesis in fibroblasts of patients with primary and secondary CoQ(1)(0) deficiency. *J. Inherit. Metab. Dis.* **2014**, *37*, 53–62. [CrossRef] [PubMed]

61. Fernandez-Ayala, D.J.; Lopez-Lluch, G.; Garcia-Valdes, M.; Arroyo, A.; Navas, P. Specificity of coenzyme Q10 for a balanced function of respiratory chain and endogenous ubiquinone biosynthesis in human cells. *Biochim. Biophys. Acta* **2005**, *1706*, 174–183. [CrossRef] [PubMed]

62. Arias, A.; Garcia-Villoria, J.; Rojo, A.; Bujan, N.; Briones, P.; Ribes, A. Analysis of coenzyme Q(10) in lymphocytes by HPLC-MS/MS. *J. Chromatogr. B Anal. Technol. Biomed. Life Sci.* **2012**, *908*, 23–26. [CrossRef] [PubMed]

MDPI

St. Alban-Anlage 66

4052 Basel

Switzerland

Tel. +41 61 683 77 34

Fax +41 61 302 89 18

www.mdpi.com

Journal of Clinical Medicine Editorial Office

E-mail: jcm@mdpi.com

www.mdpi.com/journal/jcm

www.ingramcontent.com/pod-product-compliance
Lightning Source LLC
Chambersburg PA
CBHW051731210326
41597CB00032B/5676